VLSI FABRICATION TECHNOLOGY

VLSI FABRICATION TECHNOLOGY

By

Dr. Balwinder Raj

Ph. D: IIT Roorkee,
PDF: University of Rome, Italy
Assistant Professor, Dept. of ECE, NIT Jalandhar (Punjab)

Balwinder Singh

Senior Engineer, centre for Development of Advanced Computing (C-DAC) Mohali (Punjab)

Ashish Dixit

Assistant Professor, Dept. of ECE

Amity University, Lucknow (U.P.)

UNIVERSITY SCIENCE PRESS

(An Imprint of Laxmi Publications Pvt. Ltd.)

An ISO 9001:2015 Company

BENGALURU • CHENNAI • GUWAHATI • HYDERABAD
KOCHI • KOLKATA • LUCKNOW • RANCHI
NEW DELHI

Printed and bound in India
First Edition : 2013; *Edition* : 2025
ISBN : 978-93-81159-60-6

Branches

Bengaluru	080-26 75 69 30
Chennai	044-24 34 47 26
Guwahati	0361-254 36 69
Hyderabad	040-27 55 53 83
Kochi	0484-405 13 03
Kolkata	033-40 04 77 79
Lucknow	0522-430 36 13
Ranchi	0651-224 24 64

Published in India by

UNIVERSITY SCIENCE PRESS
(An Imprint of Laxmi Publications Pvt. Ltd.)

An ISO 9001:2015 Company
113, GOLDEN HOUSE, GURUDWARA ROAD, DARYAGANJ,
NEW DELHI - 110002, INDIA

Corporate Office : B-33, Sector-5, Noida,
Gautambuddha Nagar (U.P.) - 201 301, INDIA
Telephone : +91-120-44 46 900
www.laxmipublications.com
info@laxmipublications.com

CONTENTS

FOREWORD

Since the invention of IC (Integrated Circuits) five decades ago, Microelectronics as a field has come to govern our lives, which leads with VLSI design and technology. This book has brought out the development in this field of VLSI design and technology in a systematic manner. This book begins with an introduction chapter on the history of semiconductor devices and its evolution throughout years. It also addresses the importance of VLSI design and technology right from semiconductor devices, CMOS circuits. VLSI design, and fabrication technology. The authors have covered the topics in an exhaustive manner with a thorough review. This book is excellent for a beginner/undergraduate and postgraduate student and shall be helpful to students wanting to have a conceptual grasp on the subject "VLSI design and technology".

I recommend the book for undergraduate and postgraduate of many Indian universities students especially under the PTU colleges and UPTU colleges, as it is written as per their syllabus.

I congratulate the young authors Balwinder Raj, Balwinder Singh and Ashish Dixit for this excellent book. The effort is all the more commendable as authors are teaching the subject for a number of years and the book is a product of feedback from students. I expect the book to be further fine tuned with the feedback from readers which I am sure the author shall incorporate in the next edition.

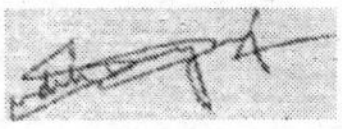

Dr. Sudeb Dasgupta
Faculty in Microelectronics and VLSI Technology (MVT) Group
Dept. of Electronics and Computer Engineering
Indian Institute of Technology Roorkee, (IIT Roorkee)
Uttrakhand-247667 (India)

1

INTRODUCTION VLSI FABRICATION

1.1 INTRODUCTION

The Thomas Edison's discovery of thermionic emission opened the door to electronic technology. The progress was slow in the beginning, but each year brought new and more amazing discoveries. The development of vacuum tubes soon leads to the simple radio. Then came more complex systems of communications. Modern systems now allows us to communicate with other parts of the world via satellite. Data is now collected from the space by probes without the presence of man because of microelectronics technology.

Sophisticated control system allows us to operate equipment by remote control in hazardous situations, such as the handling of radioactive materials. We can remotely pilot aircraft from takeoff to landing. We can make course corrections to space craft millions of miles from earth. Space flight, computers and even videogames, would not be possible without the advances made in microelectronics.

Early developments in semiconductor materials were studied in laboratories as early as 1830. The first materials were a group of elements and compounds that were usably poor conductors if heated. Shining light on some of them would generate electrical current that could pass through them in one direction only. In 1874, electricity was being used not only to carry power, but to carry information. The telegraph, telephone and later the radio were the earliest devices in an industry that would eventually be known as electronics.

Radio receivers required a device called a rectifier to detect signals. Firdinand Braun used the rectifying properties of the galena crystal, a semiconductor material composed of lead sulfide, to create the cat's whisker diode for this purpose. Thus was born the fist semiconductor.

The revolution of change in the art of electronics design brought about the development of integrated circuits is the result of invention of the transistor in 1948 by Burdeen, Brattain and Sckockley who were awarded Nobel prize in the physics for their such invention. Integrated circuits is a collection of desecrate elements (diode, resistor, capacitor and transistor) created by means of a single construction process in which all elements are formed.

The first integrated circuit (shown in Fig. 1.1) in which all of the components, both active and passive were fabricated in a single piece of semiconductor material half the size of paper clip was developed in 1998 by Jack Kilbey at Taxas instrument in USA. It was a simple device which Jack

Kilbey developed. But, it revolutize the electronics industry. New technology emerged and integrated circuit were soon developed with various components (transistors, resistor and capacitor) formed on the same clip, but interconnection of the various components still required tedious hand wiring.

Fig. 1.1 *First Integrated Circuit*

In 1959, Jean Hernia and Robert Noyce developed a new process called planner technology at Fairchild semiconductor which enabled them to diffuse various layers onto the surface of a silicon wager to make a transistor, leaving a layer of protective oxide on the functions. This process allowed metal inconnections to be evaporated onto the flat transistor surface and replaced the hand wiring. The new process used silicon instead of germanium and made commercial production of IC's possible.

The initial resistance to the new integrated circuit technology gave way to enormous popularity. By the end of the 1960's nearly 90% of all the components manufactured were integrated circuits from the first simple circuit was grown worldwide integrated circuit, market whose sales in 2004 totaled $ 179 bilions. These components supported a 2004 world wide electronics end equipment market of $ 1, 186 billion, such is the power of one idea to change the world the 2000 semiconductor industry report (shown in Fig. 1.2) shows that scales volume of the semiconductor devices based electronics industry in the past 30 years and projects sales to the year 2010. It also shows world product and sales volume of automobile, steel and semiconductor industries. Note that electronics industry crosses the automobile industry in 1998 if current trend continues the sales of electronics industry will reach 3 trillion dollor and will constitute about 10% of GWP by 2010 (source is 2000 semiconductor industry report).

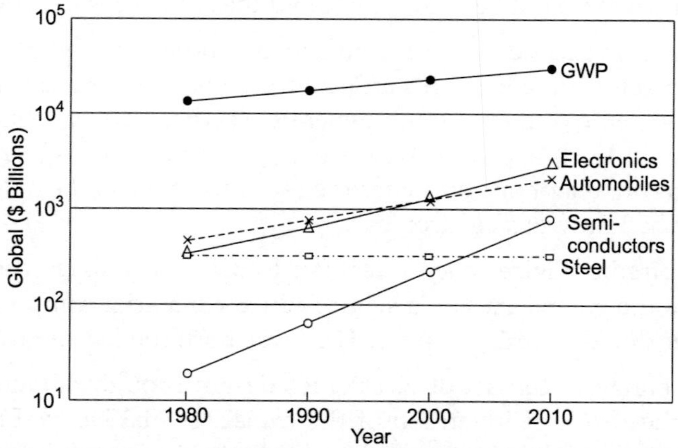

Fig. 1.2 *Sale of Electronics Automobile Semiconductor and Steel Industries and Gross World Product (GWP)*

1.2 EVOLUTION OF MICROELECTRONICS

The earliest electronics circuits were fairly simple. They were composed of a few tubes, transformers, resistors, capacitors and wiring. As more was learned by designers. They began to increase both the size and complexity of circuits component limitations were soon identified as this technology development.

Vacuum Tube Equipment: Vacuum tubes were found to have several built in problems.

1. The tubes were light weight, associated components and chassis were quite heavy. It was common for such chassis to weight 55 to 65 kg.
2. In addition, that tubes had the problem of heating thus, required a warm-up time from 1 to 2 minutes.
3. They also required power supply voltage of 300 volt dc and more.
4. No tow tubes of the same type were exactly alike in output characteristics, therefore, the designers were required to produce circuit that could work with any tube of a particular type. This meant that additional components were often required to tune the circuit to the output characteristics required for the tube used.
5. Another problem with vacuum tube circuit was the method of wiring components referred as point to point wiring. Those wiring look like rat's nest and it after caused unwanted interactions between components.
6. Point to point wiring posed a softy hazard when trouble shooting were performed on energized circuit because of usually repaired with general purpose test equipment and common hand tools.

Vacuum tube circuit proved to be reliable under many conditions still, the drawbacks of large size, heavy weight and significant power consumption made them undesirable in most situations, for example, computer system using tubes were extremely large and difficult to maintain. ENIAC, a completely electronic computer built in 1945, contained 18,000 tubes, it often required a full day just to locate and replace faulty tubes.

In some applications we are still limited to vaccum tubes. Cathode ray tubes used in radar, television and oscilloscope tics date.

Solid State Devices: The transition from vacuum tubes to solid state devices took place rapidly. As new types of transistors and diode were created, they were adapted to circuit. The reduction in size, weight and power use were impressive. Circuits that earlier weighed as much as 50 kg were reduced in the weight to just few grams by replacing bulky components with the much lighter solid state devices.

The earliest solid state circuits still relied on point to point wiring which caused many of the disadvantages mentioned earlier. Metal chassis, are used with tubes required to provide physical support for the components. The solid state chassis were still considerably small and lighter than the older, tube chassis, still greater improvements in component mounting methods were yet to come.

One of the most significant developments in circuit packaging has been the Printed Circuit Board (PCB) as shown in Fig. 1.3. The PCB is usually an epoxy board on which the circuit leads have been added by the photo etching process. This process is similar to photography in that copper clad board are exposed to controlled light in the desired circuit pattern and then etched to remove the unwanted copper. This process leaves copper strips [Lands] that are used to connect the components. In general, PCB's eliminate both the heavy, metal chasis and point to point wiring.

Although, printed circuit boards represents a major improvement over tube technology, they are not without fault, for example, number of components on each board is limited by the size and shapes of components. Also, while vacuum tubes are easily removed for testing or replacement, PCB components are soldered into place and are not as easily removed.

Fig. 1.3

Normally, each PCB contains a single circuit or subassembly of a system. All PCBs within the system are routinely interconnected through cabling Hardness's (group of wiring or ribbons of wiring) you may be confronted with problems in faulty connections that affect system reliability. Such problems are often caused by wiring errors, because of the large number of wires in a harness and by damage to those wires and connectors.

Discrete Components: Discrete components are those components which put into separate container (like a wax, paper sheet is sandwitched between aluminium foils, a wire in the insulator material etc.) for the electronics circuit implementation it required PCB on which the components are placed.

1.3 INTEGRATED CIRCUITS

In integrated circuits, all the components are placed on a single thin silicon piece of size of few hundred square mils which is called chip. When all the components (transistor, resistors, capacitor, etc.) and interconnection wire are formed on the same semiconductor (silicon) place. They cannot be separate from each other after fabrication and none of the component can be replaced.

The integrated circuits are classifiable based on fabrication technology and applications. They may be more clear by tree diagram represtantation as shown in figure.

Classification on based of fabrication techniques

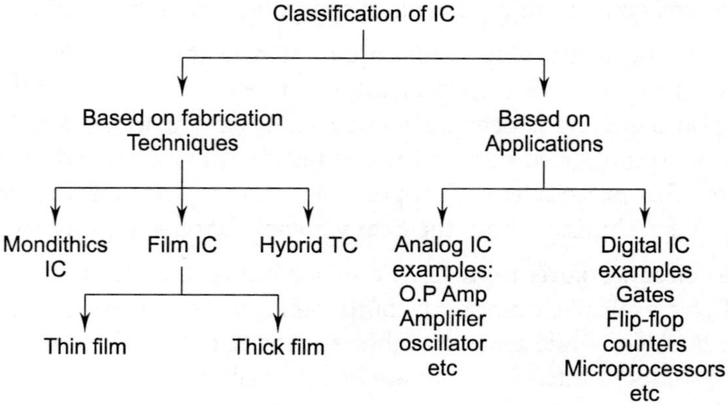

1.3.1 Monolithic Integrated Circuits (IC)

The word monolithic is originates from two Greek words mono's meaning "single" and lithics means "stone". So monolithic IC's are those that are formed completely within a semiconductor substrate (Piece). These integrated circuits are commonly referred to silicon chips a smaller chip can accommodate large no. of circuit elements or components. (detail will be discuss later) For example, IC chip measuring 1.65 mm × 1.65 mm can be normally contain 35 transistor, 30 resistors and a few capacitors and their interconnections.

Film Integrated Circuits (IC) are broken down into two categories. Thin film and thick film components are made of either conductive or non-conductive material that is deposited in desired pattern on a ceramic or glass substrate. Film can only be used as passive circuit components, such as resistors and capacitor.

1.3.2 Thick Film

Thick films are produced by screening patterns of conducting and insulating materials on ceramic substrate. A thick film is a film of material with a thickness that is at least 10 times greater than the mean free path of an electron in that material, or approximately 0.001 cm. The technique is used to produced only passive components such as resistor and capacitors.

One method which are used in fabricating a thick film is to produce a series of stencils called screens. The screens are placed on the substrate and appropriate conducting or insulating material are wiped across the screen. Once the conducting or insulating material has been applied the screens are removed and the formulations are fired a temperatures above 600°C. This process forms alloys that are permanently bonded to the insulating substrate. To a limited extent, the characteristics of the film can be controlled by the firing temperature and length of firing time.

Examples of thick film fabrication

1. **Resistors:** Thick film resistance values can be held to a tolerances of ± 10 per cent. Closer tolerance are obtained by trimming each resistor after fabrication. Hundreds of different cermets formulations are used to produce a wide range of component parameters, for example, the material used for a 10 Ω per square resistor is quite different from that used for 100 kΩ per square resistor.
2. **Capacitors:** Capacitors are formed by a sequence of screenings and firings. Capacitors in this case consist of a bottom plate, intraconnectors, a dielectric and a top plate.

1.3.3 Thin Film

Thin films are produced by deposition or sputtering of conductive, resistors and dielectric films on a insulating substrate the thickness of these films are from few hundred Armstrong (1Å – 10^{-10}) to few micrometer (1 μm to 10^{-6} m).

The various methods are used for thin film processing like vapour deposition, vacuum evaporation and sputtering etc. Here only the sputtering process is discussed other processes will be discussed later.

In sputtering process the meterial which we want to deposit as a thin layer, place on the cathode when cathode get charged by supply (d.c). The cathode is release some free electrons that collide with argon atom causing them to ionize and form the glow discharge. The positive charged atoms accelerate toward the cathode and cause some atom of cathode to sputter or breakaway then these sputtered atoms are attracted towards the anode and deposited on the substrate this process is called sputtering process a uniform film thickness is obtained over in large substrate area because of flatness of sputtering source.

1.3.4 Hybrid Integrated Circuit (IC)

Combines two or more integrated circuit types and discrete (separate) components. As shown in Fig. 1.4 hybrid IC consisting of silicon chips and film circuitry (resistance, capacitor, etc. fabricated on substrate).

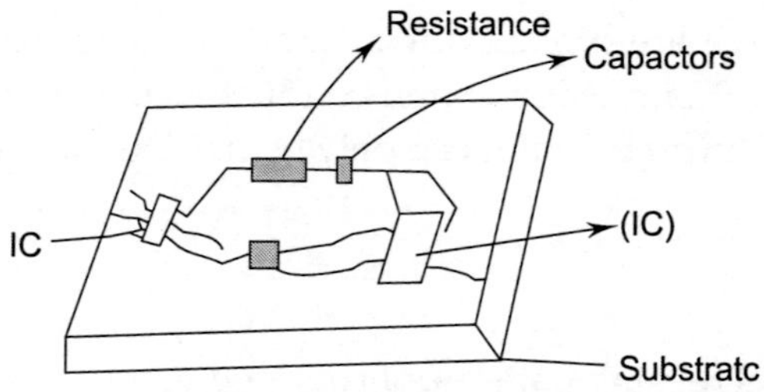

Fig. 1.4 *Hybrid IC*

The primary advantage of hybrid microcircuits is design flexibility, that is hybrid microcircuits can be designed to provide wide use in specialized applications such as low volume and high frequency circuits several components available for hybrid applications. These include discrete components that are electrically and mechanical compatible with IC's such components may be used to perform functions that are supplementary to those of ICs. They can be handled, tested and assembled with essentially the same technology and tools.

Complete circuit is available in form of (unencapsulated ICs) dice. These chips are usually identical to those sold as part of the manufacture's regular production line. They must be properly packaged and connected by the user if a high-quality final assembly is to be obtained. The circuits are usually seated in a package to protect them from mechanical and environmental stress. One mil (0.001 inch), gold wire leads are connected to the appropriate pins which are brought out of the package to allow external connections.

Some benefits of hybrid IC's technology as compared to monolithic. IC technology these are:

1. These IC's have high performance because of precise component value.
2. These IC's have high frequency and speed capability.
3. These IC's have high voltage and power capabilities.
4. They are more flexible, because it is ease to design and testing on the bread board before transfer to the substrate.
5. It takes low circuit development time, efforts and cost.
6. These are more reliable because of rigid bonding of components/Interconnection on substrate.

1.4 CLASSIFICATION BASED ON APPLICATIONS

1. Analog IC
2. Digital IC

1.4.1 Analog Integrated Circuits

Analog Integrated Circuits is one which performs linear operations and amplify the input signals this type of IC's are also called linear ICs. In this type of ICs the input signal is analog signal these input signal may be constant or can be change with time. Analog ICs are much versatile and capable of responding to many shapes and values of input signals.

Examples of analog ICs are operational amplifier (OP-AMP), amplifier, oscillators, phase locked loops (PLL), and instrumentation ICs or Process Control ICs.

1.4.2 Digital Integrated Circuits

Digital Integrated Circuits is one which perform digital operation and deals with discrete qualities or binary quantities. (Binary quantities are 0 or 1) the digital ICs requires the input signal to be in a specified form. These ICs responds to only two ranges of input voltages (0 or 1), providing only two ranges as output (0 or 1).

Examples of digital ICs are ALU (Arithmetic Logic Unit) various gate logics, memory, counters, microprocessors and computer, etc.

1.5 STATE OF THE ART

Microelectronic technology today includes thin film, thick film, hybrid and integrated circuits and combinations of these such circuits are applied in digital, switching and linear (analog) circuits. Because of the current trend of producing number of circuits on a single chip. At the same time you may expect a reduction in the size, weight and the number of connections in a individual systems. Also you may look for the farther increase in the packaging density of electronic circuits. Improvements in reliability and system capability are also expected. Thus, even as existing capabilities are being improved, new areas of microelectric use are being explored. To predict where all this use of technology will lead is impossible. However, as the demand for increasly effective electronics system continues, improvements will continue to be made in state of art microelectronics to meet the demands.

As the complexity increasing to fulfill the demands of customers like as more and more complete functions are required in various data processing and telecommunications devices, so the need to integrate these functions in a small package is also increasing. So, it needs to be define the different levels of integration of IC chips. Table shows different level and evolution of logic complexity in ICs over last three decades.

Different levels and types of integration of IC's

Integration type or level	*Year*	*Complexity (No. of logic gates per chip)*
Single transistor	1958	<1
Small Scale Integration (SSI)	1964	Less than 10
Medium Scale Integration (MSI)	1967	10 to 100
Large Scale Integration (LSI)	1972	100 to 1000
Very Large Scale Integration (VLSI)	1978	1000 to 10,000
Ultra Large Scale Integration (ULSI)	1989	10,000 and above

In LSI and VLSI, a variety of circuits can be implanted on a wafer, resulting in further size and weight reduction. ICs in modern computers, such as home computers, may contain the entire memory and processing circuits of a single substrate.

Development of a VLSI device begins with a demand from industry or as the result of research. A device that is needed by industry may be a simple diode network or a complex circuit consisting of thousands of components. No matter how complex the device, the basic steps of production are similar. Each type of device requires circuit design, component arrangement, preparation of a substrate and the depositing of proper materials on the substrate. The first consideration in the development of a new device is to determine what the device is to accomplish. Once, this has been decide, engineers can design the device. During the design phase, the engineers will determine the numbers and types of components and the interconnections, needed to complete the planned circuit.

1.6 INTEGRATED CIRCUITS (IC) MANUFACTURING FLOW

The fabrication of monolithic integrated circuits is most complex aspect of VLSI devices. This section gives an overview of processed involved in the manufacturing of IC's. The Fig. 1.5 shows the sequence of steps involved in IC manufacturing process.

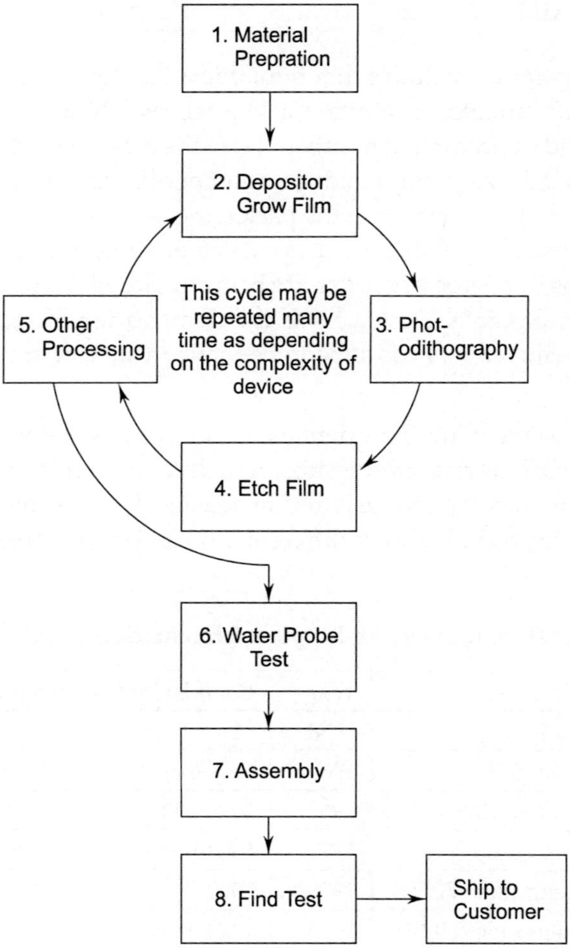

Fig. 1.5 *Sequence of Steps Involved in IC Manufacturing*

As shown in the figure above first step is material preparation for the device manufacturing. Material (sand) silicon available in natural form is not suitable for manufacturing. It needs to be purified as given in three steps in Fig. 1.6.

PREFACE

Jack Kilby and Robert Noyce are reported to have famously said at the birth of IC (Integrated Circuits) that these components would serve the needs of entire world. They were thinking of applications like in Computers, in Electronics and Mobile Communications etc. With advances in VLSI design and technology, a few Silicon chips replaced entire banks of vacuum tubes, occupied a hundredth of the volume and consumed a small fraction of the power, such that the home PC now is more powerful than a mainframe of yesteryears. The microchip industry continues its evolutionary march to the physical limits of silicon based ICs. Small Scale Integration (SSI) soon gave way to Large Scale Integration (LSI) and Very Large Scale Integration (VLSI) following Moore's Law, with a few million gates per chip. We are now at the threshold of Nanoscale Devices which operate on entirely new principles of Quantum Mechanics.

This book named as "VLSI Fabrication" is intended for the students of Electronics, VLSI and Nanotechnology. The purpose of writing this book is to present the subject matter in a most concise, compact and lucid manner. The focus of this book is basic VLSI design and technology for students, designers and scientific community. The authors have also borne in mind the syllabi of many Indian universities for the requirements of B.Tech, M.Tech and B.Sc students of Electronics and VLSI design technology. To the best of the authors' knowledge, no such text has been published in the country, though there are specialized books on Basic VLSI Design by Pucknell Eshraghian, VLSI Design by Mead, Weste and Eshragian and on VLSI Fabrication by Sze, Ghandy and others.

It is the reason of this introductory text to present that the Silicon-based NMOS and CMOS devices are at the heart of most VLSI processors as well as memories; these are the subject of Chapters 1–5. Chapter 1 introduces the basic principles of semiconductor and integrated circuits. Chapter 2 deals with the Silicon Fabrication steps and clean room Fabrication. Chapter 3 describes in detail the Etching process with its types and various etchers. Other Chapters discusses metallization process with the perfect material used for metallization, the lithography, Oxidation, Diffusion steps and its various types etc.

While preparing manuscript, we constantly kept in mind the requirements of all the students, regarding the latest as well as changing trend of examination, the contents of this book are adequate, simple language and easy to understand, in order to explain the topics in very simple way, analytical and mathematical approach is provided whenever is necessary. We hope the book will meet the requirement of students, for whom it has been designed. Authors' contributions and perspectives are critical to the completeness of this book.

We would like to thank all our friends, co-workers and seniors for their encouragement during the preparation of this book.

Although every care has been taken in printing the book, yet the very factor of human error can't be ruled out. Any errors, omissions and suggestions for the improvement of this book, brought to our notice will be acknowledged.

—Authors

ACKNOWLEDGEMENTS

This book has been evolved from the lecture notes of the course that we have taught in many collages. Throughout the years, colleagues and students have detected mistakes and suggested modifications to the text. We would like to thank all those who were willing to give us the feedback for further improvement and enhance the quality of the book.

We are very grateful to Mr. Gurinder Pal Singh, Mr. Varun Kumar, Mr. Ajay Dhaiya, as they have read large parts of the almost final manuscript.

Of course, the publication of this book would not have been possible without the support of my friends Dr. Santosh Vishvakarma, Dr. S. S. Rathod, Dr. Ramesh Vaddi, Dr. Jawar Singh, Dr. R. K. Sunkaria, Dr. Vikas Chawala, Dr. Samrjeet Rawat and Dr. Vinod. We would like to thank our mentors Dr. A. K. Saxena, Dr. S. Dasgupta, Dr. A. K. Khosla, and Dr. Sukhleen Bindra for their guidelines to make this book technically sound and complete well in time. We would like to thank Dr. Manish Hooda and Dr. Dilip Kumar for readiness to answer all our questions and their flexible interpretation of deadlines.

No words are sufficient to express our gratitude to our family for their cooperation, patience and support during preparation of this book. I would also like to thank my students Mr. Amandeep Singh and Mr. Navdeep Prashar for encouraging me for preparation of this book.

Any constructive comments, suggestions, criticisms from the students and colleagues will be highly appreciated and greatly acknowledged.

—Authors

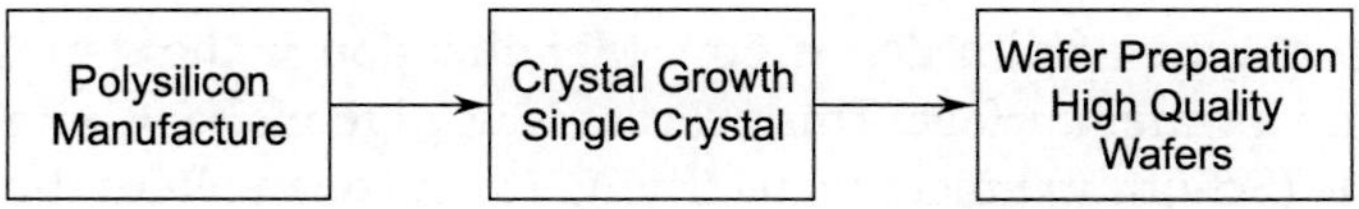

Fig. 1.6 *Purifications Steps*

1. **Material Preparation:** The initial step in the production of wafer grade silicon is to mix silicon dioxide (sand) with wood and coke at 1600°C to give 98% pure metallurgical grade silicon (MGS). Since silicon purity is essential, a second step is added to produce electronic grade silicon (EGS). EGS, which is separated using various methods like distillation fractional and chemical vapour deposition process. It is 99.9999% pure. The detail of material preparation will be discussed in chapter 3.

   Raw EGS with any dopants desired for the electrical specifications for the wafer are placed into a crucible and heated to the melting point of silicon (1200°C). The Czochralski method is used in crystal production. The crystal growth is then slowly lifted out of the melted material to form an ingot 8 to 10 feet long of a diameter up to 12 inches depending on wafer specifications.

   The ends of the ingot are removed, lathed to precise roundness and the diameter checked for exactness to specifications. The crystal orientation is checked using x-ray crystallography. The material properties of the crystal will change with different crystal orientations. The axis of orientation determines the type devices that can be fabricated from the silicon crystal. The ingot is sliced to create individual wafers, 20–30 mils thick, using a diamond-edged, inner diameter saw. These slices are tested for thickness, diameter, flatness and uniform electrical resistance. Once, these wafers are found to meet specifications The wafers are now considered to be high quality, prepared to meet customer specifications and are shipped for the Device manufacturing in Fabrication Labs.

2. **Film Deposition:** Detail given in chapter 5

   A. **Thin film deposition:** Thin films may be metals, semiconductors, or insulators. Films may be used as temporary protective layers or to construct circuit or device components.

   Methods:

   1. Physical vapour deposition
   2. Oxide growth
   3. Chemical Vapour Deposition (CVD) - forms solid films on a wafer surface by chemical reactions of gaseous reactants at the wafer surface. During CVD, only the solid product is left as film and all other gaseous products are removed.

   Types of CVD:

   - Atmospheric pressure chemical vapour deposition (APCVD)
   - Low pressure chemical vapour deposition (LPCVD)
   - Plasma-enhanced chemical vapour deposition (PECVD)
   - Photon or laser enhanced chemical vapour deposition (PHCVD of LCVD)

   B. **Oxidation:** An oxide is grown on silicon for one of the following reasons:

   1. A mask for protection during diffusion or ion implantation.
   2. A gate oxide in MOS devices.
   3. A field oxide for device isolation.
   4. A protective layer for the finished device.
   5. Interlayer dielectric (an insulating material between the plates of a capacitor).

The detail of the oxidation process discussed in chapter 6

C. **Metallization:** A type of film deposition. Metallization is the deposition of a metal or metal alloy onto a wafer surface. This process is used to make IC interconnects, bonding pads and diode (Schottky) junctions rectifying (to make an alternating current unidirectional) junctions. Schottky diodes are a two-terminal device that offers a high resistance to current flow in one direction and low resistance to current flow in the opposite direction. It is a lower voltage barrier than standard silicon diodes and is able to change operating states (on/off) much faster than ordinary *p-n* junction diodes. There are three methods in metallization:

- *Evaporation*: The metal is evaporated with electrical thermal energy in an evacuated chamber (extremely low pressure). The metal vapor condenses on all surfaces including those exposed areas on the wafers.
- *Sputtering*: Plasma energy is used to knock (sputter) fragments from the target allowing them to drift to the wafer surface and be deposited.
- *CVD*: LPCVD is the most commonly used form of CVD for metallization. LPCVD provides better step coverage and better uniformity than the other methods of CVD.

In chapter 2, the various metallization material and metallization process and discussed in detail.

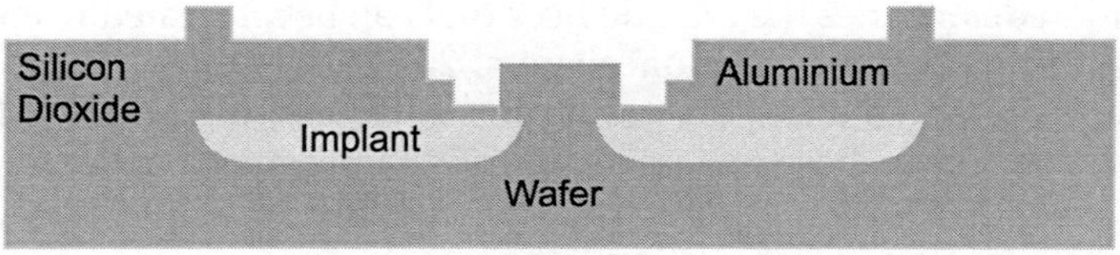

Fig. 1.7 *Metal Deposition*

3. **Photolithography (Detail in Chapter 8):** This is the third major step in the process and allows for selective processing on the wafer surface. Photolithography defines the device or circuit on the wafer by using a mask (reticle) to transfer the IC design pattern onto the wafer. This transfer is similar to the development of a photograph. The wafer is covered with photoresist, a material that is light sensitive and is soft baked to increase adhesion to the wafer.

   The photoresist is then exposed to light through the mask. There are two types of photoresists, which have opposite effects when exposed to the light passing through the translucent regions of the mask. Positive photoresist will become softened, to be rinsed away during the development stage leaving the unexposed hardened areas on the wafer. Negative photoresist will become hardened when exposed to light and the unexposed areas will be softened and be rinsed away from the wafer during the development stage. Positive photoresists usually form islands of photoresist while negative photoresist usually forms holes or "ditches.

   Methods of exposure to light:

   (*i*) Contact printing – mask in contact with the wafer.

   (*ii*) Proximity printing – mask is close to wafer, but does not come in contact with the wafer.

   (*iii*) Projection printing – using mirrors to project the actual size image on the wafer.

   (*iv*) Step and repeat – using high quality masks and image reduction lenses (5:1 or 10:1) to iniaturize the image. The wafer is sequentially moved until the entire wafer is exposed.

Major process steps:

(*i*) Coating – surface prep
photoresist application – negative = exposed areas harden
– Positive = exposed areas soluble
– Soft bake – increase adhesion

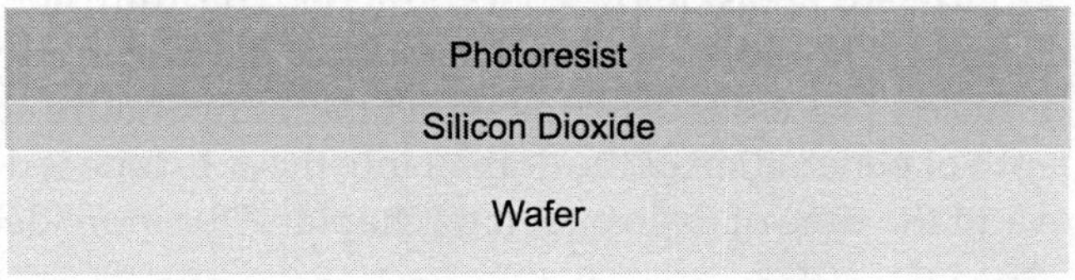

Fig. 1.8 (*a*) *Photoresist*

(*ii*) Exposure – light passes through translucent areas of mask

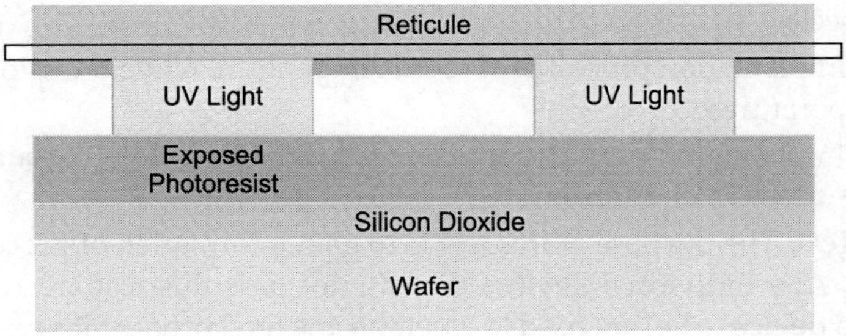

Fig. 1.8 (*b*) *Light Exposure to Photoresist*

(*iii*) Developing – soluble portions of photoresist rinsed off.

(*iv*) Inspect

4. **Etch Film:** Etch is an intermediate process step used to remove material from the water surface. Removal may be patterned (in conjunction with photolithography) or unpatterned (taking place over the entire wafer surface). Both wet and dry etch involve three steps: 1. Disperse the etchant 2. Remove etched material 3. Remove the by-products created during etching. (Detail about the process of etching and etching equipment are discussed in chapter 9)

   Dry (plasma) Etch: uses plasma to remove material by physical means, chemical reaction, or some combination of the two. Plasma is a mixture of charged and uncharged gas particles that maintain an overall electrical neutrality. It is an energetic, ionized gas sustained by RF energy.

   Three methods of dry etch:

   - *Ion milling (sputter etch)*: – An entirely physical etching process using a plasma of non-reactive ions to physically remove material.
   - *Plasma Etch*: A chemical etching process using a plasma energy driven reactive gas rather than liquid chemicals.
   - *Reactive ion Etch*: A dry etch process combining chemical and physical removal of wafer surface material.

   Wet Etch: uses acids, bases and solvents that react with the wafer material to be removed. The etch chemicals change the exposed portion of the unwanted layer into a by-product by either oxidation/reduction reaction or an exchange reaction. The undercutting due to the isotropic etching imposes limits on critical geometries.

5. **Some Other Processes Involved in Manufacturing**
   - *Diffusion*: Diffusion is the movement of particles from a region of high concentration to a region of low concentration. It is used to introduce dopants into the semiconductor wafer or films and to cause dopants already present to migrate deeper into the substrate. High concentrations of dopant particles deposited on the surface of the wafer diffuse inward toward areas of low concentration. This diffusion of *p*-type and *n*-type dopants will modify the electrical characteristics of the silicon and produce the desired effect. The diffusion process is dependent on time, temperature and the dopant species used. The depth of penetration of the dopant into the substrate is dependent on the oven temperature and the time of exposure. (See chapter 7 for more detail).
   - *Ion implantation*: Like diffusion, is used to introduce dopant atoms into the substrate. Ion implantation is used to modify the electrical characteristics of selected areas of the wafer. The ion implanter consists of an ion source, mass analyzer, accelerator and a scanning system (detail is given in chapter 10). This technique differs from diffusion in two respects:
     (*i*) Ion Implantation processes at room temperature while diffusion processes at high temperatures.
     (*ii*) Ion Implantation peak dopant concentration lies below the surface rather than at the surface as in diffusion.
6. **Wafer-Probe Test:** The purpose of this test is to gain information of process yields and final performances. Any individual devices that do not pass this test are rejected and not assembled. Good devices (die) are used to assemble the final product it saves the assembly cost of bad dies.
7. **Assembly (see chapter 15 for more detail):** The purpose of assembly is to package the die in a form that is usable to the customer and to protect the quality of the device. The following steps is involved in the assembly process:
   1. Die (chips) separation with diamond saw.
   2. Mounting process – places die onto lead frame strips.
   3. Wire bonding – a computer controlled process in which gold or aluminium wires from the pads on the die are attached to the inner leads on the lead frame. These wires connect the die (chip) integrated circuit to the outside world.
   4. Encapsulation – Die is covered or enclosed with a plastic coating using epoxy to seal.
   5. Symbolization – A laser writes the part, lot number, date and country of origin on the package for identification.
   6. Trim and form the leads.
   7. Solder tin leads and rinse.
8. **Final Test:** This final step in the manufacturing process is a series of AC and DC measurements of the performance of the die. In this process, various inputs are given to the Packed IC's and output responses are observed. This is the last check for device quality before the package is shipped to the customer.

1.7 FABRICATION OF BIPOLAR JUNCTION TRANSISTOR

The steps in the fabrication process described here and illustrated in Fig. 1.9, would produce an npn, planar-diffused transistor. But, with slight variations, the technique may also be applied to the production of a complete circuit, including diodes, resistors and capacitors. The steps are performed in the following order:

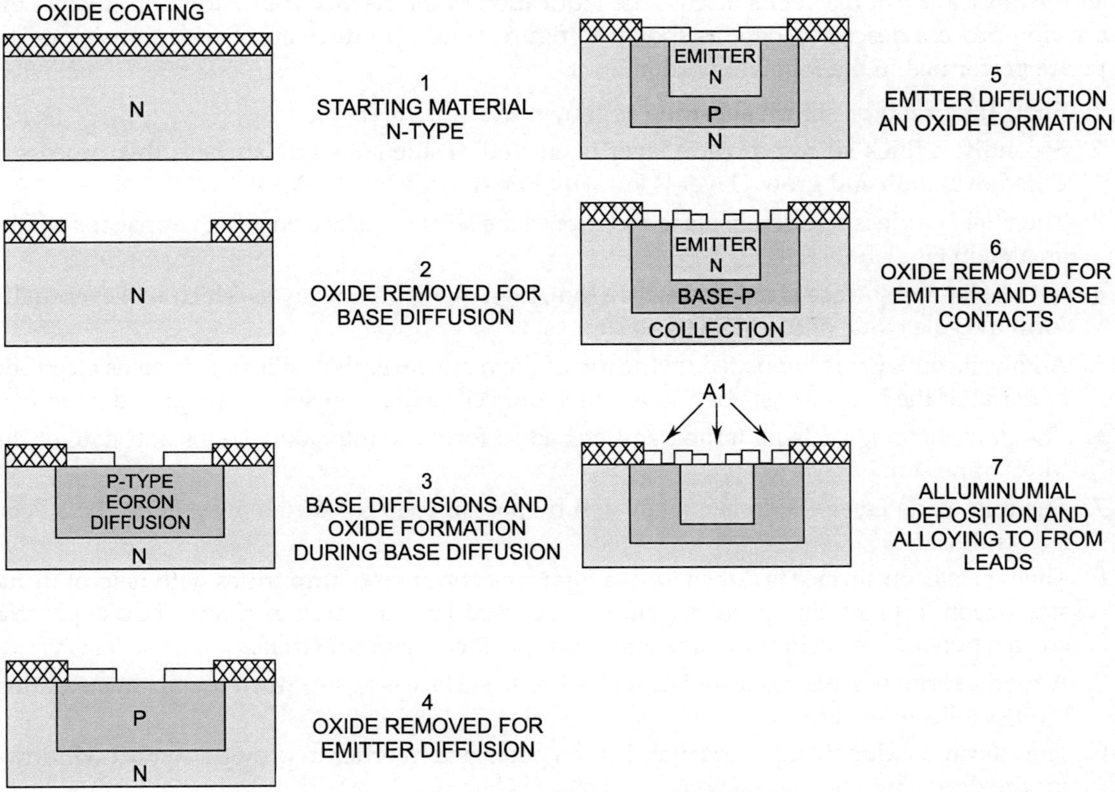

Fig. 1.9 *Planar-Diffused Transistor*

1. An oxide coating is thermally grown over the *n*-type silicon starting material.
2. By means of the photolithographic process, a window is opened through the oxide layer. This is done through the use of masks, as discussed earlier.
3. The base of the transistor is formed by placing the wafer in a diffusion furnace containing a *p*- type impurity, such as boron. By controlling the temperature of the oven and the length of time that the wafer is in the oven, you can control the amount of boron diffused through the window (the boron will actually spread slightly beyond the window opening). A new oxide layer is then allowed to form over the area exposed by the window.
4. A new window, using a different mask much smaller than the first is opened through the new oxide layer.
5. An *n*-type impurity, such as phosphorous, is diffused through the new window to form the emitter portion of the transistor. Again, the diffused material will spread slightly beyond the window opening. Still another oxide layer is then allowed to form over the window.
6. By means of precision-masking techniques, very small windows (about 0.005 inch in diameter) are opened in both the bass and emitter regions of the transistor to provide access for electrical currents.
7. Aluminium is then deposited in these windows and alloyed to form the leads of the transistor or the IC. (Note that the *pn*-junctions are covered throughout the fabrication process by an oxide layer that prevents contamination.)

1.8 N-MOS FABRICATION STEPS

After the fabrication of BJT let us discuss the fabrication of the N–Mos transistor. The steps in the fabrication process described here as shown in figure would produce an N–Mos transistor. The steps are performed in the following order.

1. First, Take a *p*-type silicon substrate as shown in Fig. 1.10(*a*)
2. Secondly, a thick silicon dioxide layer is created on the substrate surface, this process is called oxidation and grown layer is known as field oxide Fig. 1.10(*b*)
3. Then field oxide is selectively etched to expose the silicon surface on which transistor will be created in Fig. 1.10(*c*)
4. In this step the surface is covered with a thin, high quality oxide layer, which will eventually form the gate oxide of the MOS transistor.
5. A Polysilicon layer is deposited on the top of the oxide layer. Polysilicon is used as electrode material of the MOS transistor and an interconnect medium in silicon integrated circuits.
6. The polysilicon layer is patterned and etched to form the interconnections and gate of the MOS transistor.
7. The thin oxide layer which is not covered by polysilicon is also etch away upto the silicon substrate.
8. Then the silicon surface is doped with a high concentration of impurities with help of diffusion or ion Implantation process (will be discussed later in detail in chapter) these process creates two *n*-type regions (source and drain) in the *p*-type substrate.
9. After the formation of source and drain the entire surface is again covered with an insulating layer of silicon dioxide.
10. The silicon dioxide (insulating layer) is then patterned in order to provide contact windows for the drain and source junction.
11. Then interconnections will be form by covering the surface with evaporated aluminium.
12. Finally, the metal layer is patterned and etched and it completes the interconnections of the MOS transistor on the surface.

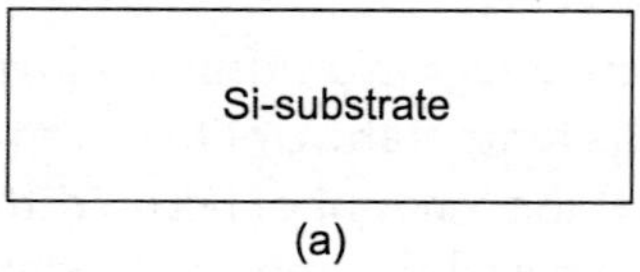

(a)

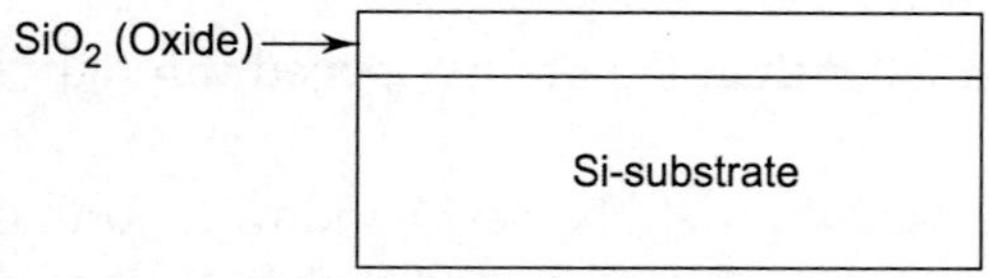

(b)

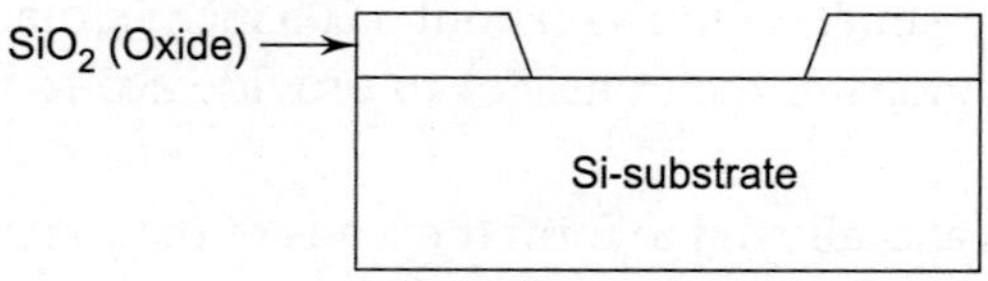

(c)

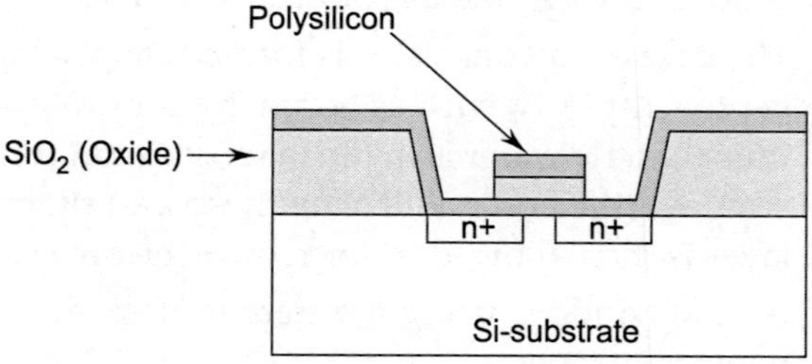

(h)

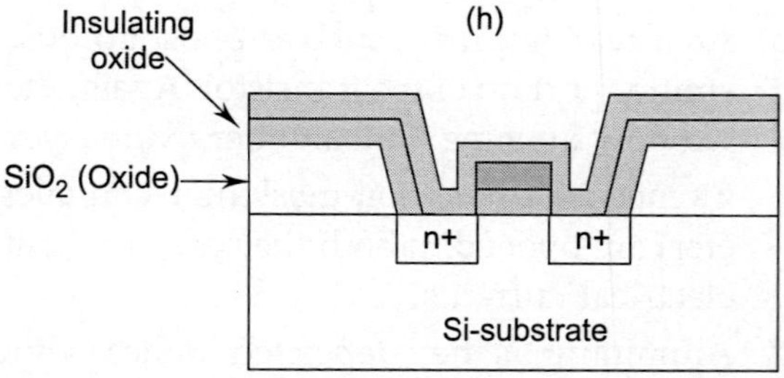

(i)

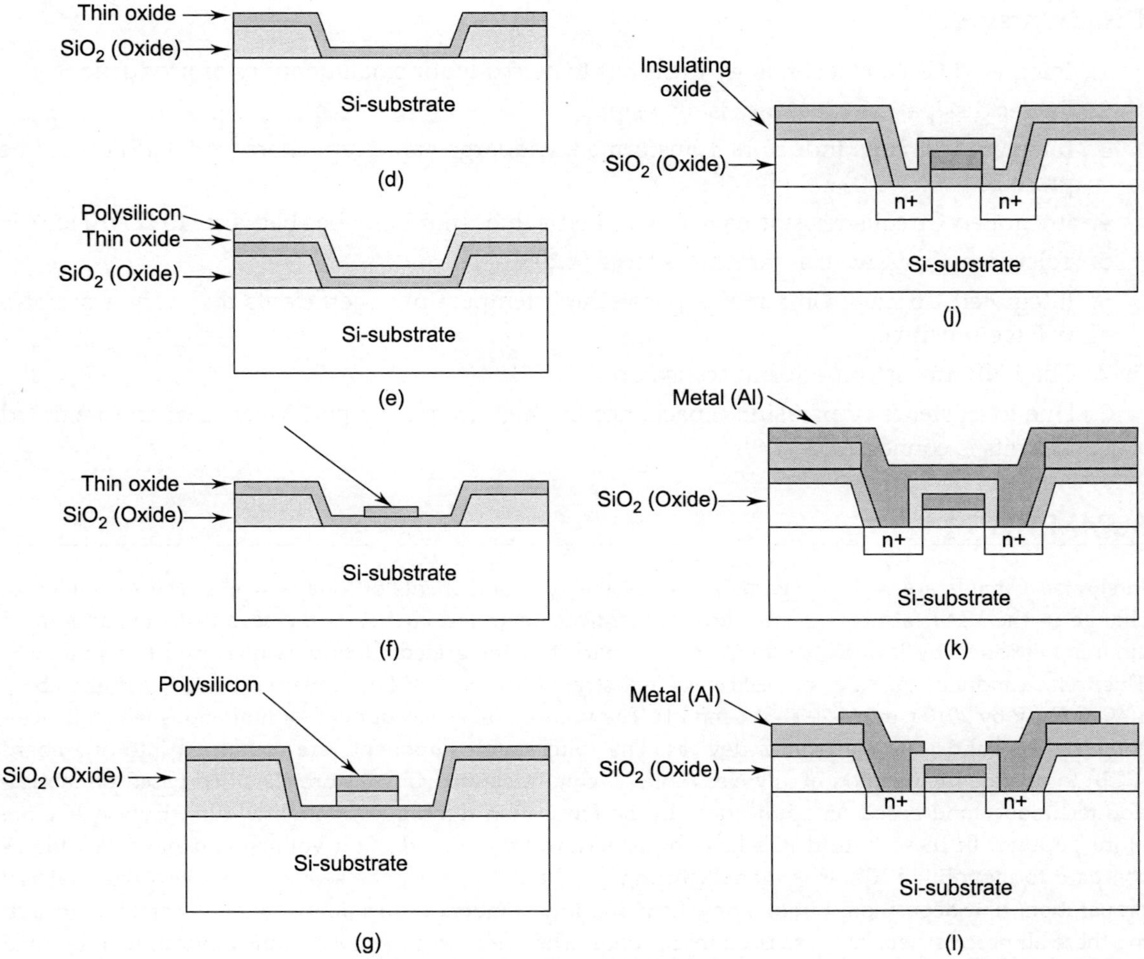

Fig. 1.10 *Fabrication Steps*

1.9 ADVANTAGES AND DISADVANTAGES OF INTEGRATED CIRCUITS

Advantages

1. An Integrated Circuits in much smaller in size and very light in weight as compare to vacuum tubes.
2. A large number of circuit components can be processes in a single operation resulting in reduced cost.
3. Integrated Circuits are more reliable because of the absence of solder joints due to improved on chip interconnections.
4. A complex circuits can be integrated with reasonable size and cost to give improved performance characteristics.
5. Integrated Circuits consumed less power for their operation.
6. Integrated Circuits required less testing at system level.
7. Integrated Circuits Amplifier or circuits provide better high frequency performance.
8. These circuits are very fast because of reduced length of the interconnecting lead.

Disadvantages

1. Integrated Circuits technology turns out to be costly for small quantity of production
2. Power dissipation capability is also small.
3. Integrated Circuits Inductors, transformers and large value capacitors (> 200 μF) cannot be produced.
4. Integrated Circuits resistor have the limited values (normally lies between 10 Ω - 10 kΩ).
5. Tolerance of passive components is large (± 20%)
6. Integrated Circuits components passes high temperature coefficients and values are also voltage sensitive.
7. The PNP transistor performance is poor.
8. Due to existence of parasitic capacitance the high frequency performance of an Integrated Circuits is connected.

SUMMARY

Integrated Circuits are widely regards as one of the key components of today's world. The revolution of change in the VLSI fabrication and development of integrated circuit, is the result of first Integrated circuits fabricated by Jack Kilbey in 1958. The market of Integrated Circuits is increased tremendously. This trend continues the sales of electronics industry will reach 3 trillion dollar and will constitute about 10% of GWP by 2010 (source 2000 SIA report). The vacuum tubes has number of limitations which is overcome by the solid state electronics devices. The solid state components are combined into one board (PCB) to operate the function of any electronics circuit. Integrated Circuits are classified based on fabrication techniques and based on application. In the fabrication techniques monoliths film, Hybrid IC's are more popular. In these hybrid IC's have high performance, speed, high voltage and more flexible as compare to monolithic IC's. The manufacturing of IC's are done by the sequence of steps like material preparation, film deposition. Lithography, Etch, ion implantation, diffusion, oxidation, assembly and testing these all process steps are discussed in this book. The basic steps involved in the fabrication of BJT and N-MOS are discussed in the last section of the chapter.

TUTORIAL SHEET

1. Who was Jack Kilbey? Discuss the contribution of Jack Kilbey in the VLSI Revolution.
2. Discuss the disadvantage of vacuum tube equipment. Why solid state devices are needed?
3. What do you understand by PCB? What is the role of PCB in evolution of electronics field?
4. What do you understand Integrated Circuits? How will you classify the IC's?
5. What is monolithic IC's?
6. Differentiate between Thick film and Thin film Integrated Circuits with help of example.
7. How Hybrid IC's are different from other IC technology. Discuss the advantages of hybrid IC's.
8. Different between digital IC's and analog IC's.
9. Draw and discuss the manufacturing flow of Integrated Circuits.
10. Discuss the fabrication steps of Bipolar junction transistor.
11. Discuss the fabrication steps of N-MOS fabrication steps.
12. Write advantage and disadvantages of Integrated Circuits.

2

ENVIRONMENT FOR VLSI TECHNOLOGY : CLEAN ROOMS

2.1 INTRODUCTION

VLSI circuits are developing in such a way that makes both production tool and the process environment, increasing critical for the manufacturing process. The issue is above all, to increase the yield. This yield can be increased by the control of wafer and surrounding environment. The environment affects the yield. Some of these environmental factors are:

- Temperature Control
- Relative Humidity
- Electrostatic Discharge
- Airborne Particles
- Chemical Contamination
- Electromagnetic field
- Oxygen and Vibrations.

To eliminate micro-contamination and reduce native oxide growth on silicon wafers, the wafer processing and loading/unloading sections of process tools are enclosed in an extremely high cleanliness min-environments are flashed with ultra pare nitrogen containing no oxygen and moisture. Reducing defect density is most important aspects of increasing yield. Defects density is defined as the number of defects per cm^3 wafer that occurs during the processing due to all kind of contamination. These particles are responsible for over 80% of defects.

2.2 TYPES OF CONTAMINATION

The industry started with a clean room technology developed by the space industry. However, these techniques were not sufficient for VLSI chips. In VLSI, many chemicals are used during fabrication. How to clean chemicals after every process is more important and some more challenges for contamination control are material and construction knowledge. The ability of the industry to grow has been dependent on keeping up solving the contamination presented by each generation of chips. If industry ignores these minor problems of contamination that will become the killer defects of tomorrow's chip. These contaminants are divided into four types.

1. **Particles Contaminants:** Semiconductor devices are sensitive to all kind of contaminants, the sensitivity is due to the small feature size and the thinness of deposited layer on the wafer surface. One micron is very small value. 1 cm = 10,000 μm, for more clearing let us take the example of human hair, a human hair diameter is about 100 μm as shown in Fig. 2.1.

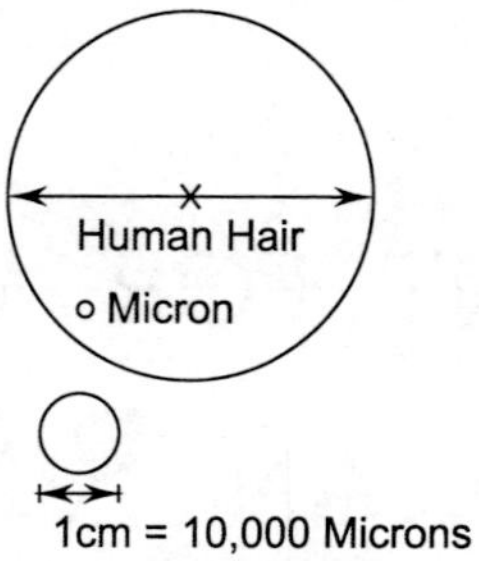

Fig. 2.1 *Relative Size of One Micron*

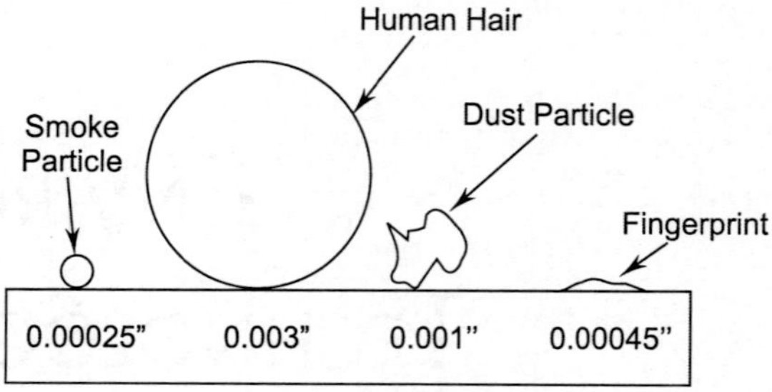

Fig. 2.2 *Size of Contamination*

As shown in Fig. 2.2, some more examples of contaminants smoke particle, dust particle, finger print which effects the performance of the devices. These particles come with the workers, generated by the equipment and present in processing steps. As feature size and films become smaller, the allowable particle size must be controlled to small dimensions as shown in Fig. 2.3. Two metallic lines are passing in parallel and some particle dust particles are in between these metallic lime, it may short circuit. If it falls on the oxide layer then it also effects the performance of the devices. Particles that locate in a critical part of the devices and destroy its functioning are called killer defects. Killer defects also include crystal defects and other process induced problems.

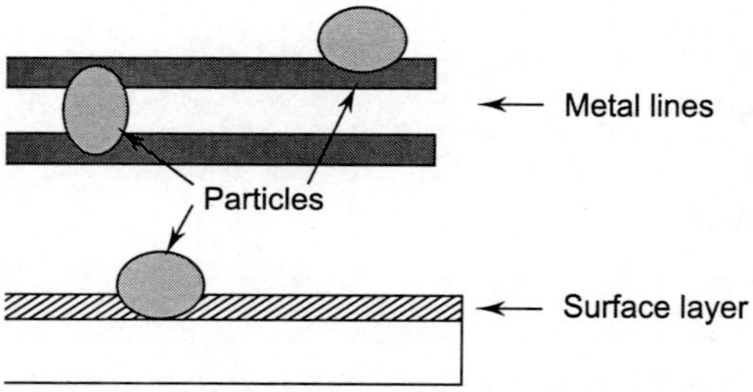

Fig. 2.3 *Size of Airborne Particles*

2. **Metallic Ion Contaminants:** In semiconductor devices has very important parameter is controlled resistivity of wafer. For example, in the case of P-N-Junction. Resistivity is controlled at 3 points : at doped N-region, P-region and in P-N Junction. These resistivities can effect with small amount of dopent. If a very small amount of electrically active contaminants present in the wafer.

   It can change the performance and reliability parameters of the devices. Contaminants causing these types of problems are known as Mobile Ionic Contaminants (MIC). They are atoms of metal exist in the ionic form in the material. These ionic atoms are highly mobile in the semiconductor material, metallic ions moves in the device. It causing the device to fail, even after passing through the various electrical testing this problem is most serious in MOS devices. Some metallic ions are : Sodium, Potassium, Iron, Copper etc.
3. **Chemicals Contaminants:** During the preparation of wafer unwanted chemical effects the area of wafers. Process water and process chemicals can be contaminated with trace chemicals that interface with the water processing. Due to these chemicals unwanted etching will

take place and some compounds are formed during process reaction that cannot be removed from the devices or cause non-uniform processes. For example, chlorine is such a contaminant which causes non-uniform processes.

4. **Bacteria Contaminants:** This type of contamination take place when surface of wafer or devices contain water particles and are not cleaned properly. Because bacteria are organisms that grow in water systems. If once the bacteria grows on the surface, it may contribute unwanted metallic ions to the devices surface.

2.3 PROBLEMS DUE TO CONTAMINATION

The contaminants discussed above affects the processing and performance of the devices. The three specific performance areas are:

Yield: The dimensions of the device part may change due to contamination and some time dust particle adhere to the surface of a photo mask, they behave as opaque patterns on the mask. These patterns will be transferred to the underlying layer along with circuit patterns on the mask. When the high level of contamination are checked it results only a few wafers complete the process. It results only good devices are very few so yield become poor and it becomes the reasons of high cost.

Reliability: Reliability is one of the important issue in the VLSI Technology. Specially, when the devices are designed and fabricated for the space and defence industry. Sometimes, small amount of metallic ions can get into the wafer during processing and these contaminants cannot be detected during the normal testing. However, in the field, these ions can travel inside the device and end up in the electrically sensitive area. It may cause failure of the device.

Performance: At the post fabrication level there is one wafer sort electrical test, in which the device failure is detected. This failure effects the performance of the device. It is due to the undetected particulates, unwanted chemical and high level of mobile ionic contaminants in the wafer, which can change the electrical performance of the device.

2.4 CLEAN ROOM

A clean room is an environment, typically used in manufacturing or scientific research, that has a low level of environmental pollutants such as dust, airborne microbes, aerosol particles and chemical vapours. More accurately, a cleanroom has a *controlled* level of contamination that is specified by the number of particles per cubic meter at a specified particle size. To give a perspective, the ambient air outside in a typical urban environment might contain as many as 35,000,000 particles per cubic meter, 0.5 um and larger in diameter, corresponding to an ISO class 9 clean room. It solves the above-mentioned problems.

Clean rooms can be very large. Entire manufacturing facilities can be contained within a clean room with factory floors covering thousands of square meters. They are used extensively in semiconductor manufacturing, biotechnology, the life sciences and other fields that are very sensitive to environmental contamination.

2.4.1 Clean Room Features

The main features of clean rooms are:

- It provides filtered air (99.9% at 0.15 mm particle size)
- It provides overpressure (50 p.a.) for kept the particle outside.
- It minimizes the mechanical and electrical interface.
- It provides laminar (unidirectional) air flow in working area.
- It controls the heating and cooling of the incoming air.
- It also provides feasibility to control humidity of incoming air from outside.

2.4.2 Clean Room Standards

Clean rooms are classified on the basis of particle counts. Some ISO standards are given in table (2.1 & 2.2) these standards employs units of particle per cubic meter (conversation factor $1\,m^3 = 35.5\,ft^3$) according to US FED STD 209 E clean room standards. And ISO standard of cleanliness class N with particle concentration C_n (particle/m^3) is calculated as

$$C_n = 10^N \times (0.1\ \text{mm/D})^{2.08}$$

where, D is particle size in micro meters.

The proper way to specify clean room is therefore class x (at y mm particle size)

The example shows in Table 2.2 that there are a magnitude of clean room feature in addition to particle specification. These are related to air quality plus mechanical and electrical environment.

The clean room standards tables indicate the maximum number of allowable particles of the indicated size per cubic foot (for US FED) or cubic meter (for ISO) for each class.

Table 2.1: US FED STD 209E Clean Room Standards Particle/ft^3

Class	*0.1 μm*	*0.2 μm*	*0.3 μm*	*0.5 μm*	*5 μm*
1	35	7	3	1	
10	350	75	30	10	
100		750	300	100	
1,000				1,000	7
10,000				10,000	70
100,000				100,000	700

The above mention standards are not valid in these days:

Table 2.2: ISO 14644-1 Clean Room Standards Particle/m^3

Class	*0.1 μm*	*0.2 μm*	*0.3 μm*	*0.5 μm*	*1 μm*	*5 μm*
ISO 1	10	2				
ISO 2	100	24	10	4		
ISO 3	1,000	237	102	35	8	
ISO 4	10,000	2,370	1,020	352	83	
ISO 5	100,000	23,700	10,200	3,520	832	29
ISO 6	1,000,000	237,000	102,000	35,200	8,320	293
ISO 7				352,000	83,200	2,930
ISO 8				3,520,000	832,000	29,300
ISO 9				35,200,000	8,320,000	293,000

Table 2.3: Clean Room Class Comparison

ISO 14644-1	*FED STD 209E*
ISO 3	1
ISO 4	10
ISO 5	100
ISO 6	1,000
ISO 7	10,000
ISO 8	100,000

2.5 DESIGN AND CONSTRUCTION OF CLEAN ROOMS

Design and construction of clean rooms is very tedious job. Some special equipments and contamination control strategies are discussed and installed in the clean rooms and placement of equipments, flow of the air, humidity, temperature, ESD (Electrical Static Discharge) are decided and installed with very technical skills. In the latest versions, air filtering is consist HEPA (High Efficiency Particulates Efficiency) in the ceilings and returns in the floor to give a continuous flow of clean air. The work stations are table tops with perforations to allow the filtered air to pass through the table top. One typical clean room is shown in Fig. 2.4. Some contamination control systems are:

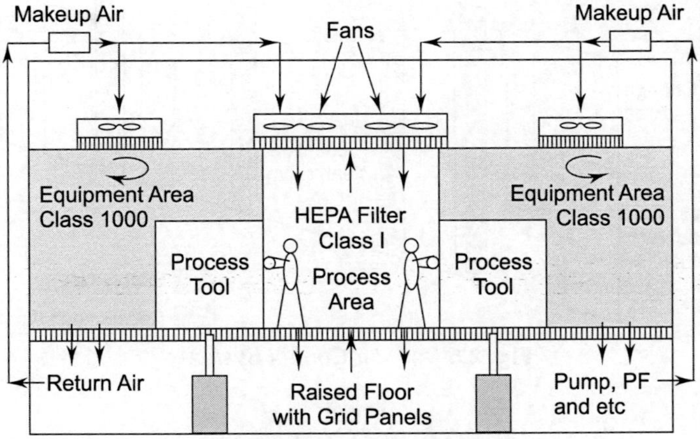

Fig. 2.4 *Clean Room Design*

(*i*) Air control system
- → HEPA Filters
- → TUNNEL / Bays
- → VLF Hoods

(*ii*) Di-Water system
(*iii*) Gas systems
(*iv*) Temperature, Humidity and smog control
(*v*) Personal generated contamination control
(*vi*) Electro Static Discharge (ESD) Control.

2.5.1 Air Control System

The design of a clean room is integral to its ability to produce contaminant free wafers. The early semiconductor industry evolved from ceilings and wall filters. Concept to clean room work stations. In this strategies focussed on individual work stations with air filters as shown in Fig. 2.5(*a*).

2.5.2 High Efficiency Particulate Attenuation (HEPA) Filters

The fabrication laboratory is consist of large rooms with the work stations called Hoods. These hoods are arranged in rows so that the wafer could move sequentially through the different processes without exposed to dirty air. The filters which keep clean these work stations by filling the outer air are called high efficiency particulate attenuation (HEPA). These filters are constructed with fragile fibres with many small holes (as shown in Fig. 2.5(*b*)) and folded into the filter holder. The high density of small holes and large area of the filter allow the large volume of air to pass at low velocity. Operator also feel comfortable at low velocity of air when they are working on these work stations.

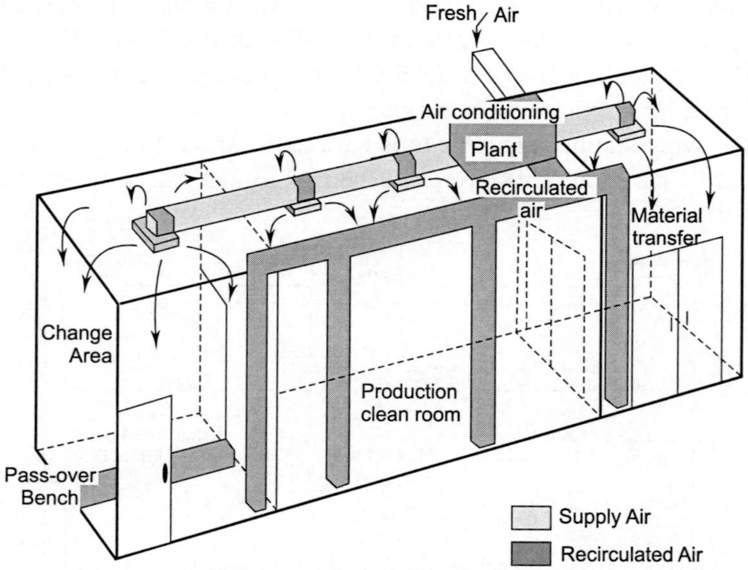

Fig. 2.5 (*a*) *Air Control System*

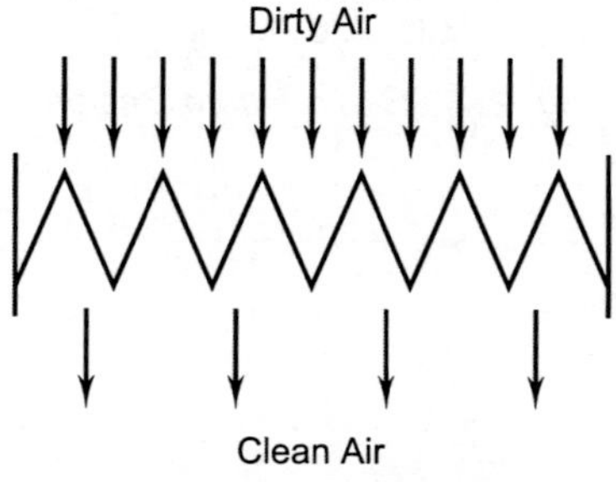

Fig. 2.5 (*b*) *HEPA Filter*

2.5.3 Verticle Laminar Flow (VLF)

The air is taken inside the clean room through a pre filter fan and forced through the HEPA filters. The air leaves the filters in the laminar pattern and at the work surface turns and exit the work station. These type of work stations are called verticle laminar flow stations. This type of VLF station is shown in Fig. 2.6. Some more special type of VLF stations are designed specially for the chemical processes. These stations are connected to an exhaust system to remove the chemical fumes. The fumes can pose both safety and contamination problems, but care must be taken to balance VLF air and exhaust to maintain the required class number in the station.

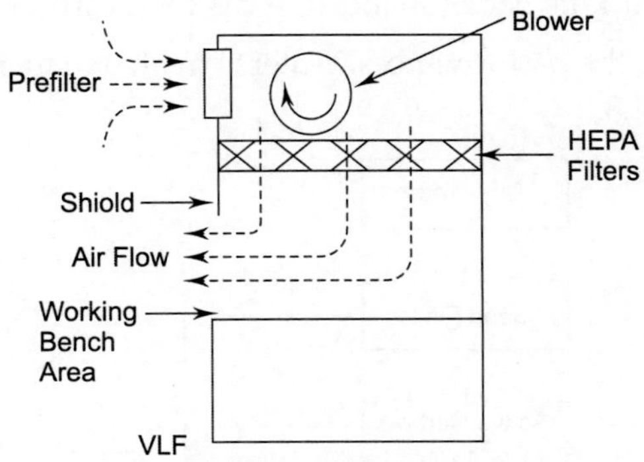

Fig. 2.6 *Verticle Laminar Flow*

2.5.4 Tunnels/Bay Concept

When people entering and existing can contamination, all the processes in the process station. This problem is solved by dividing the fabrication area into separate tunnel or bays as shown in Fig. 2.7. In this type of arrangement, the filters are fitted in the ceiling and serve the same objective which is served by individual VLF station and one more advantage of it is that the filtered air from the ceiling kept the wafers clean and less vulnerable to the worker-generated contaminations. This is more costly and not too much versatile when the process is changed.

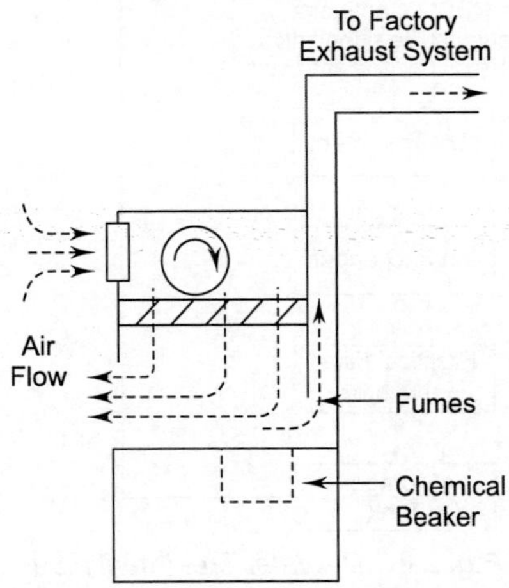

Fig. 2.7 *Tunnels/Bays Concept*

2.5.5 Di-Water System

During the fabrication processes a wafer will etched and clean many times with chemicals. After chemical cleaning, the etching wafers are cleaned with water and rinse with water many times. In modern integrated circuits fabrication laboratory uses a million cubic metres of water in a day, but the water coming from the city is not fit for these cleaning processes because it contains unacceptable amount of contaminations. These contaminants are particulates, dissolved minerals, bacteria, organism, dissolve oxygen and silicon.

In the normal water contain some salts like NaCl, this salt breaks into Na^+ and Cl^- ions. These ions are very harmful for the device. They are removed from the water by Reverse Osmasis (RO) and Ion Exchange System. The RO is a process in which water molecules, diffuse through a porous membrane, while micro-organisms, particles and ions are rejected. Some steps for the water treatments : Ultra violet treatment kills bacteria and reduces total carbon contents.

Both UV and RO improves the performances. Water is treated in many step as represented in the flow diagram shown in Fig. 2.8.

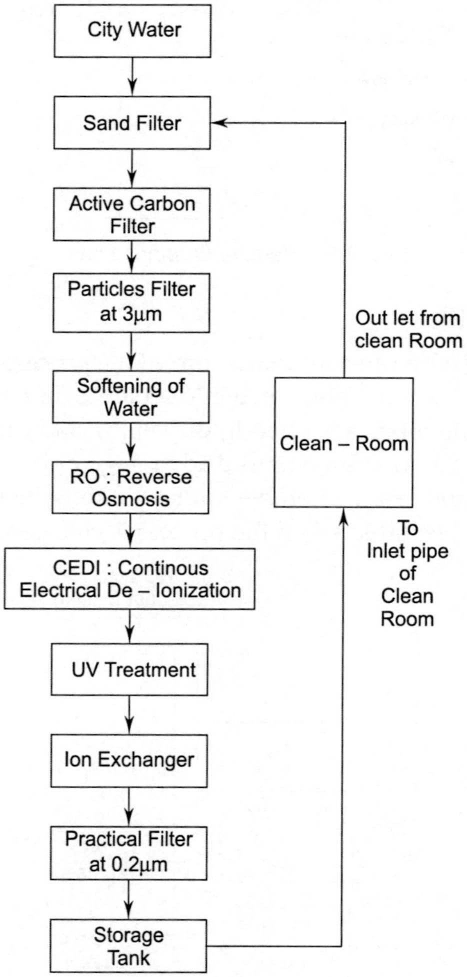

Fig. 2.8 *Di-Water Treatment Flow*

The process of removing the electrically, active ions changes the water from conductive medium to resistive one. This fact is used to improve the quality of de-ionized or ultra pure water (UPW), (DI) water. DI water quality is monitored by resistivity measurement 18 M ohm-cm is required. Regular bacteria checks as well as particle test are performed.

2.5.6 Gas Systems

The semiconductor wafer is processed with many gases at different stages like in oxidations. Sputtering, plasma, etch, chemical vapour deposition etc., These gases are nitrogen, oxygen, arsine and carbon tetra flouride.

To maintain the purity gas in the fabrication labs is very challenging job. The gas is mostly pass through the piping system' a gas panel containing valve and flow meters and connections to the tool: Leakage at any part is disastours. Outside air can enter into process gas and react with it and change the composition and the desired reaction in the process gas and react with it and changes the composition and the desired reaction in the process.

A typical system designed with the features stainless steel, piping valve along with some polymer components such as connectors and seals. For ultra clean systems, the stainless steel surface have electro polished and double vacuum melt inside surface to reduce.

2.5.7 Temperature, Humidity and Smog Control

Temperature Control: It is necessary to control for process for operator comfort. In many wet processes of etching and cleaning take place is non-temperature control baths and rely on the room temperature for control, the typical temperature range is (72° F ± 2° F).

Humidity Control: It is a critical process parameter. In the pattering area, thin films of a polymer are put on the wafer to act etch stencil. If the humidity is too high, the wafer collect moisture, preventing the polymer from sticking, the situation is the same as applying point on a wet surface.

Smog: It is another airborne contaminant in clean room. Again the problem is most critical in the pattering areas. Ozone, is filtered out by installing carbon filters in the incoming air ducts.

2.5.8 Personal Generated Contamination

Clean room personal are the biggest source contamination. A person can give off 100,000 and 1,000,000 particles even after showering. The particles come from died hair and normal skin flaking and additional particle sources are hair sprays, cosmetic, facial hair and expressed clothing like wooler and cotton clothes. Even human's breath also contain high level of contaminates, the breath of smoker carries million of particles for long time after a cigarette is finished (as shown in Table 2.4).

The material of clothing selected for clean room depends on the level of cleanliness required. The material is non-shedding and have ability to draw off the static charge of the body. Mostly the polyster and GORETEX material is preferred to make dresses. The dresses are dry-cleaned after wearing once.

Table 2.4

Normal Breath	*No Particles*
Smoker's Breath after smoking	500%
Sneezing	2000%
Sitting Quietly	20%
Rubbing Hands on Face	200%
Walking	200%
Stamping Foot on Floor	5000%

Some steps are taken in consideration to control the contamination produced by personal contacts.

1. Every part of the body is covered up. The head will have an inner cap that keeps the hair in place. This is covered by an outer shell that is designed to fit close to the face and has snaps or a tail for security the head gear under the body covering smock.
2. The Eye, which are also major source of fluid particles and for the contamination control the covering glasses (usually safety glasses) with side shidds are used.

3. Body covered with over suit also legs, arms and necks well designed suit will cover over zippers and no pockets.
4. Hands are covered with at least one pair of gloves, most favoured PVC gloves are medically type, that permits good PVC factile feeling ; gloves should be pulled up over the sleeves to prevent contamination from travelling down the arm and into the clean room.
5. Skin Flacking can also be controlled with use of special type of lotions that moisturise the skin. These lotions must be free from sodium (Na) and chlorine (Cl).

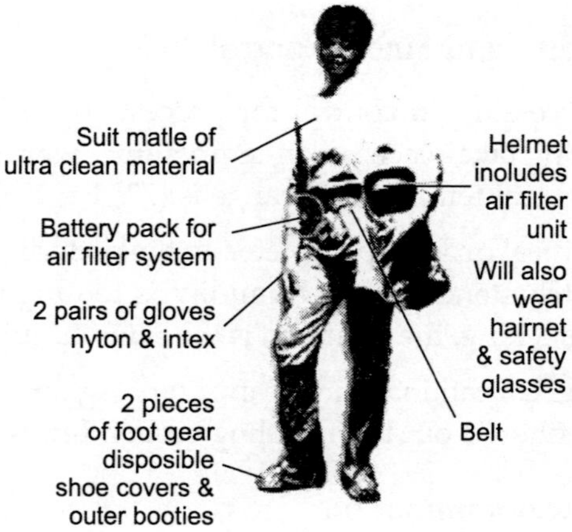

Fig. 2.9 *Dress for Fabrication Lab*

The figure shown in 2.10 shows layout of a typical fabrication laboratory. During the construction some methods are used to control dirt from the outside of the laboratory. Some of these methods are:

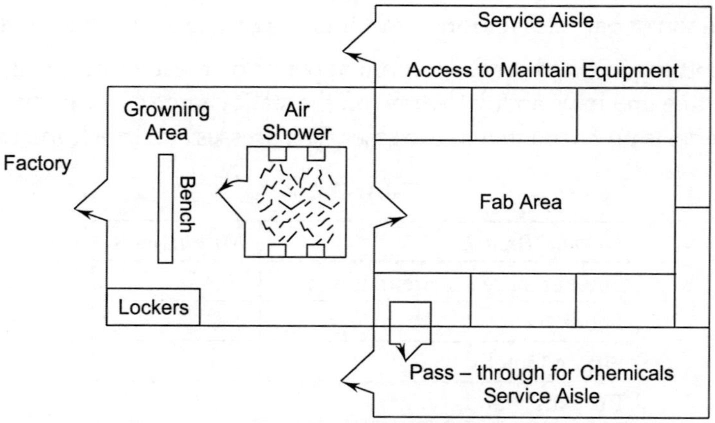

Fig. 2.10 *Typical Fabrication Lab Area*

1. **Adhesive Floor Mats:** Adhesive floor mats are used at the main entrance of laboratory. These mats pulls and hold dirt adhering to the bottom of shoes.
2. **Change Room (Growning Room):** This area consist of HEPA filters in ceiling which provide filtered air. This area also consist of lockers to store the operators shoes and cloths which are not permitable inside the fabrication area.
3. **Air Showers:** These air showers located between growning room and a clean rooms. Operators of the fabrication laboratory enter through the air shower cabin, where high velocity air jets off particles on the outside of garments.

4. **Double Door System:** This door system is used to protect the cleanliness of the clean room. When some material is supplying from outside to the laboratory. It will kept at this place whose HEPA filters and air showers clean the material. Both the doors should not be opened at same time.
5. **Shoe Cleaner:** To remove the dirt from the sides of shoes, a rotating brushes are at the entrance which remove the dirt and vacuum system captures the loosened dirt.
6. **Gloves Cleaner:** After every shift the gloves used are discards. But these gloves are very costly. So in some fabrication labs gloves cleaners are used to clean the gloves.

After the all these precautions and contamination controls methods. There are some more areas where the contaminants can affect the performance of the device. These areas are during the fabrication processes. So, it is also necessary for the performance of devices to clean the fabrication equipments ; time to time and clean the wafer at different stages of fabrication. So, let us discuss some wafer cleaning techniques in next section.

2.5.9 ESD Control Systems

Static electricity is an electrical charge that is at rest. This is mainly created by an imbalance of electrons that stay on a specific surface, or in the environmental air. The imbalance of electrons (in all cases, is caused by absence or surplus of electrons) thus causes an electrical field that is capable of influencing other objects at a distance.

The level of charge is affected by material type, speed of contact and separation, humidity and several other factors. The **ESD** effect can be seen easily in everyday life, while it could always been hardly detected

Sources of static electricity

Triboelectric charging occurs when two materials are separated after coming into contact with each other or any rubbing activities. The charging thus takes place during the transfer of electrons from one material to another. Triboelectric charging can happen between any materials such as solids, liquids and air particles.

Field Induction: Whenever an object is electrostatically charged, an electrical field associated with that charge is created around it. Once an ungrounded sensitive device enters the electrical field, a charge is induced on the device, causing a surging transfer of charges between the two bodies. This transfer of charges thus resulted in catastrophic failures, which leaves the device destroyed permanently.

Electrostatic discharge can change, degrade or destroy the electrical characteristics of electronic devices such as integrated circuits and components. Therefore, effective static control is crucial, in order to protect products from undesirable damages

There are three predominant ESD models for IC's:

1. Human Body Model (HBM);
2. Charged Device Model (CDM); and
3. Machine Model (MM).

The HBM simulates the ESD event when a person charged either to a positive or negative potential touches an IC that is at another potential. The CDM simulates the ESD event wherein a device charges to a certain potential and then gets into contact with a conductive surface at a different potential. The MM simulates the ESD event that occurs when a part of an equipment or tool comes into contact with a device at a different potential. HBM and CDM are considered to be more 'real world' models than the MM.

Methods to control ESD in Clean Room

Static Control: Many instances, people at work are one of the key generators of static electricity. The simple act of walking or repairing a PCB is sufficient to generate thousands volts on the body. It is obvious that personal grounding is the first step to effective static control.

Packaging: Packaging materials such as Static Shielding bags, Conductive bags, ESD containers and boxes provide direct protection of devices from electrostatic discharge. The principal use of these packaging materials is to protect the product when it leaves the ESD protected facility. The main ESD function of these packaging materials is to eliminate or minimize the possible impact of electrostatic discharge created from triboelectric charging, direct discharge and electrostatic induction fields.

Wrist Straps: When proper worn and grounded, a functioning wrist strap keeps the human body near ground potential, thus preventing hazardous discharge between bodies and objects. Wrist straps allow safe dissipation of charges from the body to ground.

ESD Flooring and Footwear: A good combination of ESD floor materials and proper footwear provides a grounding path for dissipating electrostatic charges generated during walking. The use of ESD floor materials is particularly appropriate in those areas where increased personnel mobility is essential.

ESD-Safe Garments: Clothing is an important consideration in most ESD protective areas, particularly in dry environments. Grounded static control garments are recommended to minimize the effects of electrostatic fields or charges that may exist on a person's clothing.

Ionizers: Air ionization can neutralize the static charges on insulated and isolated objects by charging the molecules of the gases in the surrounding air. Static charges that exist on any surface can be neutralized by attracting opposite polarity charges from the air.

ESD-Safe Workstations: Proper ground of workstations plays an important role in protecting devices from electrostatically induced damages. ESD table mats, grounding cords and awareness signs are key elements in an electrostatic protective workstation shown in Fig. 2.11.

Fig. 2.11 *ESD Controlled Work Station*

2.6 WAFER CLEANING TECHNIQUES

There are a number of wafer cleaning techniques or steps employed to ensure that a semiconductor wafer is always free of contaminants and foreign materials as it undergoes the wafer fabrication process. Different contaminants have different properties and therefore, have different requirements for removal from the wafer. Below are some examples of commonly-used methods for wafer cleaning.

1. **Photoresist Stripping:** Photoresist stripping, or simply 'resist stripping', is the removal of unwanted photoresist layers from the wafer. Its objective is to eliminate the photoresist material from the wafer as quickly as possible, without allowing any surface materials under the resist to get attacked by the chemicals used. Resist stripping can be classified into : (*a*) Organic stripping ; (*b*) Inorganic stripping and (*c*) Dry stripping.
   (*a*) Organic stripping employs organic strippers, which are chemicals that break down the structure of the resist layer. The most widely-used commercially available organic strippers used to be the phenol-based ones, but their short pot life and difficulties with phenol disposal made low-phenol or phenol-free organic strippers the more popular choice now-a-days.
   (*b*) Wet inorganic strippers, which are also known as oxidizing-type strippers, are used for inorganic stripping, usually to remove photoresist from non-metallized wafers, as well as post-baked and other hard-to-remove resists. Inorganic strippers are solutions of sulphuric acid and an oxidant (such as ammonium persulfate), heated to about 125 degree C.
   (*c*) Dry stripping pertains to the removal of photoresist by dry etching using plasma etching equipment. Its advantages over wet etching with organic or inorganic strippers include better safety, absence of metal ion contamination, decreased pollution issues and less tendency to attach underlying substrate layers.
2. **Chemical Removal of Film Contaminants:** Chemically bonded films of contaminant material may be removed from a wafer surface by chemical cleaning. Chemical cleaning comes in various forms, depending on the nature of the film contaminants that need to be removed and from which surface. In general, however, chemical cleaning is performed with a series of acid and rinse baths. As an example, removal of film contaminants from a wafer with nothing but thermally grown oxide may consist of the following steps : preliminary cleaning, removal of residual organic contaminants and some metals, stripping of the hydrous oxide film created by the previous step, desorption of atomic and ionic contaminants and drying. Storage of cleaned wafers must be avoided but, if necessary, must be done using closed glass containers inside a nitrogen dry box.
3. **Sputter Etching of Native Oxide Films:** A thin oxide layer grows over silicon or aluminium when these are exposed to air, forming SiO_2 and Al_2O_3, respectively. These 'native' oxide layers need to be removed in places where they should n't be, because they exhibit adverse effects such as higher contact resistance or hampered interfacial reactions. In-situ sputter or plasma etching are the techniques commonly utilized to remove these native oxides from the wafer. "In-situ" means performing the sputter or plasma etching in the same vacuum environment, where the overlying layer will be deposited.

4. **Elimination of Particulates:** The contamination of wafers with insoluble particulates is also a common problem. There are two frequently utilized techniques for removing particulates from a wafer: (*a*) ultrasonic scrubbing (*b*) a combination of mechanical scrubbing and high-pressure spraying.
   (*a*) **Ultrasonic scrubbing** consists of immersing the wafer in a liquid medium which is supplied by ultrasonic energy. The sonic agitation causes microscopic bubbles to form and collapse, creating shock waves that loosen and displace particles. Ultrasonic scrubbing requires a filtration system that removes the particles from the bath as they get detached. One drawback of ultrasonic scrubbing is that they can cause mechanical damage to substrate layers.
   (*b*) **Mechanical scrubbing** employs a brush that rotates and hydroplanes over a solvent applied on the wafer surface. This means that the brush does not actually contact the wafer, but the solvent moved by the rotating brush dislodges particles from the wafer surface. As this happens, high-pressure spraying of a D/I water jet over the wafer surface is done to help in clearing the wafer surface of particulate contamination.
5. **Water Rinsing:** Rinsing performs the functions of removing the cleaning chemicals from the surface and ending action of oxide etch. De-Ionized (DI) water is used to achieve the goal.
   (*a*) **Over flow:** A box containing the wafers, put into the water tank. DI water is brought into the bottom of the box and flow through and around the wafers and existing over a dam into the drain system. It cleans the oxide and other contaminations from wafers as shown in Fig. 2.12.

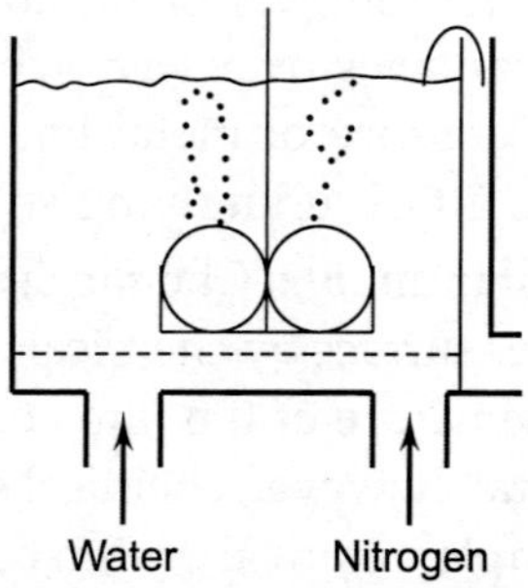

Fig. 2.12 *Rinse Systems Flow*

   (*b*) **Spray rinsing:** By a dilution mechanism, flowing water removes a water soluble chemical from the surface of wafers. When the water flows from the top layer of the wafer chemical dissolves into it and carried away to outlet of the furnace. The rinsing rate can increase with water flow rate, which can remove the dissolved chemical more quickly.
   (*c*) **Dump rinsers:** This type of rinser have the characteristics of both dump and spray rinser, means an overflow rinser, but with spray capability. The wafer is placed into the dry rinser and immediately sprayed with di-oxized water. While they are sprayed, the cavity of the rinser is rapidly filled with water. As the water overflow the top, a trap door is the bottom swing and open then the water is instantly exits to the drain system. This process is repeated several times until the wafers are entirely rinsed.

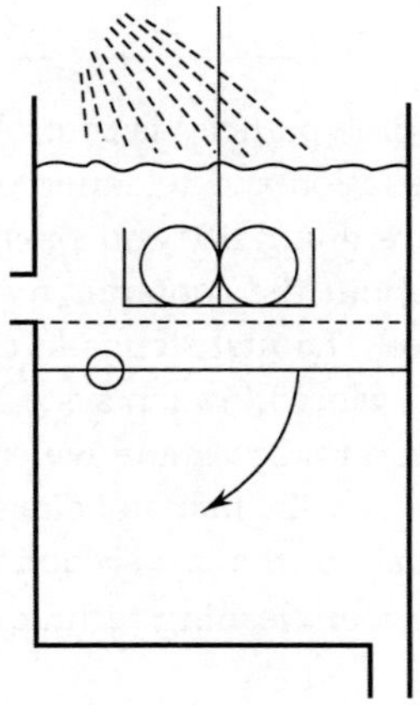

Fig. 2.13 *Spray-Dump Rinser*

(*d*) **Ultrasonics cleaning and rinsing:** A UV waves are generated from the outside of the tank with help of transducer as shown in the figure. The range is (20,000 –50,000 Hz) UV assists rinsing through cavitation, the wave pass through the liquid causing microscopic scrubbing action and disloges particles. The phenomenon is called cavitation. The velocity of rinse channel increases the cleaning efficiency.

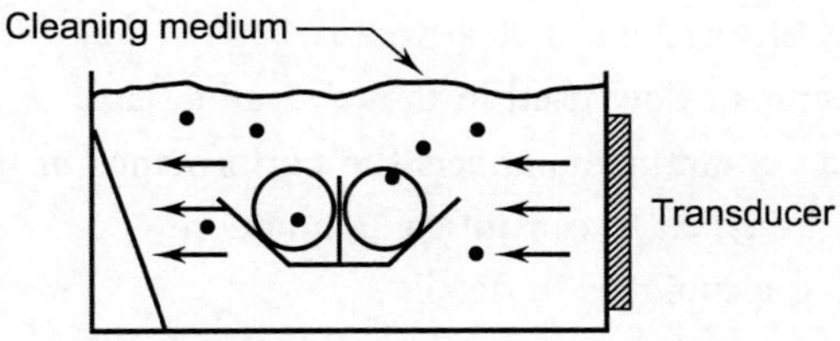

Fig. 2.14 *Ultrasonics Cleaning*

6. **Spin Rinse Dryers (SRD):** The processes starts with a rinse of the wafers and also after the rinse, it is necessary to remove the small droplets of water from the wafer, for this purpose some drying methods are used these type of dryers, the wafer boats are put in holders around the inside the drum. In the center of the drum is a pipe with holes that is connected to a source of deionized water and hot nitrogen as shown in Fig. 2.15.

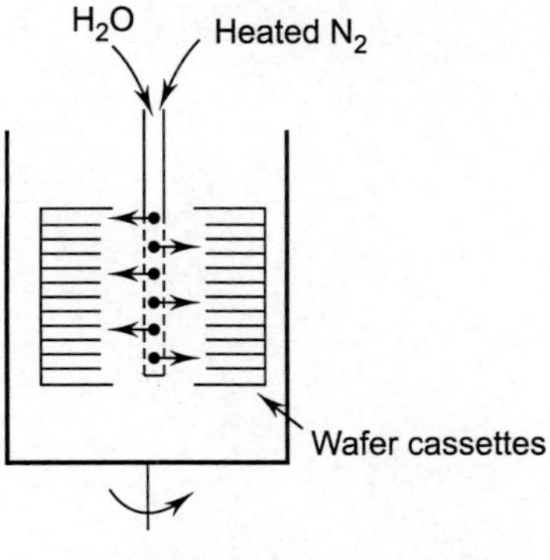

Fig. 2.15 *Spin Rinse Dryers*

SUMMARY

A room in which the concentration of airborne particles is controlled and which is constricted and used in manner to minimize the introduction, generation and retention of particles inside the room and in which other relevant parameters *e.g.*, temperature, humidity and pressure are controlled as necessary. In first part of the chapter, we have discussed about the contamination types and sources of contaminations which affect the performance of the devices. The first step is to clean the environment around the wafers. Most commonly this is done through the advance filtration systems to remove particle like HEPA, VLF and other filters. Second the contamination is control by careful wafer cleaning and third is the human contact control systems like Gloves, shoes, and dress of the human beings used in fab labs can be useful to control the contamination flow into the clean rooms. In the last section of the chapter, we have discussed about perpetual generated contamination and wafer cleaning techniques.

TUTORIAL SHEET

1. What is contaminants? How it effects the processing of Integrated Circuit fabrication?
2. Discuss the types of contaminants. What are the main sources of contamination?
3. Discuss the three specific performance areas in contamination control.
4. What do you understand the clean room? Why it is needed in fabrication of IC's?
5. Explain the standards of clear rooms.
6. Explain the Design and Construction of clear rooms.
7. Discuss the Di-water treatment flow used in the fabrication labs.
8. How personal generated contaminants effects the performance of the device under fabrication?
9. What precaution should be taken to control contamination?
10. Explain the wafer cleaning techniques in detail.
11. Write a short note on:
    (*i*) Chemical Contaminants
    (*ii*) Bacteria Contaminants
    (*iii*) HEPA Filter
    (*iv*) Vertical Laminar flow.

3

SILICON GROWTH

3.1 INTRODUCTION

The silicon, occuring naturally in the form of silica (Si) and silicate is the important semiconductor for electronics industry, another important semiconductor is germanium (Ge). The electron mobility in germanium was much higher than silicon. However, silicon with its 1.12 eV bandgap, was better sloted to higher operating temperature and small reverse currents. Germanium proved unsuitable in certain applications because of its property to exhibits high junction current leakage. The current results from germanium is relatively narrow bandgap (0.66 eV), for this region silicon is more preferable for fabrication. Silicon devices operates upto 150°C but germanium operates upto 100°C only.

The other reasons could have ultimately lead to same material substitution. The planner processing technology derives its successes from the properties of thermally grown SiO_2. SiO_2 is not suitable for device applications. The undopped (internisic) resistivity at germanium is about 47 Ω cm. which would have produced the fabrication of rectifying devices with high breakdown voltages. The undoped resistivity of silicon is about 2,30,000 Ω cm.

Thus, high voltage rectifying devices and certain infrared sensing devices are practical and Si is less costly than germanium.

3.2 SILICON MATERIAL PROPERTIES

Silicon material properties are an excellent for performance related parameters. Energy gap of 1.12 eV makes silicon devices less prone to thermal noise than germanium devices with a 0.67 eV gap. Silicon source gases can be purified to extremely high degree of purity, means that high resistivity material can be made. Taken together with the high solubility of dopants upto 10^{21} cm^{-3} for the common dopants are boron, phosphorous Arsenic. This translate to eight orders at magnitude resistivity, tailoring opportunites as shown in Fig. 3.1. Optical absorption in the visible makes silicon suitable for photodetector and solar cells and transpirancy above (1.1 μm) as used in IR microsystem. There are some more properties of silicon are shown in Table 3.1.

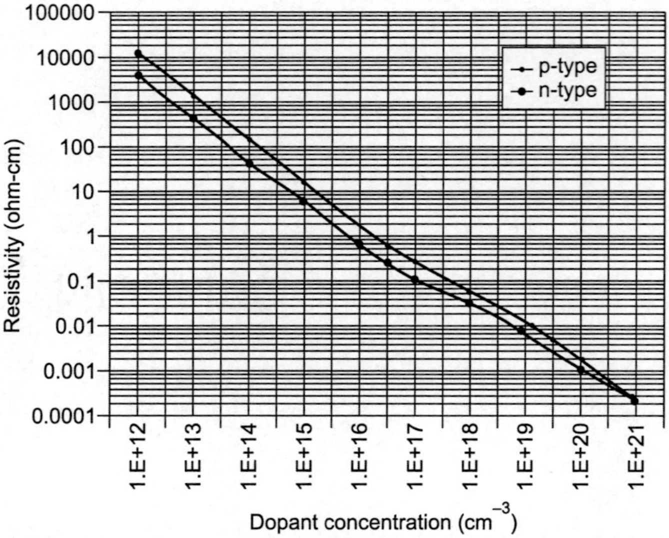

Fig. 3.1: *Silicon Resistivity can be Varied Over Eight Orders of Magnitude By Doping*

Table 3.1 Properties of Silicon at 300 K

Structual and mechanical		
Atomic weight	28.09	
Atoms, total (cm^{-3})	4.995×10^{22}	
Crystal structure	Diamond (FCC)	
Lattice constant (Å)	5.43	
Density (g/cm^3)	2.33	
Density of surface atoms (cm^{-2})	(100) 6.78×10^{14}	
	(110) 9.59×10^{14}	
	(111) 7.83×10^{14}	
Young's modulus (GPa)	190	(111) Crystal orientation
Yield strength (GPa)	7	
Fracture strain	4%	
Poisson ratio, ν	0.27	
Knoop hardness (kg/mm^2)	850	
Electrical		
Energy gap (eV)	1.12	
Intrinsic carrier concentration (cm^{-3})	1.38×10^{10}	
Intrinsic resistivity (Ω-cm)	2.3×10^5	
Dielectric constant	11.8	
Intrinsic Debye length (nm)	24	
Mobility (drift) (cm^2/Vs)	1500 (electrons)	
	475 (holes)	
Temperature coefficient of resistivity (K^{-1})	0.0017	

Thermal		
Coefficient of thermal expansion (°C^{-1})	2.6×10^{-6}	
Melting point (°C)	1414	
Specific heat (J/kg K)	700	
Thermal conductivity (W/m K)	150	
Thermal diffusivity	0.8 cm^2/s	
Optical		
Index of refraction	3.42	$\lambda = 632$ nm
	3.48	$\lambda = 1550$ nm
Energy gap wavelength	1.1 μm	(Transparent at larger wavelengths)
Absorption		
	$> 10^6$ cm^{-1}	$\lambda = 200 - 360$ nm
	10^5 cm^{-1}	$\lambda = 420$ nm
	10^4 cm^{-1}	$\lambda = 550$ nm
	10^3 cm^{-1}	$\lambda = 800$ nm
	< 0.01 cm^{-1}	$\lambda = 1550$ nm

Silicon is strong semiconductor. Its youngs modulus can be as high as 190 GPa (for <111> orientation) stress and strain (longitudaral) is represented by

$$\sigma = \varepsilon E$$

σ = stress

ε = strain

E = Const of proportionality or young modulus.

Strain $\varepsilon = \dfrac{\Delta L}{L}$; stress is force per unit area $\dfrac{F}{A}$ which gives the most familiar expression of Hook's law:

$$\frac{F}{A} = E\frac{\Delta L}{L}$$

When a piece of material is tensile stress it elongation leads also a lateral shrinkage of its diameter

$$\varepsilon_{lateral} = \frac{\Delta D}{D}$$

Poisson ratio is defined as $\nu = \dfrac{-\varepsilon_{lateral}}{\varepsilon_{tensile}}$.

Silicon poissons ratio 0.27 is among the lowest of the all solids so, silicon is strong as steel but fact is disguised by two factors.

1. Most of us do not have experience with 0.5 mm thick steel plate.
2. Silicon is brittle and breakage pattern is therefore, different from the ductile fracture of multicrystalline.

3.3 SILICON CRYSTAL STRUCTURE

The atoms are arranged spatially in a periodic form in the crystalline material, a basic unit cell can be defined, which repeats in all the three dimensions. The electrical properties of the devices depend on the crystal orientation of the material (like silicon) used for starting wafer. Bulk properties of the silicon are generally isotropic (independent of direction) because of the cubic symmetry of the silicon crystal.

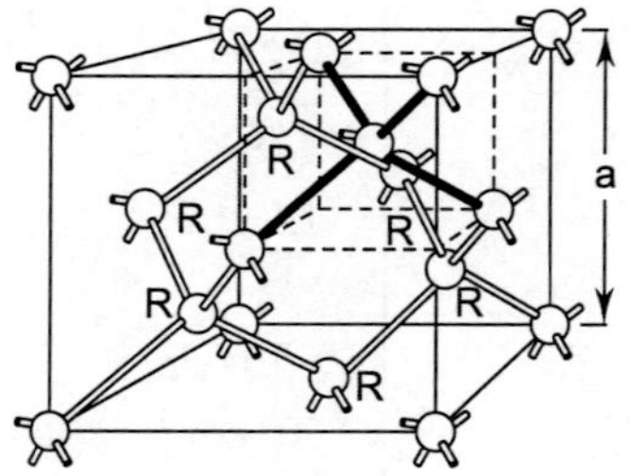

Fig. 3.2 *Unit Cell*

Silicon has a cubic diamond lattice structure as shown in Fig. 3.2. The unit cell can be thought of as two inter leaved FCC (face central cubic) lattices with their origin in (0, 0, 0) and $\left(\frac{1}{4}, \frac{1}{4}, \frac{1}{4}\right)$. The distance between two atoms is $\frac{\sqrt{3}}{4}$ a, and radius is $\frac{\sqrt{3}}{8}$ a, where a is unit cell edge length (1 unit = 5.43095 Å). There is 18 atoms in a unit cell of Si, 8 at vertices. They are shared between 8 unit cell, 6 face atom are shared between two neighbouring unit cells and contribute 3 atoms and 4 atoms are fully inside unit cell.

These crystal lattices of Si are bounded by specific force, which arise from the un-ion pairs of valance electrons of neighbouring atom. This bond is called covalent bond *i.e.*, valance bond.

The millar indices define the plane of a crystal. The plane that defines the faces of the cubes as shown in Fig. 3.3. Millar indices use brackets to represent it, this index enclosed in bracket *e.g.*, (111) (100). The intersect axes 1, 2, 3 at (1, 8, 8) respectively. The Miller index of a plane is given by the reciprocal of these intersects *i.e.*, (1, 0, 0). The edges of tie planes are designated (1, 1, 0) and diagonals of planes are (1, 1, 1). The angles between the planes can be calculated from the scalar product of normal vectors.

$$\vec{a} \cdot \vec{b} = |\vec{a}||\vec{b}| \cos(\vec{a}, \vec{b})$$

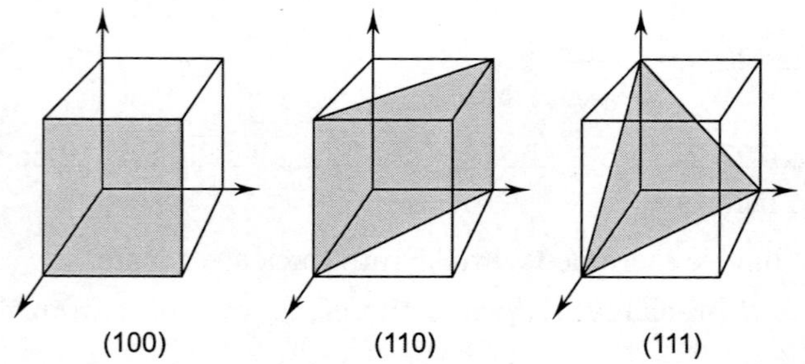

Fig. 3.3 *Some Important Silicon Crystal Planes with Their Miller Indices*

As Fig. 3.3, shows that (1, 0, 0) and (1, 1, 0) planes meets 45° and all other angles can be calculated easily, when the negative unit vector are accounted for (1, 1, 0) is (– 1, 1, 0) and the angle between (1,1,1) and (1, 1, 0) planes is calculated from $\frac{1}{\sqrt{3}} = \cos\alpha$, giving $\alpha = 54.7°$.

Example. *Calculate the angle between the (1, 0, 0) and (1, 1, 0) plane ?*

Solution. The angle between the planes can be calculated from the scalar product of the normals value:

$$\vec{a} \cdot \vec{b} = \left|\vec{a}\right|\left|\vec{b}\right| \cos\left(\vec{a}, \vec{b}\right)$$

First, we calculate the direction cosines of (1, 0, 0) and (1, 1, 0)

$$\left|\vec{a}\right| = \sqrt{1^2 + 0^2 + 0^2} = 1$$

$$\left|\vec{b}\right| = \sqrt{1^2 + 1^2 + 0^2} = \sqrt{2}$$

Let the normals be $\overrightarrow{n_1}$ and $\overrightarrow{n_2}$

i.e.,

$$\overrightarrow{n_1} = \hat{i} + 0\hat{j} + 0\hat{k}$$

$$\overrightarrow{n_2} = \frac{1}{\sqrt{2}}\hat{i} + \frac{1}{\sqrt{2}}\hat{j} + 0\hat{k}$$

then,

$$\overrightarrow{n_1} \cdot \overrightarrow{n_2} = \left|\overrightarrow{n_1}\right|\left|\overrightarrow{n_2}\right| \cos\alpha$$

$$\frac{1}{\sqrt{2}} = \cos\alpha \quad \Rightarrow \quad \alpha = 45°$$

Wafers of two crystal orientations which are widely used for fabrication are <1 0 0> and <1 1 1>. The <1 1 1> planes have the highest density of atoms on the surface, so that crystal grows most easily on these planes and mechanical properties such as tensile strength is highest in <1 1 1> direction. <1 1 1> plane oxidize faster than <1 1 0> planes, because they have more atoms per unit surface area, available for oxidatiion reaction to occur. Bipolar circuits have preferred <1 1 1> oriented material and MOS <1 0 0>.

3.4 OTHER SEMICONDUCTOR COMPOUND MATERIALS

There are many semiconductor compounds like Gallium Arsenide (GaAs) and Gallium Arsenide - Phosphide (GaAsP), Indiam Phosphide (In P), Indium Gallium Phosphide (InGaP)

GaAs: This material is used for diodes, these diodes give laser light when electrical supply is given to it. These diodes are called Light Emitting Diodes (LED). These GaAs devices operate two or three times faster than normal silicon devices, thats why they are used in special real time circuits such as airplanes control panels of superfast computers.

GaAs is also semi-insulating. It is helpful to minimize leakage between the adjacent devices, and allows higher packaging density. It has many more advantages, but it is not expected to replace the silicon in industry because of the trade offs between the performance and processing difficulty. The steps to fabricate GaAs is more critical than silicon. Arsenic evaporates from the compound at normal process temperature, requiring the additional suppression layers. More detail of semiconductor compound materials with properly is given in Table 3.2.

Table 3.2: Physical Properties of Semiconductor Materials

	Ge	*Si*	*GaAs*	*SiO_2*
Atomic Weight	72.6	28.09	144.63	60.08
Atoms/cm^3 or Molecules	4.42×10^{22}	5.00×10^{22}	2.21×10^{22}	2.3×10^{22}
Crystal Structure	Diamond	Diamond	Zinc-Blends	Amorphous
Atoms/Unit Cell	8	8	8	—
Density	5.32	2.33	5.65	2.27
Energy Gap	0.67	1.11	1.40	8 (approx.)
Dielectric Constant	16.3	11.7	12.0	3.9
Melting Point (°C)	937°	1415°	1238°	1700° (approx.)
Breakdown Field (V)	8 (approx.)	30 (approx.)	35 (approx.)	600 (approx.)
Linear Coefficient of Thermal Expansion $\frac{\Delta L}{LT}\frac{1}{C}$	5.8×10^{6}	2.5×10^{6}	5.9×10^{6}	0.5×10^{-6}

3.5 CRYSTALLINE DEFECTS IN SILICON

Like anything else in this world, crystals inherently possess imperfections, or what we often refer to as **'crystalline defects'**. The presence of most of these **'crystalline defects'** is undesirable in silicon wafers, although certain types of 'defects' are essential in semiconductor manufacturing. Engineers in the semiconductor industry must be aware of, if not knowledgeable on the various types of silicon crystal defects, since these defects can affect various aspects of semiconductor manufacturing — from production yields to product reliability.

Crystalline defects may be classified into four categories according to their geometry. These categories are:

1. Zero-dimensional or 'point' defects:;
2. One-dimensional or 'line' defects;
3. Two-dimensional or 'area' defects; and
4. Three-dimensional or 'volume' defects.

Table 3.3 presents the commonly-encountered defects under each of these categories.

Table 3.3: Examples of Crystalline Defects

Defect Type	*Examples*
Point of Zero-Dimensional Defects	Vacancy Defects Interstitial Defects Frenkel Defects Extrinsic Defects

Line or One-Dimensional Defects	Straight Dislocations (edge or screw) Dislocation Loops
Area of Two-Dimensional Defects	Stacking Faults Twins Grain Boundaries
Volume or Three-Dimensional Defects	Precipitates Voids

There are many forms of crystal point defects. A defect where in a silicon atom is missing from one of these sites is known as a 'vacancy' defect. If an atom is located in a non-lattice site within the crystal, then it is said to be an 'interstitial' defect. If the interstitial defect involves a silicon atom at an interstitial site within a silicon crystal, then it is referred to as a 'self-interstitial' defect. Vacancies and self-interstitial defects are classified as intrinsic point defects as shown in Fig. 3.4.

If an atom leaves its site in the lattice (thereby creating a vacancy) and then moves to the surface of the crystal, then it becomes a 'Schottky' defect. On the other hand, an atom that vacates its position in the lattice and transfers to an interstitial position in the crystal is known as a 'Frenkel' defect. The formation of a Frenkel defect therefore, produces two defects within the lattice – a vacancy and the interstitial defect, while the formation of a Schottky defect leaves only one defect within the lattice, *i.e.,* a vacancy. Aside from the formation of Schottky and Frenkel defects, there's a third mechanism by which an intrinsic point defect may be formed, *i.e.,* the movement of a surface atom into an interstitial site.

Extrinsic point defects, which are point defects involving foreign atoms, are even more critical than intrinsic point defects. When a non-silicon atom moves into a lattice site normally occupied by a silicon atom, then it becomes a 'substitutional impurity.' If a non-silicon atom occupies a non-lattice site, then it is referred to as an 'interstitial impurity.' Foreign atoms involved in the formation of extrinsic defects usually come from dopants, oxygen, carbon, and metals.

The presence of point defects is important in the kinetics of diffusion and oxidation. The rate at which diffusion of dopants occurs is dependent on the concentration of vacancies. This is also true for oxidation of silicon.

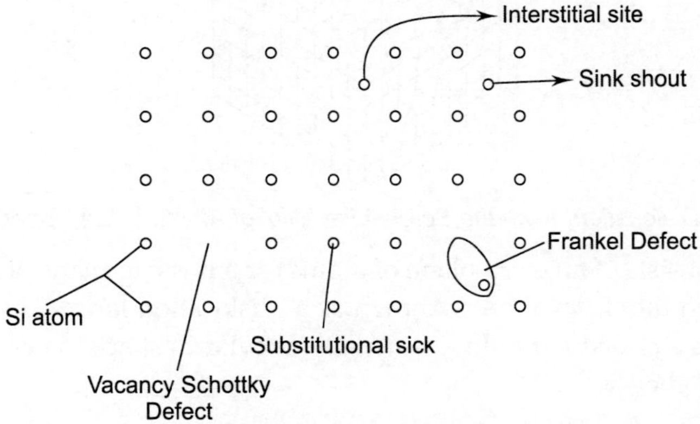

Fig. 3.4 *Point Defects*

Crystal **line defects** are also known as 'dislocations', which can be classified as following:

1. Edge dislocation ;
2. Screw dislocation ; or
3. Mixed dislocation, which contains both edge and screw dislocation components.

An edge dislocation may be described as an extra plane of atoms squeezed into a part of the crystal lattice, resulting in that part of the lattice containing extra atoms and the rest of the lattice containing the correct number of atoms. The part with extra atoms would therefore, be under compressive stresses, while the part with the correct number of atoms would be under tensile stresses. The dislocation line of an edge dislocation is the line connecting all the atoms at the end of the extra plane.

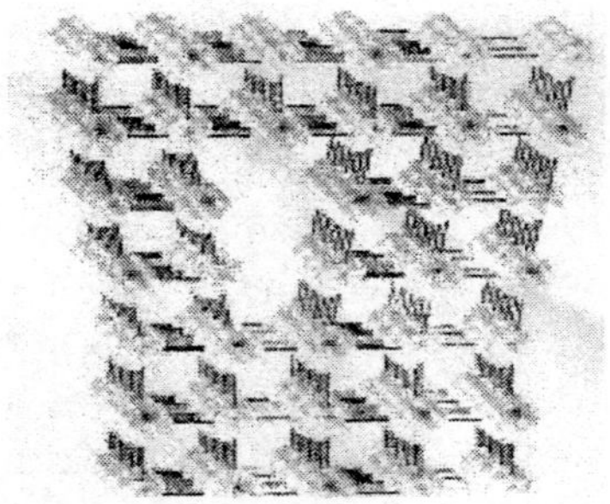

Fig. 3.5 *An Edge Dislocation ; Note the Insertion of Atoms in the Upper Part of the Lattice*

If the dislocation is such that a step or ramp is formed by the displacement of atoms in a plane in the crystal, then it is referred to as a 'screw dislocation.' The screw basically forms the boundary between the slipped and unslipped atoms in the crystal. Thus, if one were to trace the periphery of a crystal with a screw dislocation, the end point would be displaced from the starting point by one lattice space. The dislocation line of a screw dislocation is the axis of the screw.

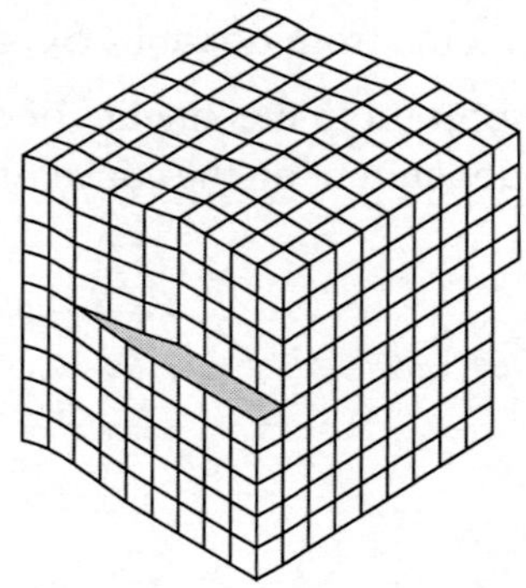

Fig. 3.6 *A Screw Dislocation, Note the Screw-Like 'Slip' of Atoms in the Upper Part of the Lattice*

If the dislocation consists of an extra plane of atoms (or a missing plane of atoms) lying entirely within the crystal, then the dislocation is known as a 'dislocation loop.' The dislocation line of a dislocation loop forms a closed curve that is usually circular in shape, since, this shape results in the lowest dislocation energy.

Dislocations are generally undesirable in silicon wafers, because they serve as sinks for metallic impurities as well as disrupt diffusion profiles. However, the ability of dislocations to sink impurities may be engineered into a wafer fabrication advantage, *i.e.*, it may be used in the removal of impurities from the wafer, a technique known as 'gettering.' It will be discussed later in this chapter.

Area defects in crystals consist of stacking faults, grain boundaries, and twin boundries. A 'stacking fault' pertains to a disturbance in the regularity of the stacking of planes of atoms in a crystal lattice. This usually occurs when a plane is inserted into or removed from the lattice. The insertion of an extra plane in the stacking is known as an 'extrinsic' stacking fault, while the removal of a plane is referred to as an 'intrinsic' stacking fault.

Stacking faults can become electrically active when decorated by impurity atoms. Electrically active stacking faults can cause device degradation, examples of which are higher reverse bias currents in *p-n* junctions and storage time reduction in MOS circuits.

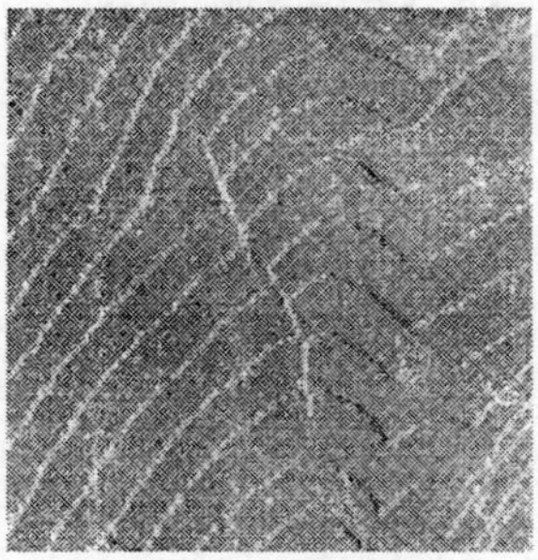

Fig. 3.7 *Photo of a Stacking Fault*

A 'twin' is an area defect wherein a mirror image of the regular lattice is formed during the growth of the silicon ingot, usually caused by a perturbation. The 'twin boundary' is the mirror plane of the twin formation.

A 'grain boundary' refers to the transition or interface between crystals whose atomic arrangements are different in orientation with respect to each other.

Volume defects in a crystal are also known as 'bulk' defects, which include voids and precipitates of extrinsic and intrinsic point defects.

Every impurity introduced into a crystal has a certain level of solubility, which defines the concentration of that impurity that the solid solution of the host crystal can accommodate. Impurity solubility usually decreases with decreasing temperature. If an impurity is introduced into a crystal at the maximum concentration allowed by its solubility at a high temperature, the crystal will become supersaturated with that impurity, once it is cooled down. A crystal under such supersaturated conditions seeks and achieves equilibrium by precipitating the excess impurity atoms into another phase of different composition or structure.

Precipitates are considered undesirable, because they have been known to act as sites for the generation of dislocations. Dislocations arise as a means of relieving stress generated by the strain exerted by precipitates on the lattice. Precipitates induced during silicon wafer processing come from oxygen, metallic impurities, and dopants like boron (as already discuss).

3.6 SINGLE CRYSTAL GROWING FOR WAFER PRODUCTION

Integrated circuits are built on **single-crystal silicon** substrates that possess a high level of purity and perfection. Single-crystal silicon is used in VLSI fabrication instead of polycrystalline silicon since, the former does not have defects associated with grain boundaries found in polysilicon. Such defects have been known to limit the lifetimes of minority carriers.

Aside from the need to be single-crystalline in nature, silicon substrates must also have a high degree of chemical purity, a high degree of crystalline perfection, and high structure uniformity. The acquisition of such high-grade starting silicon material involves two major steps:

1. Refinement of raw material (such as quartzite, a type of sand) into electronic grade polycrystalline silicon (EGS) using a complex multi-stage process; and
2. Growing of single-crystal silicon from this EGS either by Czochralski or Float Zone process.

3.7 PRODUCTION OF EGS (ELECTRONIC GRADE SILICON) FROM MGS (METALLIC GRADE SILICON)

The raw material for the preparation of single crystal silicon is the EGS (Electronic Grade Silicon) which is a polycrystalline material of high purity. The major impurities are boron, carbon and residual doner. Pure EGS generally requires that doping elements be in parts per billion (ppb) range and carbon be less than two parts per million. These properties are usually estimated on test Ingots capabilites of most laboratory method. Due to this reason, the claping level can be measured by resistivity measurements on the test ingot. Alternatively, low temperature infrared absorption measurements or photaluminece can be used.

To obtain EGS from MGS requires a same steps. These steps are given below:

1. MGS (Mattalurgical-Grade-Silicon) is produced by submerged electrode in a furnace as shown in Fig. 3.8.

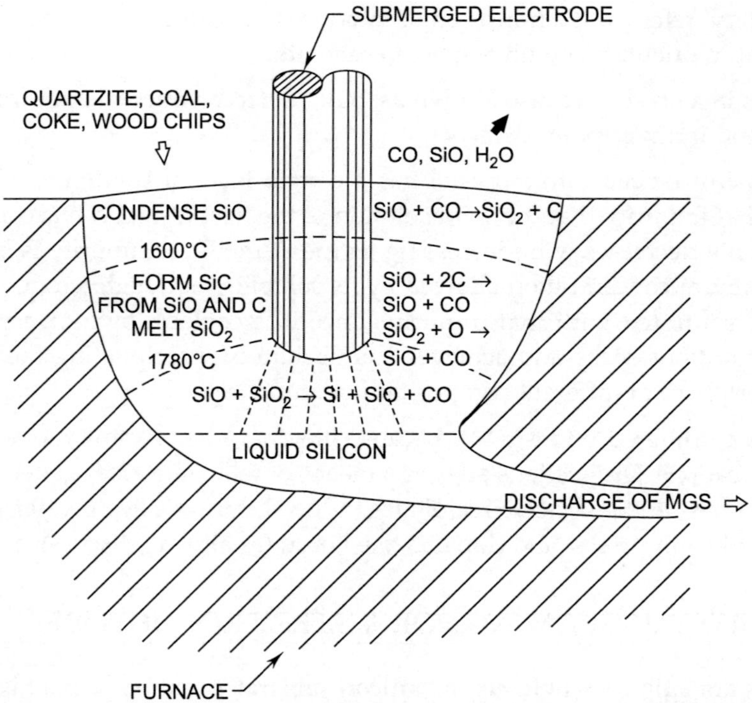

Fig. 3.8 *Submerged-Electrode Arc Furnace for the Production of Metallurgical-Grade Silicon.*

The furnace is charged with quartizite, a relatively pure form of sand (SiO_2) and Carbon in the form of coal, coke and wood chip. In the furnace, a number of reactions take place, the overall reactions are as follows:

$$SiC_{(solid)} + SiO_{2\,(solid)} \rightarrow Si_{(liquid)} + SiO_{gas} + CO_{(gas)}$$

In this process, MGS of 98% purity can be achieved, but this is not suitable for manufacturing semiconductor devices.

2. In second step, the silicon is pulverized mechanically and reacted with an hydrous hydrogen chloride (HCl) to form trichlorosiline ($SiHCl_3$), the chemical reactions take place is:

$$Si_{(solid)} + 3HCl_{(gas)} \rightarrow SiHCl_{3\,(gas)} + H_{2\,(gas)} + \text{Heat.}$$

The reaction take place at nominal temperature of 300°C using a catalyst, $SiHCl_3$ (silicon tetra chloride ($SiCl_4$) and impurities are formed at this point and purification process occurs. Trichlorosiliane is liquid at room temperature, because its boiling point is 32°C, and other unwanted chloride are liquid at room temperature, therefore, purification is done with help of fractional distillation process.

3. EGS is prepared from the purified $SiHCl_3$ in chemical vapour deposition (CVD) process. This process is similar to epitaxial deposition method (to be discuss in next chapter). The chemical reaction is a hydrogen reduction of $SiHCl_3$

$$2SiHCl_{3\,(gas)} + 2H_{2\,(gas)} \rightarrow 2Si + 6HCl$$

The above mention reaction take place in CVD reactor shown in figure. In which a resistance-heated

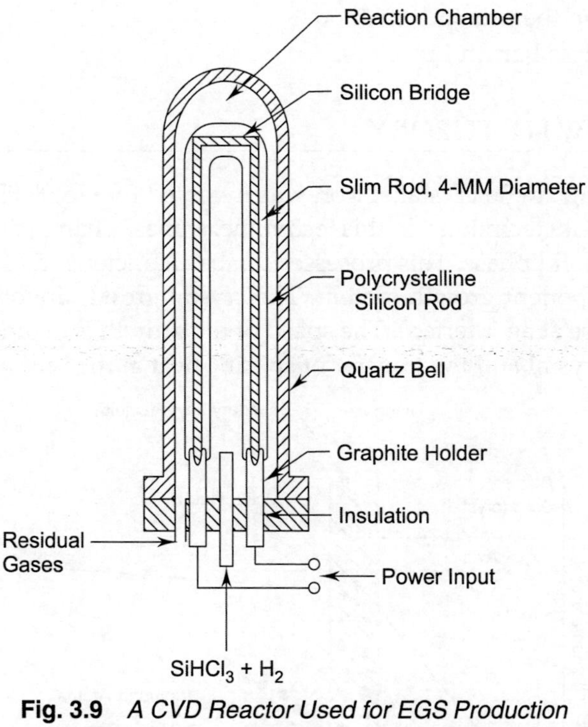

Fig. 3.9 *A CVD Reactor Used for EGS Production*

Si rod (4 mm diameter) called slim rod serves as nucleation point for deposition of silicon. The complete process cycles take many hours and a poly crystalline structured rod of EGS is formed which is 0.2 m in diameter and few meter in length. This EGS can be cut from these rods as single chunks or crushed into nuggist. To achieve high overall efficiency of the process feedback or recycling of reaction by product is used as shown in Fig. 3.10.

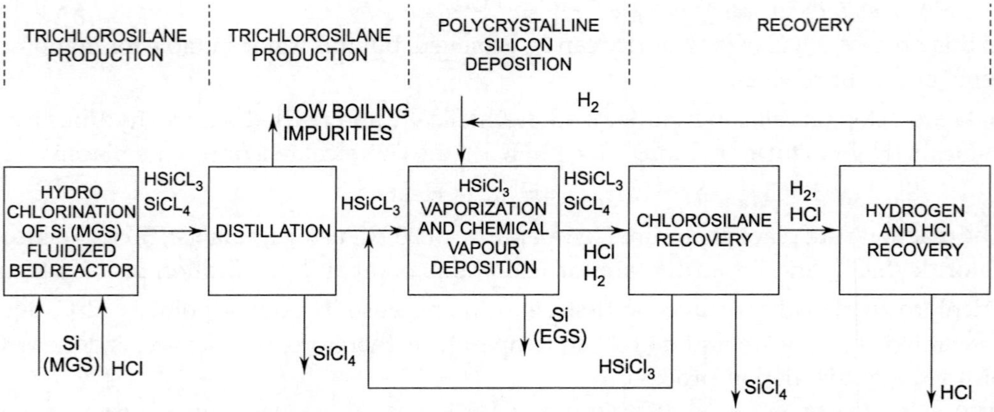

Fig. 3.10 *Schematic Diagram of the Production of Electronic-Grade Silicon from the Hydrogen Reduction of Trichlorosilane*

Another method of producing EGS from pyrolysis, the apparatus used here is same as in previous method. But, this reactor is operated at 900°C and supported with siline instead of trichlorosiline.

$$SiH_{4\,(gas)} + Heat \rightarrow Si_{(solid)} + 2H_{2\,(s)}$$

Advantages of producing EGS from siline are:

→ It is less costly from the previous method.

→ It by product are less harmful reactors.

3.8 CRYSTAL GROWTH THEORY

The technique of growing single crystal silicon was invented by CZOCHARALSKI, so this technique is called Czocharalski technique. In this technique, a phase change discuss from solid, liquid, gas phase to crystalline solid phase. This processes can be characterized as applied to silicon, to be a liquid solid monocomponent growth system. The crystal growth involves the solidification of atoms from a liquid phase at an interface. The speed of the growth is determined by the number of sites on the face of the crystal and the transfer of specific heat at the interface shown in Fig. 3.11.

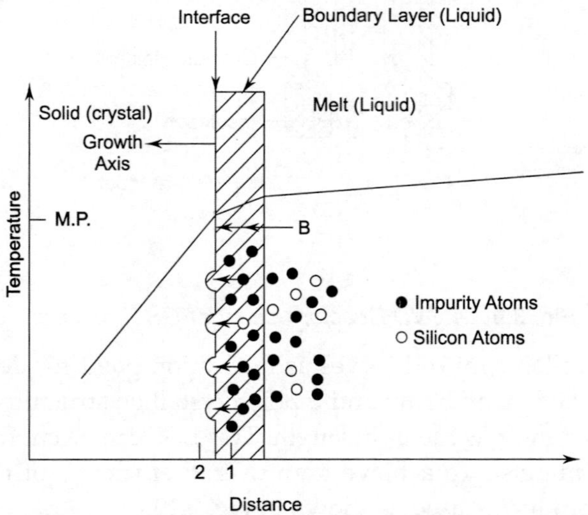

Fig. 3.11 *Temperature Gradients, Solidification, and Transport Phenomena Involved in Czochralski Growth*

Positions 1 and 2 represent the locations of isotherms and the crystal solidification at the interface. Impurity atoms are transported across the boundary layer (B) and incorporated into the growing crystal interface. M.P. is the melting point.

It shows the transport process and temperature gradient involved in the heat transfer. Condition about the interface can be modeled by the following equation:

$$h\frac{dm}{dt} + K_i \frac{dT}{dx_i} A_1 = K_s \frac{dT}{dx_2} A_2 \qquad ...(1)$$

Where, h is the latent heat of fussion,

$\frac{dm}{dt}$ is the mass rate of solidification

T is the temperature.

K_l and K_s are the thermal conductivities of liquid and solid respectively.

$\frac{dT}{dx_1}$ and $\frac{dT}{dx_2}$ are thermal gradients at points 1 and 2 which are near the interface in the liquid and solid respectively.

A_1 and A_2 are the areas of isotherms at positions M and N respectively.

From equation (1), the max pull rate of crystal under the condition of zero thermal gradients in the melt, *i.e.*, $\frac{dT}{dK_1} = 0$, can obtained, the max pull rate is given by

$$P_{max} = \frac{K_s\, dT}{h\, d_i\, dx}$$

where, P_{max} is maximum pull rate of pull speed.

d is the density of solid silicon.

The pull rate influence the incorporation of impurities into the crystal and is a factor in defect generation. Generally, when temperature gradient in the melt is small, the heat transferred to the crystal is the latent heat of fusion. The pull rate generally varies inversely with diameter. In actual practice, the pull rate is obtained 30% to 50% slower than the maximum theoritical value.

The growth rate of growth velocity of the crystal is very important growth parameter and is the instantaneous solidification rate. Note that pull rate is the macroscopic indication of net solidification rate. These two rates are differ because of the temperature fluctuations near the interface. The growth rate can be more than the pull rate or even be negative at the given time. When the growth rate is negative, remelting occurs, that is the crystal dissolves back into the melt.

Remelting must be eliminated to remove crystal defect. The growth rate affects the defect structure and dopant distribution in the crystal on a macroscopic scale. Every impurity has a solid solubility in the silicon, the impuring has different equilibrium solubility in melt. During the growth, the impurities at the interface are left in the liquid (melt), thus as the crystal grows the melt becomes progressively enriched with impurity due to temperature variation, the thickness of boundary layer fluctuates with time, the net result of the thermal convection effects is an inhomogeneous distribution of dopants in the crystal on a microscale, the pull rate speed is also a factor indetermining the shape of the growing interface.

3.9 DIFFERENT METHODS OF CRYSTAL GROWTH

There are mainly three methods used for the crystal growth are:

1. Czocharalski Method
2. Float zone method
3. Bridgeman method

Let us discuss these three methods one by one in detail:

3.9.1 Crystal Growth by Czohalki Method

This is named in honour of its inventor. It involves the crystalline solidification of atoms from a liquid phase at an interface (as discussed in previous section). The crystal growth Czochralskai Apparatus is shown in Fig 3.12. This apparatus is 6.5 m tall and 17,600 kg in weight, the puller is capable to hold a melt charge of 60 kg of silicon, which can be transferred into a crystal of 100 mm diameter and 30 m length and it is required several hour to pull the ingot. The crystal pulling is done in an inert gas atmosphere and sometime a vacuum is used. This is done to prevent oxidation.

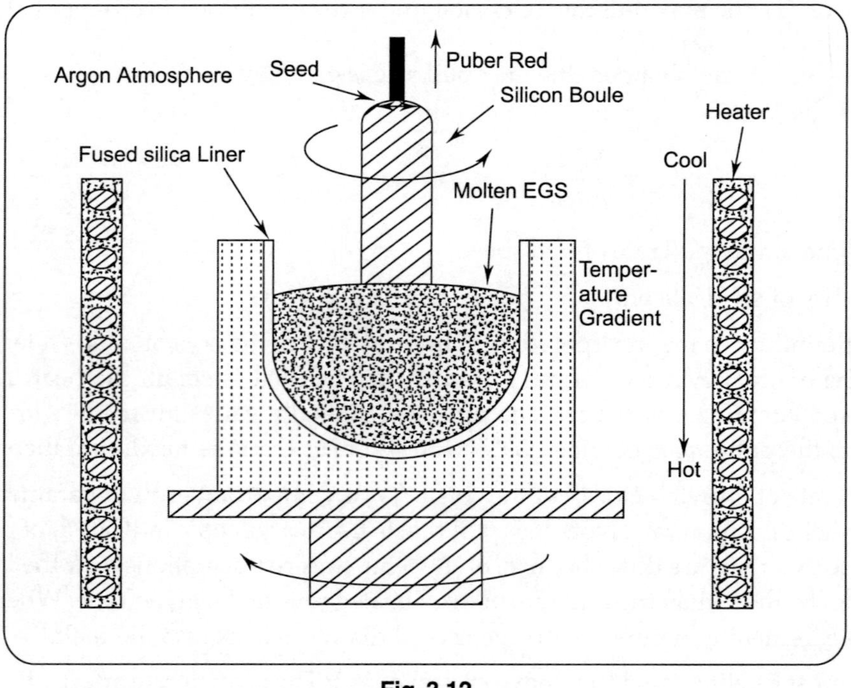

Fig. 3.12

The apparatus is divided into four parts these are (shown in Fig. 3.13)

(*a*) *Furnace*: This part consist of crucible, susceptor and rotation mechanism, heating element, power supply and chamber.

(*b*) *Ambient control*: It consist of gas sources, flow control, exhaust or vacuum system and purge tube.

(*c*) *Pulling mechanism*: It consist of seed shaft or chain, rotation mechanism and seed chuck.

(*d*) *Control system*: It consist of microprocessor, sensors and display devices or output meters to show readings.

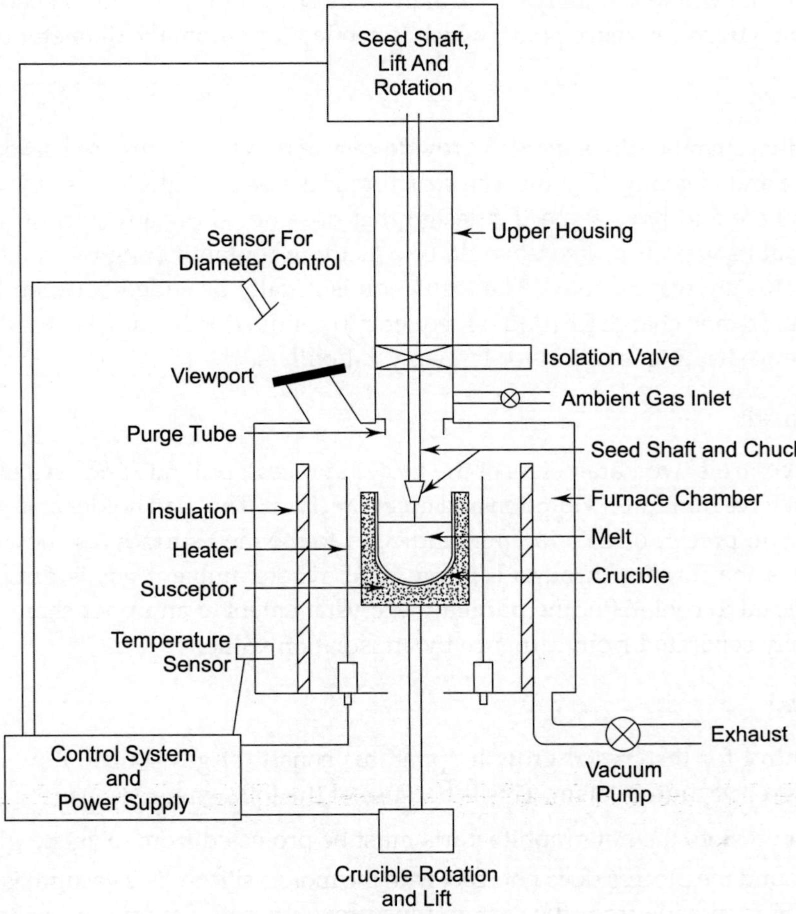

Fig. 3.13

Furnace

The crucible is the most important component of the growth apparatus. It contains the melted silicon. So, the crucible material should be chemically unreactive with silicon. The material should have high melting point, thermal stability and hardness. Carbon and silicon crucibles are unsuitable. As we know carbon is electrically inactive. In silicon, high quality crystals cannot be grown with carbon structure metals fused silica however, reacts with silicon, releasing silicon and oxygen into the melt. In this process, the crucible undergoes erosion. Crystals grown with thus crucible also contain substantial amounts of interstitial, oxygen that can be either beneficial or determental.

Susceptor

Susceptor is used to support the silica crucible. It also provide better thermal conditions. The material like graphite is first choice because of its high temperature properties. The graphite should be pure to prevent contamination of the crystal from impurities that would volatilized from graphite at temperature involved. The susceptor rests on a pedestal whose shaft is connected to a motor that provide rotation. The whole assembly can usually be raised and lower end to keep the melt level equidistants from a fixed reference point, which is needed for automatic diameter control.

Chamber Housing

The function of this chamber housing is to provide easy across to the furnace component to facilitate maintenance and cleaning. The furnace structure must be air tight to prevent contamination from the atmosphere and have a specific design that does not allow any part of the chamber to become so hot that its vapour pressure would be a factor in contaminating the crystal. The hottest part of the appratus are water cooled. The insulation is usually provided between the heater and the chamber wall. To melt charge RF (Radio Frequency) or induction heating is useful for small melt sizes. But, resistance heating is useful for large crystal pullers.

Pulling Mechanism

The mechanism controls two parameters of the growth process, pull rate and crystal rotation. This mechanism must have minimum vibration and great precision. The seed holder and pulling mechanism must maintain precise orientation perpendicular to the melt surface (as shown in Fig. 3.14). The crystal leaves the furnace through a purge tube, where ambient gas is directed along the surface of the crystal to cool. After the purge tube, crystal enters to an upper chamber. The upper chamber is usually separated from furnance by an isolation value.

Ambient Control

The ambient control for the crystal growth apparatus consist of gas source, flow control, purge tube, and exhaust or vacuum system. This is because of the following reasons:

1. To prevent erosion, the hot graphite parts must be protected from oxygen.
2. The gas around the process does not react with the molten silicon. The vacuum is also helpful in removing Si monoxide from the system thus preventing its deposition inside the furnance chamber. Growth in a gaseous atmosphere ; generally used on large growers must use an inert gas such as helium and argon. The pressure of the inert gas may be atmospheric or reduced.

Control System

It is used to provide control of the process parameters like temperature, crystal diameter, rotation speed, pull rate etc., they are controlled with help of microprocessor of sensors. The microprocessor or microcontrollers are programmed according to wafers specifications and sensors are used to judge the change. These all are included in digital control. These methods are more popular, because they are less depended on operators interaction.

3.9.2 Bridgeman Techniques

This technique is very popular for the growth of single crystal gallium arsenide. The left hand zone is kept at the temperature of 610°C approximately. To maintain the required over pressure of Arsenic, whereas, the right hand zone is kept at the temperature just above the melting point of Gallium arsenide (1240°C) the sealed tube is made of quartz and a boat is made of graphite.

In operation, firstly, the boat is located with the charge of polycrystalline Gallium arsenide, at right side of the tube. Secondly, load arsenic at the other end of the tube. After loading the furnance is moved towards the right, the melt cools at one end. Usually, there is a seed placed at the left end of the boat to establish a specific crystal orientation. The gradial freezing (solidification) of the melt allows a single crystal to propagate at liquid solid interface and after this crystal of GaAs is grown.

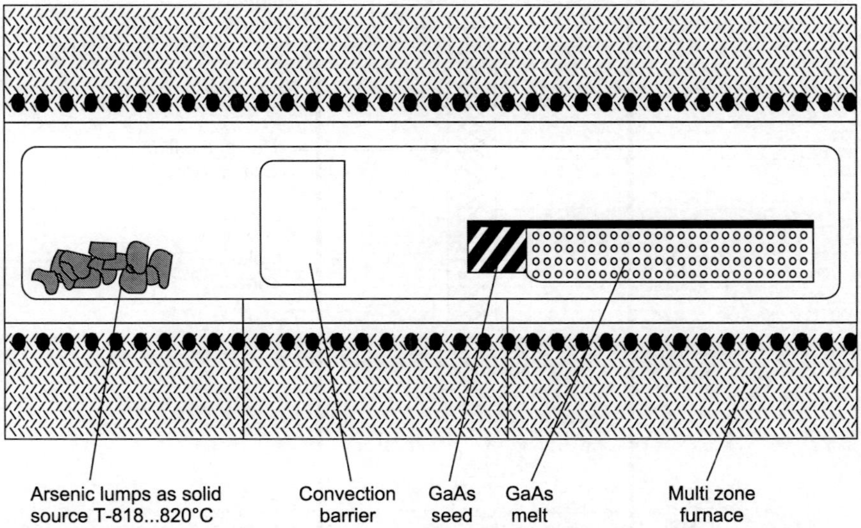

Fig. 3.14 *Schematic of a Horizontal Bridgman Growth System (After Sell).*

3.9.3 Float Zone Crystal Growth

The Float Zone (FZ) process is another method for growing single-crystal silicon. It involves the passing of a molten zone through a polysilicon rod that approximately has the same dimensions as the final ingot. The purity of an ingot produced by the FZ process is higher than that of an ingot produced by the CZ process. As such, devices that require ultrapure starting silicon substrates should use wafers produced using the FZ method.

The FZ processes consists of the following steps.

1. A polysilicon rod is mounted vertically inside a chamber, which may be under vacuum or filled with an inert gas.
2. A needle-eye coil that can run through the rod is activated to provide RF power to the rod, melting a 2-cm long zone in the rod. This molten zone can be maintained in stable liquid form by the coil.
3. The coil is then moved through the rod, and the molten zone moves along with it.
4. The movement of the molten zone through the entire length of the rod purifies the rod and forms the near-perfect single crystal.

FZ growing equipment can also use a stationary coil, coupled with a mechanism that can move the silicon rod through it.

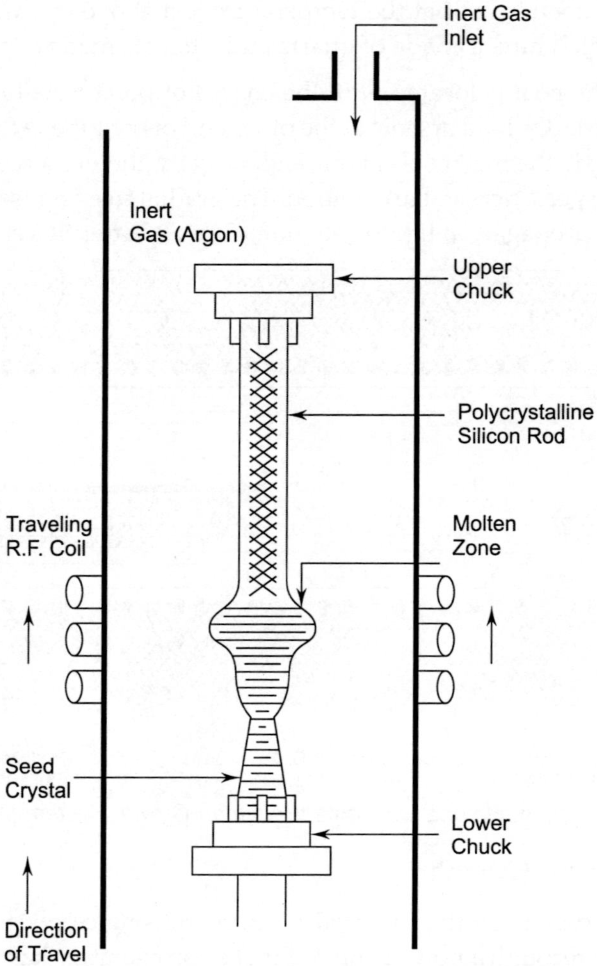

Fig. 3.15 *Float Zone Crystal Growing System*

Table 3.4: Comparison of CZ and Float Crystal-Growing Methods

Parameter	*CZ*	*Float Zone*
Large Crystal	Yes	Difficult
Cost	Lower	
Dislocations	$0 - 10^4/cm^2$	$10^3 - 10^5/cm^2$
Resistively	Up to 100 ohm-cm	2000 ohm-cm Max.
Radial		
Resistively	5 – 10%	5 – 10%
Oxygen Content	$10^{16} - 10^{18}$ atoms/cm^2	0 – Very Low

3.10 WAFER PREPARATION

The silicon crystal obtained after growth process (CZ or FZ) is known as ingot. It undergoes a routine evaluation if its resistivity impurity content, crystal perfection, size or weight. It is then grinded by using diamond wheels to make it a perfect cylinder that the right diameter, then it undergoes an etching process to remove the mechanical left by the grinding process. These processes are discussed in more detail as follows:

(*i*) **INGOT Trimming:** The extreme top and bottom portion of the ingot are cut-off using circular saw. The next operation is a surface griding and during this the diameter of the wafer decided according to that grinding take place. Silicon ingots are grown slightly oversize because the automatic diameter control in crystal growing cannot maintain the needed diameter tolerance and the crystal cannot be grown perfectly round. So, the oversized ingot have to grinded on the surface so as to have uniform diameter throughout the ingot surface. The lathe like machine tool (as shown Fig. 3.16) is used to grind the ingot using rotating diamond grinding tool. The machine tools move along the ingot surface and grind the ingot to the required diameter.

After surface grinding ingot in Fig. 3.17, one or more flats are grounded along the length of ingot. There are usually flats called primary (major) and secondary (minor) flats as shown in figure. below. The functions of the primary flat are: first, it is used as mechanical locator in automated processing equipment to position the wafer. Secondly, it helps to form an array of square die, thirdly, it is used to orient the integrated circuits devices relative to crystal.

The secondary flat is used to identity orientation and type of conducting of the material as shown in Fig. 3.18.

Surface grinding Mark crist. orient

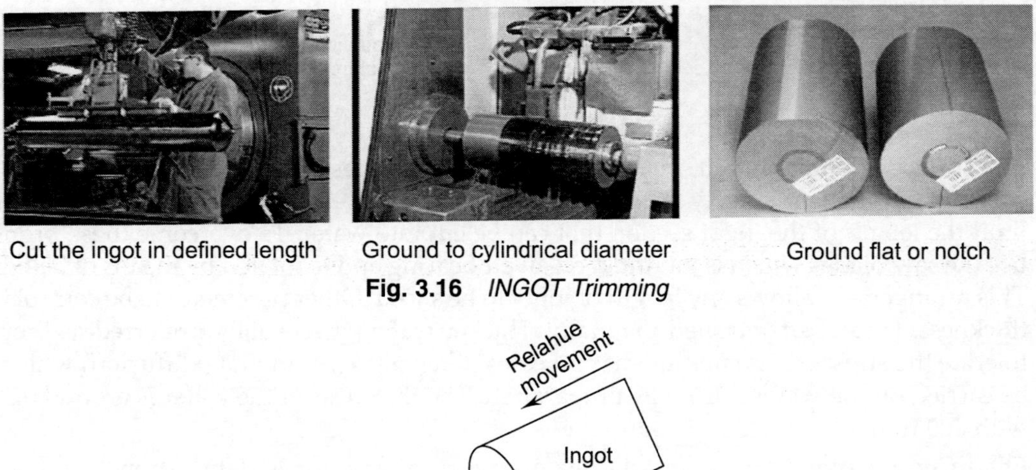

Cut the ingot in defined length Ground to cylindrical diameter Ground flat or notch

Fig. 3.16 *INGOT Trimming*

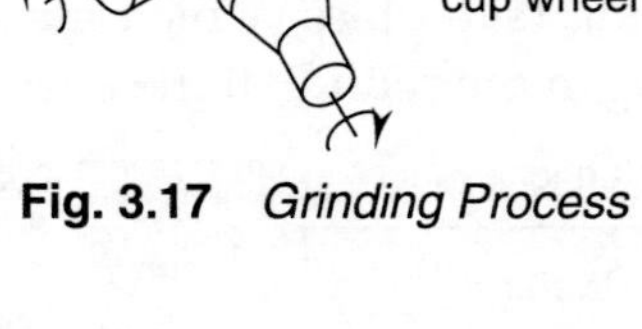

Fig. 3.17 *Grinding Process*

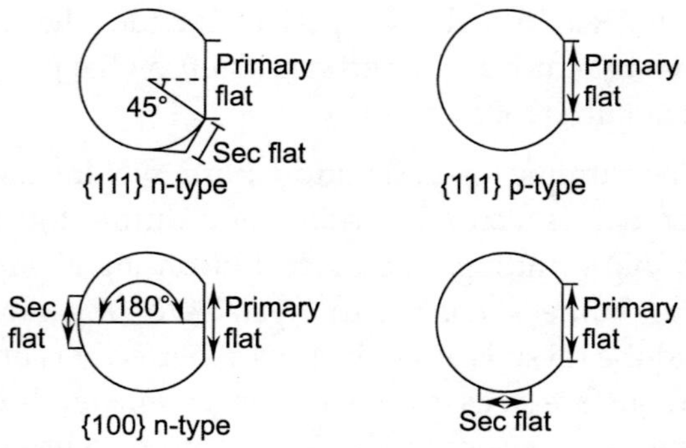

Fig. 3.18 *Flats of the Wafer*

(*ii*) **Wafer Slicing:** Slicing determine the surface orientation thickness taper, and bow, slicing is done by inner diameter sawing using a diamond saw (shown in Fig. 3.19) at high speed and the moved across the ingot to obtain wafer slice, slicing is done very slowly, because it determine.

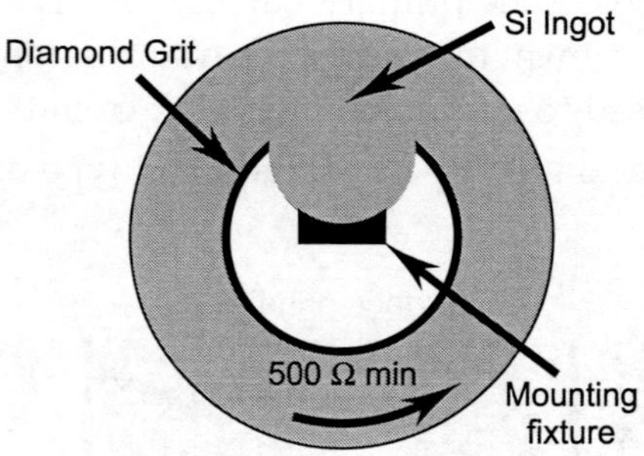

Fig. 3.19 *Wafer Sawing*

The surface orientation. But, there are some disadvantages of inner diameter slicing (*i*) The speed of cutting is relatively slow and result in slow process. (*ii*) Drum is of finite depth, it limit the length of the ingot section that can be cut into wafer. To overcome these problems use the saw blades which is mounted on an air bearing and is rotated by means of belt drive. This arrangement allows any length of ingot to be sliced. Other parameter to be control is the thickness. It can be controlled manually. Thicker wafers are usually preferred as they can tolerate the stress of the subsequent processes (like epitaxy, oxidation, diffusion) which will be carried out on wafer during IC preparation. The thickness of the wafer is around 0.5 mm with 100 mm in diameter.

(*iii*) **Edge Countoring:** After slicing the wafer is not suitable for IC fabrication. The wafer is mechanically lapped on the both side. This is performed under pressure using a mixture of Al_2O_3 and Glycerine. It remove 2.0 mm of the surface material on both sides (as shown in Fig. 3.20) this edge countouring take place, which means that wafer is rotated against a concave diamond tool resulting in the round off the edge of the wafer.

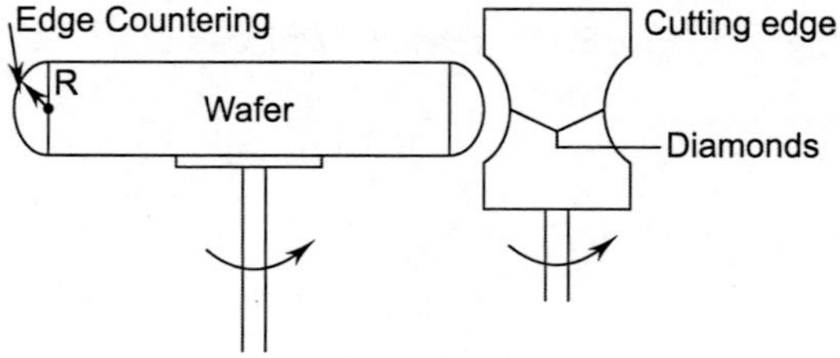

Fig. 3.20

(*iv*) **Wafer Etching :** The surface of the lapped wafers are etched to remove any remaining micro cracks or surface damage introduced by the alumina abrasive in the previous lap-stage. Etching is done chemically using a corrosive mixture of nitric acid and acetic acid solutiion. Edge beveling smooths out the peripheral edges of the wafer. A chemical/mechanical (chemomechanical) beveling technique leads to a marked improvement in water losses (shinning etc.)

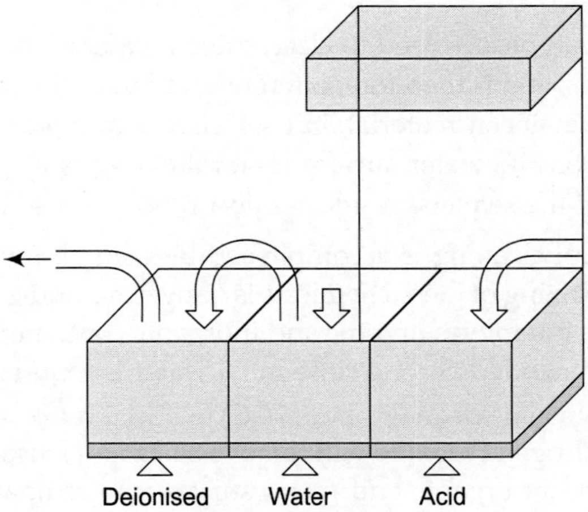

Fig. 3.21 *Etching Process*

(*v*) **Wafer Polishing :** The polish provide smoothness and specular surface on which device can be fabricated. The figure shows the polishing process. The polishing process can be done on single wafer or batch wafer process depending on apparatus. The wafers are mounted on a fixture, processed against the pad under high pressure and are rotated. The polish pad is made up of artificial fabric. A mixture of polishing slurry consist of colliodal suspension of fine SiO_2 particles in a aqueous solution of sodium hydral Na $(OH)_2$ sodium hydroxide oxidize the silicon and the silica particals in the slurry arbude the oxidized silicon away. The polishing rate depends on the properties of pad, applied pressure, composition of slurry, rotation speed and pH of the polishing solutiion.

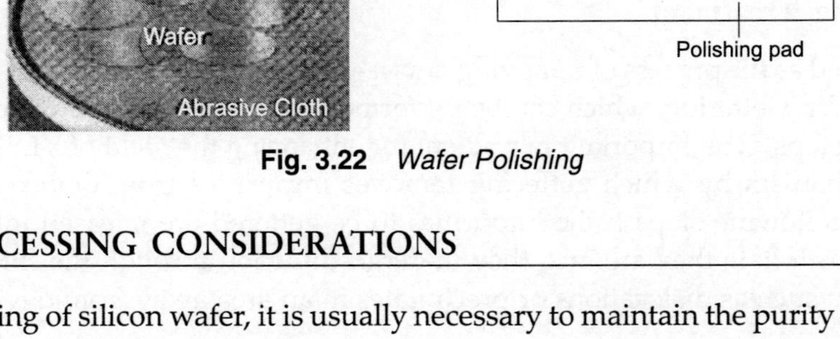

Fig. 3.22 *Wafer Polishing*

3.11 IC PROCESSING CONSIDERATIONS

In the IC processing of silicon wafer, it is usually necessary to maintain the purity and perfection of material.

Some of these processes are:

1. Chemical cleaning
2. Gettering treatment
3. Thermal stress factors

3.11.1 Wafer Cleaning

A sequence of chemistries is typically used to clean silicon wafers. This sequence was first developed at the RCA laboratories, and is therefore, often referred to as the RCA process. This chemical sequence does not attack the silicon material, but selectively removes the organic and inorganic contaminations that reside on the wafer surface. The following is a typical RCA process; many variations to the ordering of the sequence and chemical ratios are used throughout the industry.

- *General clean*: A general cleaning is accomplished by using a mixture of Sulfuric Acid and Hydrogen Peroxide. Mixing of these chemicals is dangerous and generates extreme heat. This industry standard clean removes organic and inorganic contamination from the wafer. 2–10 minute clean is recommended. Strong rinse in DI water is required after this cleaning step.
- *Particle removal*: A Megasonic clean (at about 70 C) in a 5:1:1 ratio mixture of DI water: Ammonium Hydroxide: Hydrogen Peroxide will remove silica and silicon particles from the wafer, as well as remove certain organic and metal surface contamination. 2–10 minute clean is recommended. Strong rinse in DI water is required after this cleaning step.
- *Oxide removal*: A 15–60 second dip in 1:20 HF:DI water will remove the native oxide layer and any contamination in the oxide from the wafer surface. HF is extremely dangerous and must be handled with great care. Strong rinse in DI water is required after this cleaning step.
- *Metal contamination removal*: A Megasonic clean (at about 70° C) in a 6:1:1 ratio mixture of DI water: HCL: Hydrogen Peroxide will remove certain ionic and metal surface contamination. 2–10 minute clean is recommended. Strong rinse in DI water is required after this cleaning step.
- *Spin rinse dry*: Wafers should be rinsed and dried in a standard spin-rinse dryer.

Megasonic agitation is commonly used with the chemcial bath and most commonly with the particle removal step. Also, heavy DI rinse steps are used between each chemical treatment. DI rinsing may use dump-baths, over-flow baths, and spray-dump baths, as well as combinations. Proper removal of all cleaning chemistry with 18 MegaOhm DI water is critical and needed after each chemical bath.

There are commercially available premixed cleaning solutions that can be used directly to clean wafers and serve the same purpose of the RCA cleaning process. These chemicals typically achieve the function of several cleaning steps with one solution.

3.11.2 Gettering Treatment

Gettering is defined as the process of removing device-degrading impurities from the active circuit regions of the wafer. Gettering, which can be performed during crystal growth or in subsequent wafer fabrication steps, is an important ingredient for enhancing the yield of VLSI manufacturing. The general mechanism by which gettering removes impurities from device regions may be described by the following steps:1. the impurities to be gettered are released into solid solution from whatever precipitate they are in:2. they undergo diffusion through the silicon ; 3. they are trapped by defects such as dislocations or precipitates in an area away from device regions.

There are two general classifications of gettering, namely, extrinsic, and intrinsic. Extrinsic gettering refers to gettering that employs external means to create the damage or stress in the silicon lattice in such a way that extended defects needed for trapping impurities are formed. These chemically reactive trapping sites are usually located at the wafer backside.

Several methods have been used to accomplish external gettering. For instance, the introduction of mechanical damage by abrasion, grooving, or sandblasting can produce stresses at the backside of a wafer, which when annealed create dislocations that tend to relieve these stresses. These locations can then serve as gettering sites. The main drawback of this method, of course, is its tendency to initiate and propagate wafer backside microcracks that may compromise the mechanical strength of the wafer.

Diffusing phosphorous into the wafer backside is another technique used for external gettering. P diffusion into silicon result in phosphorus vacancies or dislocations that serve as trapping sites for impurity atoms, such as gold. Another effect of P diffusioin is the creation of Si-P precipitates, which have been shown to be capable of removing Ni impurities through interactions between Si self-interstitials and Ni atoms, nucleating $NiSi_2$ particles in the process.

Introduction of damage by laser is another external gettering method. Scanning a laser beam across the wafer surface induce damage that is very similar to mechanical damage with the exception that the laser damage is 'cleaner.' Laser subjects the irradiated areas to thermal shock, forming dislocation nests that serve as gettering sites.

Ion bombardment to produce wafer backside damage is another method for external gettering, this time using high-energy ions to induce the necessary stress within the lattices of the wafer backside. Deposition of a polysilicon layer on the wafer backside has also been used for external gettering. Polysilicon layers introduce grain boundaries and lattice disorder that can act as traps for mobile impurities.

Intrinsic gettering refers to gettering that involves impurity trapping sites created by precipitating supersaturated oxygen out of the silicon wafer. The precipitation of supersaturated oxygen creates clusters that continuously grow, introducing stress to the wafer as this happens.

Eventually these stresses reach the point where they need to be relieved. Dislocation loops or stacking faults are thus formed to provide the necessary stress relief. These dislocations and faults subsequently serve as trapping sites for impurities.

A basic requirement of intrinsic gettering is, of course, starting wafers that have sufficient, but not excessive oxygen levels (15–19 ppma).

The advantages of intrinsic gettering over extrinsic gettering are:1. it does not require subjecting the wafer to any treatment except for heating; 2. its volume of impurity sink is significantly larger than that of external gettering on the wafer backside; 3. its gettering regions are much closer to the device regions.

3.11.3 Thermal Stress Factors

It is generally required to maintain the crystal perfection of wafers throughout the device fabrication process. The wafer should not be deformed during various process. To protect from deformation, the wafers are processed in furnance using racks with high wafer diameter to spacing ratio, in which a high temperature is maintain the wafer after taking out furnance, the edge of wafers cool rapidity by radiation to the surrounding, but the wafer centers remain relatively lot. The temperature gradient create the thermal stress S and can be estimated as

$$S = \alpha E \Delta T$$

where, α is the coefficient of thermal expansion.

E is young's modulus.

ΔT is the thermal difference across the wafer.

If these stresses exceed the yield-strength of material, dislocation in the wafer, to avoid dislocation. Before removing the wafer, the furnance temperature is lowered to the point where the yield strength at this temperature, increases the stress as shown in Fig 3.23.

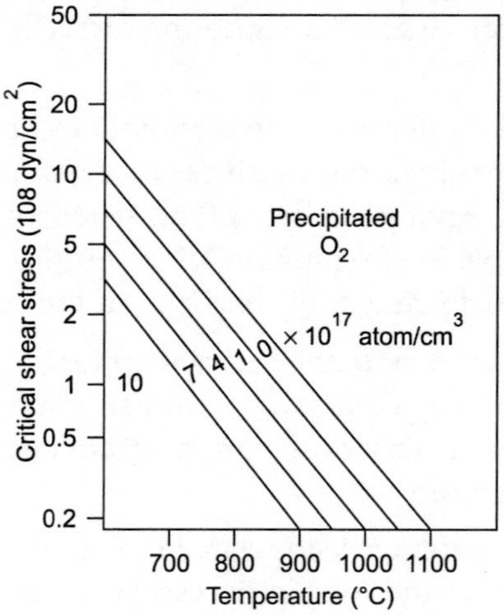

Fig. 3.23

3.12 REQUIRED PROPERTIES OF SILICON WAFERS

The wafers used for VLSI manufacturing must meet a multitude of stringent requirements to be able to serve as reliable starting substrate onto which integrated circuit will be built. These requirements include:

1. Electrical characteristics
2. Mechanical characteristics
3. Chemical characteristics
4. Surface characteristics

3.12.1 Common Electrical Specifications for Silicon Wafers

1. **Conductivity type**
   - specifies whether the wafers are *p*- or *n*-type.
   - specifies which element was used as dopant for the wafer.
2. **Resistivity or resistivity ranges**
   - usually measured in ohm-cm.
   - specifies the average resistivities or range of resistivities of the wafers.
   - the resistivity of the wafer is related to the doping density or concentration of dopants in the wafer.
   - wafer resistivity is often measured using the 4-point probe method defined.

3. **Radial resistivity gradient**
   - usually measured in % variation.
   - specifies the variation of the radial resistivity exhibited by the wafers as measured from the center to selected points in the outer regions of the wafers.
   - minimum radial resistivity gradient is desired to keep device characteristics in control.
   - radial resistivity is a function of the process used in growing the crystal as well as the dopants used.
   - radial resistivity gradient is often measured using the 4-point probe method.
4. **Local resistivity variations**
   - provides a measure of localized variations in resistivity exhibited by the wafers.
   - local resistivity variations are often due to remelt phenomena, and are usually more pronounced in FZ than CZ wafers.
   - local resistivity variation is often measured using the 4-point probe method.

3.12.2 Mechanical Specification for Silicon Wafers

1. **Crystal growth and doping process**
   - specifies the wafer's single crystal growing process, *i.e.*, whether CZ or FZ.
   - specifies the wafer's doping process, *e.g.*, ion implantation, etc.
2. **Wafer diameter**
   - specifies the diameter of the wafer (excluding flats).
   - the diameter specification is given with high precision to facilitate automated wafer handing; example:150+/-0.2 mm.
   - it is usually measured using a ring template.
3. **Wafer thickness**
   - specifies the thickness of the wafer.
   - wafer thickness is measured at the center of the wafer with a wafer thickness gauge.
   - wafer thickness is also an important parameter since a very thin wafer is prone to warp age or breakage while a very thick wafer may not allow processing in all wafer fab equipment.
4. **Total thickness variation(TTV)**
   - specifies the difference between the maximum and minimum values of wafer thickness.
   - excessive thickness variations can result in automated processing problems, especially with respect to mechanical handling and photolithographic processes.
   - TTV is measured using a non-contact thickness gauge.
5. **Bow**
   - specifies the concavity of the wafer due to sawing or the deformation of the wafer centerline due to thermal processing.
   - bow is a bulk property (and not just a surface property) expressed in microns, and can be measured.

6. **Warp**
   - specifies the difference between the minimum and maximum distance of the wafer center line from a reference plane.
   - warp which is also a bulk property of the wafer, takes into account both concave and convex deviations of the wafer.
   - warp can be measured ASTM Std F-657-80 thank.
7. **Flat dimensions**
   - specifies the precise lengths and orientations of the wafer flats.
8. **Edge contour**
   - specifies the rounding of wafer edges.
   - wafer edges are contoured to reduce chipping and film build- up.
   - edge contour is measured by comparison with a template.
9. **Laser markings**
   - specifies any markings made by laser on the surface of the wafer.
10. **Other mechanical attributes**
    - specifies the presence of attributes like chip-outs and indents on the wafer.
    - chip-outs are regions of missing silicon that do not go all the way through the slice while indents are regions of missing silicon that go all the way through the slice.

3.12.3 Chemical Specifications for Silicon Wafers

1. **Bulk structural defects**
   - a starting wafer must not have any structural defect as revealed by x-ray topography or any suitable inspection process.
2. **Surface orientation**
   - specifies the orientation of the surface of the wafer as well as the tolerance for Misorientation.
   - generally, <100> material is cut-on orientation-/-0.5 deg.
   - generally, <111> material is cut-off orientation 3+/-0.5 deg.
3. **Oxygen content**
   - specifies the amount of oxygen within the silicon wafer.
   - oxygen content needs to be specified because it affects wafer strength, impurity gettering mechanisms, thermal donors, etc.
   - oxygen content is measured.
   - interstitial oxygen content specifications may be tailored to customer requirements.
4. **Carbon content**
   - specifies the amount of carbon within the silicon wafer.
   - carbon content is measured.
   - carbon content measurement can be difficult due to low levels of carbon in modern wafers (in the order of 1 ppma).

5. **Swirl defects**
   - swirl defects, which exhibit a spiral or swirl-like pattern, should not be present in wafers before processing.
   - swirl defects are the result of the agglomeration of point defects when the crystal cooled down.
   - swirl defects consist of two variants: 1) the A- swirls, which can be delineated as hillocks; and 2) the B- swirls, which can be delineated as small and shallow pits.
   - wafer crystals can now be grown without swirl defects.

3.12.4 Surface Specifications that Must be Checked in Wafers Include

1. *Flatness*: the maximum deviation from peak to valley exhibited by a wafer as measured with respect to a reference plane. Flatness is usually express in micrometers (microns) and measured using an industry-approved method. Depending on the application of the wafer, a wafer flatness of 1–3 microns is typically acceptable.
2. *Particle density*: the number of particles present per unit area of the wafer. Detection of particles can be done by visual means, optical microscopy, or with automatic surface scanners. ASTM Std F-154 and F-24 are examples of standards used for this requirement. Particle density must be minimized in wafers.
3. *Cracks*: fracture that extends to the wafer surface caused by impact of sharp features of fixtures against the wafer Crow's feet which is a special form of crack, is a fracture that resembles a crow's foot in appearance in <111> material and a cross pattern in <100> material. A wafer must not have any cracks.
4. *Haze*: non-localized scattering of light caused by microscopic irregularities or roughness on the surface of the wafer. Such surface irregularities are often caused by crystalline defects and surface imperfections in the wafer or by mechanical damage produced by external factors. Haze is usually detected with a narrow bean of tungsten light per
5. *Saucer pits*: etch pits that are small and shallow. These microscopic defects are associated with wafer fab processes like chemical etching, oxide stripping, etc. Also known as S-pits, their presence is usually expressed as a defect density. S - pit densities of less than 100 per sq. cm is tolerable in many applications.
6. *Saw marks*: mechanical damage caused by the slicing operation of the saw blade. Saw marks may be removed by lapping.
7. *Surface defects*: other defects on the wafer surface due to various causes, e.g., stains, streaks, smudges, residues, dimples, mounds, orange peel, scratches etc.

SUMMARY

In this Chapter, first we have given the properties of silicon, Crystal Structure of silicon. The silicon crystals used in the electronic industry, crystal perfection, purity and uniformly are not necessarily highest on the list of desirable attributes for crystalline S Incorporated into commercial PV modules. The defects like point line, Area and volume is also discussed. The silicon is available MGS form it is necessary to convert it into EGS for the device fabrication, then raw material is then used to grow single crystal principally using Czochralski method float zone and bridgeman method is also discussed, then we have discussed about the wafer preparation techniques like grinding, calting, etching and polishing etc. The wafers then form the starting material for IC manufacturer. It is remarkable that such wafer's can be manufactured in large volumes and sold in about $50 per wafer for IC manufacturing.

TUTORIAL SHEET

1. Discuss the silicon material properties at 300 K.
2. Explain the Crystal structure of silicon.
3. How the silicon is different from germanium?
4. Discuss the following crystalline defects in silicon
   (*a*) Point defect
   (*b*) Line defect
   (*c*) Arco defect
   (*d*) Volume defect.
5. How will you prepare EGS from MGS?
6. Explain Crystal growth theory of silicon.
7. Explain the Czocharlski method for crystal growth.
8. Explain Bridge man method for crystal growth.
9. At 300 K the lattice constant for the silicon is 543 A°. Calculate number of silicon atoms per centimeter and density of the silicon at room temperature?
10. What are the steps for the wafer preparation after crystal growth?
11. Why are wafer edge rounded?
12. Given two reasons why semiconductor wafers require a flat surface.
13. Draw a Cubic unit cell and identify the <1 0 0> plane.
14. What is the orientation of a semiconductor crystal if the seed has a <1 0 0> orientation?

4

EPITAXIAL GROWTH

4.1 INTRODUCTION

Epitaxy is a kind interface between a thin film and a substrate. The term epitaxy (Greek ; epi "above" and taxis "in ordered manner") describes an ordered crystalline growth on a monocrystalline substrate.

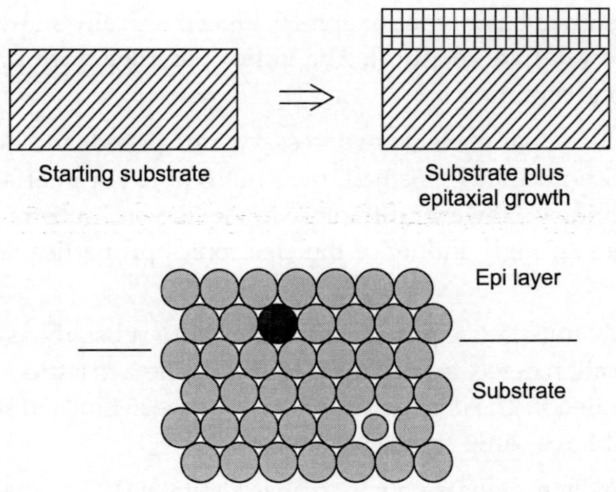

Fig. 4.1 *Epitaxy Basic*

Epitaxy is used in silicon based manufacturing processes for BJTs and modern CMOS, but it is particularly important for compound semiconductors such as gallium arsenide. Manufacturing issues include control of the amount and uniformity of the deposition's resistivity and thickness, the cleaniness and purity of the surface and the chamber atmosphere, the prevention of the typically much more highly doped substrate wafer's diffusion of dopant to the new layers imperfections of the growth process, and protecting the surface during the manufacture and handling.

Epitaxical wafers account for some 20% of all wafer sold. A lightly doped epitaxial *p*-type layer (10 ohm – cm) can be grown on a heavily *p*-doped substrate, wafer (0.2 ohm – cm). This is the material for advanced microprocessors and other high performance logic circuits. *n*-silicon on *p*-substrate is cased in many micro mechanical device because electrochemical etching are not used these days. The number and thickness of layer is practically unlimited for IGBT (Insulated Gate Bipolar Transistor) is power transistor in which a *n*-type layer is grown and then a thicker lightly doped layer. Epitaxy is generally enhanced the performance of these power transistor. It also control and the doping profile in a device structure beyond that available with diffusion and ion implantation.

Epitaxial films may be grown from gaseous or liquid procedures. Because the substrate acts as a seed crystal, the deposited film taken on a lattice structure and orientation identical to those of the substrate. This is different from other thin-film deposition methods which deposit polycrystalline or amorphous films, even on single-crystal substrates. In a film is deposited on a substrate of the same composition, the process is called homoepitaxy, otherwise it is called heteroepitaxy.

Homoepitaxy is a kind of epitaxy performed with only one material. In homoepitaxy, acrystalline film is grown on a substrate or film of the same material. This technology is applied to growing a more purified film than the substrate and fabricating layers with different doping levels.

Heteroepitaxy is a kind of epitaxy performed with material that are different from each other. In heteroepitaxy, a crystalline film grows on a crystalline substrate or film of material of which single crystals cannot be obtained end to fabricating integrated crystalline layers of different material. Examples include gallium nitride (Ga N) or sipphire or aluminium gallium indium phosphide (Ai Ga In P) or gallium arsenide (Ga As).

The difference between two basic types of epitaxy are simitarcally shown in Fig. 4.2. When the epilayer and substrate crystal are identical. The lattics parameter are preferably matched and there is no interfacial bond straining.

In heteroepitaxial the lattic parameters are necessary unmatched, and depending on the extent of mis-match, if the lattice mismatch is small, then hetro-junction interfaced structure is essentially like that for homoepitaxy. However difference in the film and substrate chemistry, coefficient of thermal expansion can strongly influence the electronics properties and performance of the interface.

When the film and substrate lattice parameters differ more substatly as shown in other case in Fig. 4.2 either edge dislocation defect from at the interface or the two lattices strain to accommodate their crystallographic difference. The phenomenon occurs between film and substate pair compossed of different material that have same crystal structure.

Epitaxial silicon is usually grown using vapor-phase epitaxy (VPE), a modification of chemical vapor deposition. Molecular-beam and liquid phase epitaxy (MBE and LPE) are also used, mainly for compound semi-conductors.

4.2 PRINCIPLE OF EPITAXIAL GROWTH

The process of epitaxial growth is taken place in the chamber same as in the case of chemical vapour deposition (CVD) but with much greater control. The basic concept of vapour phase epitaxy (VPE) is explained with help of figure.

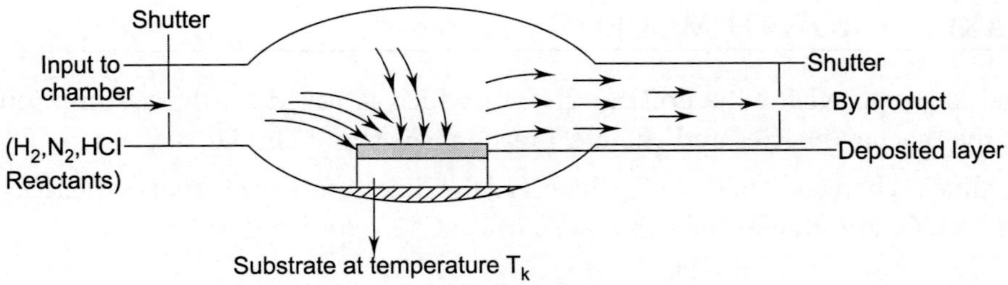

Fig. 4.2 *Growth Reaction Chamber*

First the reactants like H_2, N_2 HCl allowed to flow inside the chamber through the input side of the chamber for the short interval of time whic cleans the chamber, HCl etches the unwanted chemicals and deposited material inside the chamber. After that the substrate is placed inside the chamber and raise the temperature of substrate heater as soon as gases are fed into the growth chamber, these chemical species undergoes a series of physical and chemical reaction, which results in the layer deposition, when the reactants are passes through the chamber. It reacts with surface and absorbed on the surface of substrate then flow of gasous remove these reactants and by products from the chamber. In this process the temperature of substrate plays very important role because the single crystal epitaxial material is deposited data parameter temperature this process is explained in more accurate manner with this flow diagram as shown in Fig. 4.3.

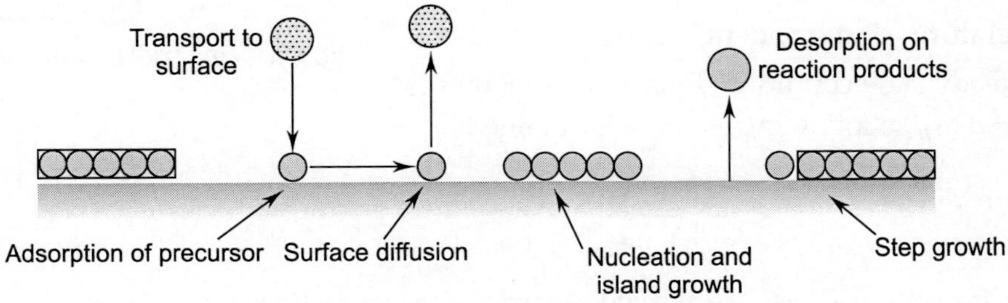

Fig. 4.3 *Reaction During Epitaxical Growth*

We now must have tremendous process control, we require:

- *A precise continuation of the substate lattice.* There should be no way whatsoever to identify the interface after the epitaxial layer has been deposited. This means that no lattice defects whatsoever should be generated.
- *Doping* of the epitaxial layer with high precision (*e.g.*, 5 Ω cm ± 5%) and doping is usually very different from that of the substrate.
- *Precise-thickness control e.g.*, 1.2 μm ± 10% over the entire wafer, from wafer to wafer and from day today. Now, there is challang if you met the first point and thus can't tell where the interface is – How do you measure the thickness ? (The answer : only electronically *e.g.*, by finding the position of the *p-n* junction produced).
- *Cleanliness.* No contaminants diffusing into the substrate and epitaxial are allowed.

This looks tough and it is indeed fairly difficult to make good epitaxial layers, it is also quite expensive and is therefore avoided whenever possible (*e.g.*, mainstream CMOS tecnology). It is, however, a must in bipolar and some other technologies.

4.3 EPITAXIAL GROWTH MODEL

Grove model is proposed for the epitaxy growth which is based on the assumption that only transport of the reactant species and chemical reactions is shown in Fig. 4.4.

Assuming linear change from C_g to C_s, the F_i *i.e.*, the flux difference between the reactants in the input gas stream, C_g and that at the substrate surface C_s is expressed as:

$$F_i = H_g (C_g - C_s) \quad \text{... }(i)$$

H_g – gas phase mass transfer coefficient and it is relatively not dependent on temperature.

C_g – concentration of reactants species in gas.

C_s – concentration on the substrate surface.

To make calculation easier, let us assume F_o is proportional to C_s.

So, F_o becomes

$$F_o = R_s C_s \quad \text{... }(ii)$$

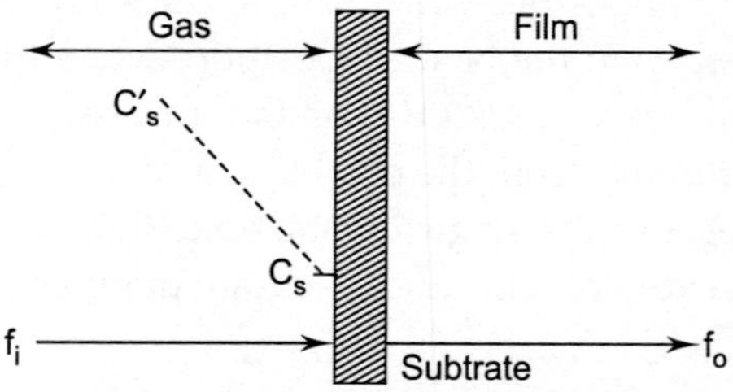

Fig. 4.4 *Epitaxial Growth Grove Model*

In the equation (*ii*), R_s is the chemical surface reaction constant.

$$R_s = R_o \exp (E_a/kT)$$

whereas, R_s strongly depends on temperature

E_a – The activation energy

R_o – temperature – Independent – Const

Since, at steady state condition, F_1 is equal to F on the equation (*i*) and (*ii*), so following equality is obtained

$$C_s = \left(\frac{C_g}{1 + R_s/H_g} \right) \quad \text{... }(iii)$$

If $H_g >> R_s$, it result in $C_s \approx C_g$ and called as surface-contracted process. Whereas,

if $H_g << R_s$, $C_s \approx O$, At this condition, the process is called mass transport-controlled. With simple mathematics, the growth rate G of vapour phase epitaxy process may be obtained and it is given as:

$$\text{Surface Controlled} - G = C_t R_s M/N \quad \text{... }(iv)$$

$$\text{Mass transport controlled} - G = C_t H_g M/N \quad \text{... }(v)$$

In the above two equations, C_t is the total number of molecules per cm^3 in the gas stream M the mole fraction of reactant species in gas stream. N the number of atoms incorporated into unit volume of the layer.

4.4 METHODS FOR DEPOSITING EPITAXIAL FILMS

The three methods of depositing epitaxial layer on the substrate are:

(*i*) Liquid Phase Epitaxy (LPE)
- Semiconductor solid from liquid solution
- An equilibrium process using liquidces – solidies equilibrium.

(*ii*) Vapour Phase Epitaxy (VPE)
- Semiconductor solid from gas sources
- A special case of chemical vapour deposition (CVD).

(*iii*) Molecular Beam Epitaxy (MBE)

- Semiconductor solid from atomic or molecular beam.
- Beam arrive directly on the growth surface without any prior interferance or interaction (feasible in an ultra high vacuum environment).

4.4.1 Liquid Phase Epitaxy (LPE)

We start with LPE, the process in which melt rather than vapour are in contact with growing film, introduced in the early 1960s. LPE still used to produce hetrojunction devices. LPE grow semiconductor crystal layer from melt on solid substrate. This happens at the temperature well below the melting point of the deposited semiconductor. The semiconductor is dissolved in the melt of another material. At conditions that are close to the equilibrium between dissolution and deposition, the deposition of the semiconductor crystal on the substrate is slow and uniform. The equilibrium condition depend vary much on the temperature and on the concentration of the dissolved semiconductor in the melt. The growth of the layer from the liquid phase can be controlled by a forced cooling of the melt. Impurity introduction can be strongly reduced. Doping can be achieved by the addition of dopants.

The method is mainly used for the growth of compound semiconductors. Very thin, uniform and high quality layers can be produced. A typical example for the liquid phase epitaxy method is growth of ternery and quartery. III-V compound on Ga As (Gallium arsinide) substrate. As a solvent quite often gallium is used in this case. Another frequently used substrate is In P (indium Phosphide).

The process of LPE can be understood by Ga As binary phase diagram as shown in Fig. 4.5.

Consider a Ga-rich melt containing 10% at As. When heated above 950°C, all of the As dissolves, if the melt is cooled below the liquidus temperature into the two phase field. It become supersaturated with respect to As. Only a melt of lower than the original. As contents can now be in equilibrium with Ga As. The excess As, is therefore rejected from solution in the form of Ga As that grows epitaxially on a suitable placed substrate.

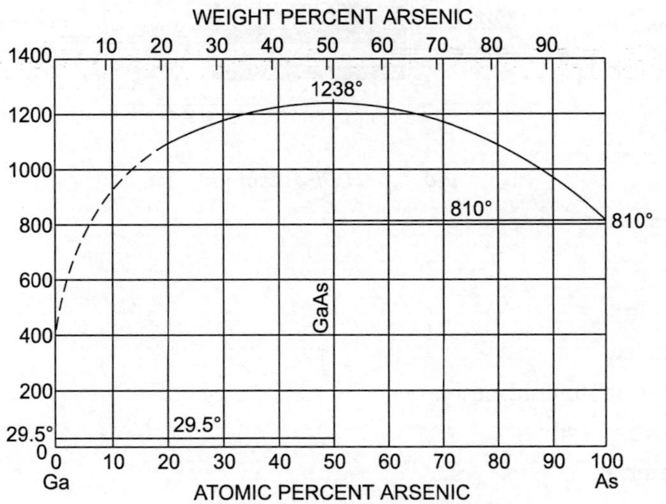

Fig. 4.5 *Binary Phase Diagram*

The various kind of reactors used for the LPE process are discussed below in Fig 4.6. The first type of LPE reactor is 'Tapper type or boat type, a graphite seed holder is used to hold substrate and solution. Quartz furnace tube is used to provide isolation from the outer side or environment. The graphite slider holding the substrate is moved so as to locate then under well this is placced in a furnace, in a neutral carrier hydrogen gas ambient. The substrate is moved under the first well at a required temperature, then the temperature is decreased. The substrate is moved out from the under the solution to stop the growth and then remove the substrate other two reactors also works in the same manner and also observe the temperature cycle for LPE growth process.

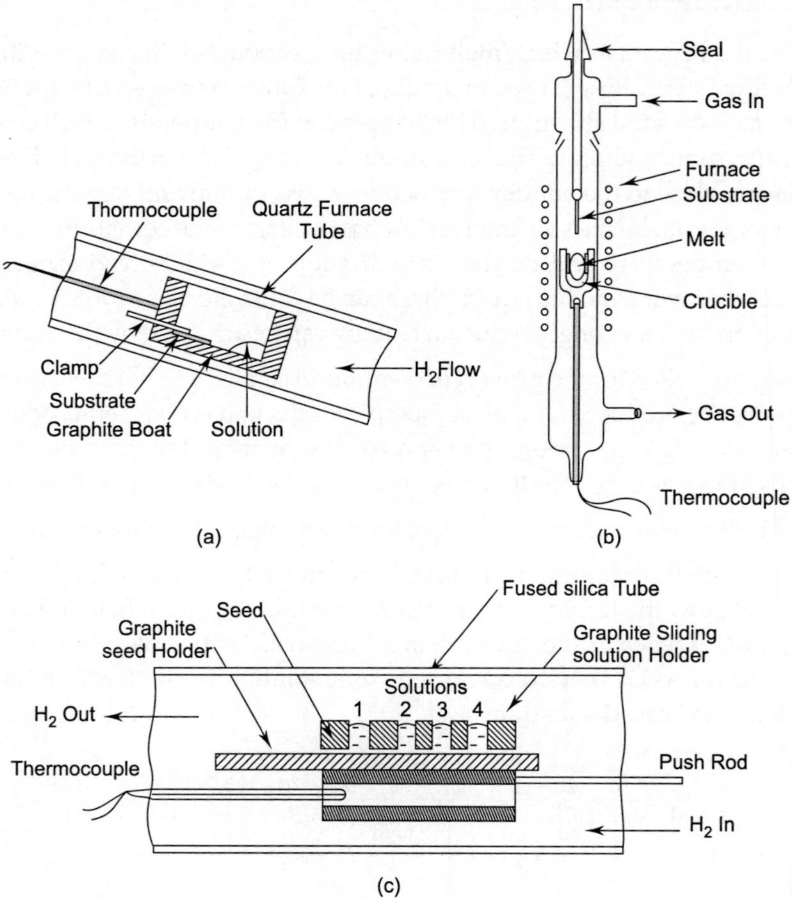

Fig. 4.6 *LPE Reactors*

Advantages of LPE

(*a*) Less expensive.
(*b*) Fast deposition rate.
(*c*) Low defect density.
(*d*) Excellent control of stoichiometry.
(*e*) Low temperature possible.
(*f*) Easy to use material.
(*g*) Suitable for selective growth.

Disadvantages of LPE

(*a*) Solubility consideration limit the number of materials that can be grown by LPE.
(*b*) Morphology (crystal orientation) control is difficult.

(*c*) Surface quality is poor.
(*d*) Difficult to control thickness.
(*e*) Difficult to grow strained layers.
(*f*) High growth temperature for certain compounds (*e.g.*, GaAs at apprax 800 – 900° C, but InP at 600°C).

4.4.2 Vapour Phase Epitaxy (VPE)

Vapour phase epitaxy method for growing epitaxial layers of GaAs, as well as advanced form of VPE, including metallorganic VPE (MOCVD) and Rapid thermal VPE (RTCVD).

The prototype of VPE systems are shown in figure, these are low pressure CVD, the wafers are loaded in the vacuum chamber on the heated susceptor. This heating is done with filament lamps (radiations) or heating the graphite susceptor (inductivity). The gas flows are controlled with mass flow controllers and pneumatic valves.

The common reactors used for the epitaxial growth are horizontal, pancage, barrel type, the reactors a consist of various reactants as shown in Fig. 4.7.

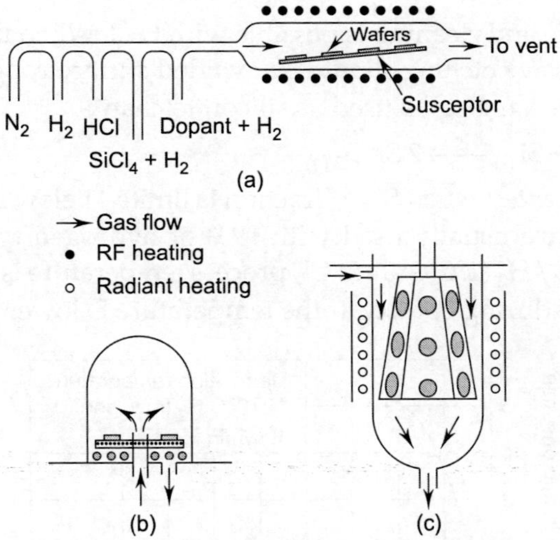

Fig. 4.7 *Vapour Phase Epitaxy Reactors*

Input valves, susceptor which not only mechanically support the wafers. They also serve as source of thermal energy for the reaction. This susceptor is made from the graphite, and outlet valve to exit the reactants and by-products of the reactions. The various kind of susceptors are used in the epitaxial process, as shown in Fig. 4.8. Each design has its own advantages and disadvantages. Horizontal reactor offers lower cost of construction, but controlling the deposition process through out the length of the susceptor is a problem. In pankage reactor, are capable of very uniform deposition but suffers from mechanical complexity. Barrel type reactors are also used for uniform deposition, but are not suitable at temperature above 1200 °C.

Working of VPE Growth

A hydrogen carrier gas purges reactor of air. The reactor is then heated to a temperature. After thermal equilibrium is established in chamber, HCl is allowed to flow through the chamber which etch the chambers walls for unwanted material and clean the chamber, this etching is take place at temperature of 1150 °C to 1200 °C for 3 min, then the temperature is reduced to T_k. The silicon sources and depand flows are turned on and growth proceeds at rates of 0.2 to 2.0 μm/min. When growth is done the silicon flow and dopants flow stopped and temperature of the chamber reduced by turn-off the power. After this to cool the reactor the flow of the hydrogen is replaced with nitrogen flow so that the reactor may be opened safely, the capacity of the susceptor is depends on the diameter of wafer and reactor type. It may varies from 10 to 50 wafers per batch.

Chemical Reactions During Epitaxial Growth

The most common silicon sources used for epitaxial growth are : silicon tetrachloride ($SiCl_4$), trichloro siline ($SiHCl_3$), dicholorosiline ($Si\,H_2\,Cl_2$) and siliane ($Si\,H_4$). Silicon tetrachloride is mostly used in industry. In chemical vapour epitaxy, the reactor is heated to 1200 °C under:

the hydrogen flow which reduce native oxide

$$Si\,O_2\,(s) + H_2 \xleftrightarrow[1150°C-1200°C]{} SiO_{(vapour)} + H_2O_{(vapour)}$$

growth commences when siliane gases of the type $Si\,H_x\,Cl_{4-x}\,(0 \le x \le 4)$ are introduced into the reactor.

$$Si\,H_{4\,(g)} \longrightarrow Si_{\,(s)} + 2H_{2\,(g)}\quad T = 1000\ °C$$

$$Si\,Cl_{4\,(g)} + 2H_2 \longrightarrow Si_{\,(s)} + 4HCl_{(g)}\quad T = 1250\ °C$$

This reaction is reversable and cleaning is possible with HCl. When the reaction takes place, the HCl etching of silicon excessive etching should be avoided surface roughness tends to increase in etching silicon tetrachloride can also be used as silicon tetchant.

$$Si\,Cl_{4\,(g)} + Si_{\,(s)} \longrightarrow 2\,Si\,Cl_{2\,(g)}$$

This reaction can be prevented when $Si\,Cl_4$ fraction is limited below 27% (shown in Fig. 4.8) but much more dilute silanes are usually used with 99% of hydrogen typical. The typical growth condition is 1 μm/min. $Si\,Cl_4\,/\,H_2$ (1% 99%). $Si\,Cl_4$ process temperature is very high and undesirable dopant diffusion take place during epitaxy. If the temperature is lower the rate is also lower.

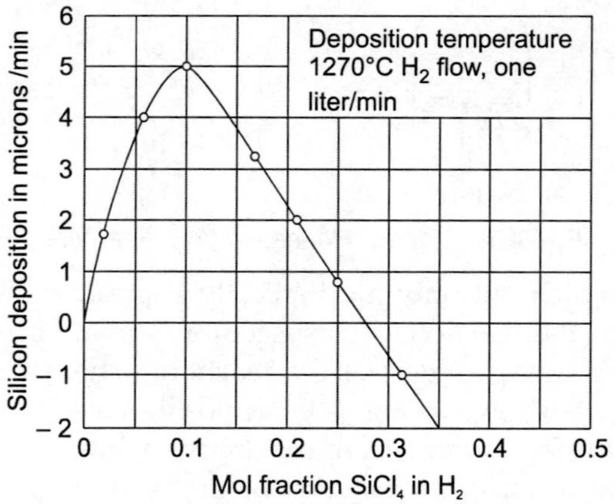

Fig. 4.8 *Epitaxial Growth Rate as a Function of $SiCl_4/H_2$ Flow Ratio*

The tricholorosiline ($SiHCl_3$) and dicholorisiline (SiH_2Cl_2) are good compromise between deposition rate and operating temperature as in reaction:

$$Si + 3HCl \longrightarrow SiHCl_3 + H_{2\,(g)}$$

$$SiH_2Cl_{2\,(g)} \longrightarrow Si_{(s)} + 2HCl_{(g)},\ 1150\ g$$

The typical epitaxial growth rates are 1 to 5 μm/min they depends on the silane gas choosen, temperature and on flows. The main advantages and disadvantages of $SiCl_4$ as a source of Si epitaxy are as follows:

Advantages:

(*i*) $SiCl_4$ is non toxic, inexpensive and easy to purity.

(*ii*) The reaction making silicon from $SiCl_4$ takes place only at the surface and not on the boat or reaction chamber walls.

Disadvantages:

(*i*) $SiCl_4$ process requires higher temperature than silane process and also has slower growth rate.

(*ii*) The growth process is accompolished by diffusion phenomenon which exchange of impurities between silicon wafer and growing films.

4.4.3 Metal Organic Chemical Vapour Deposition (MOCVD) or (MOVPE)

Metal organic chemical vapour deposition (MOCVD) or (MOVPE) began in the late 1960s. MOCVD processes developed using group III organic compounds and group V hydrides. MOCVD system is capable of the controlled growth of thin, atomically abrupt, extremely perfect epitaxial layer. The system may be divided into three subsection. These are (*a*) gas delivery — the gas delivery system can be specified easily. In this system, it is desired that growth chamber should be provided well mixed, and well controlled flow of the number of gases. (*b*) Reaction Chamber where all the reaction or process take place. (*c*) gas exhaust is used to exhaust the used gases and by-products of the process. This system is shown in Fig. 4.9.

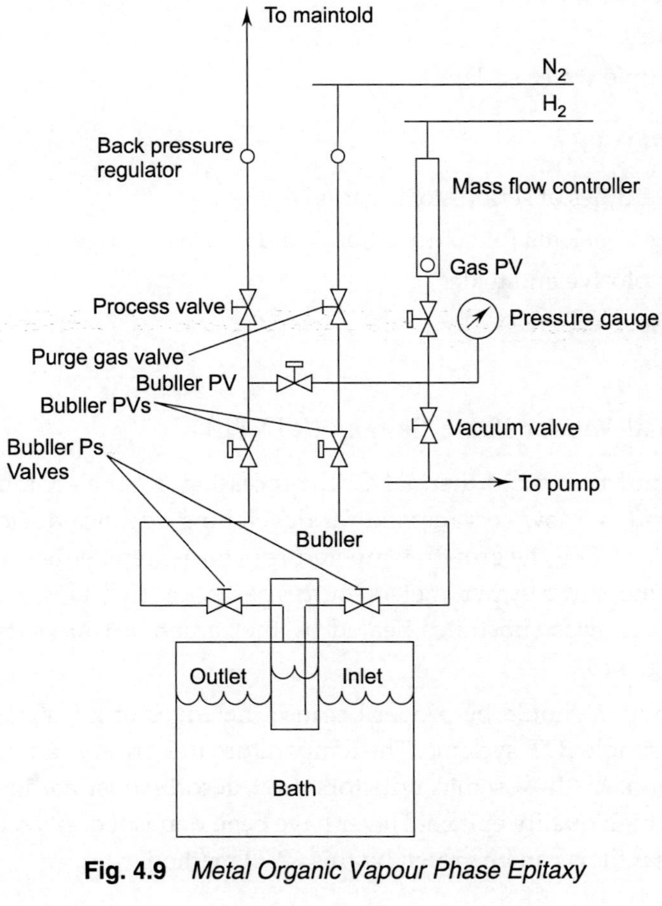

Fig. 4.9 *Metal Organic Vapour Phase Epitaxy*

Organic metal sources are GaAs this is typically H_2, TMG or TEG, AsH_3 and a dopant typically DMZ (dimethylzinc) for *p*-type doping or SiH_4 for *n*-type doping. The OM source consist of a bubbler immersed in a controlled temperature bath of glycerine and water. The OM vapours is obtained by heating the liquid in a constant temperature bath and bubbling H_2 through the bath. Temperature of the bubbler must be carefully controlled. The back pressure regulator is used along with temperature of the bath to established the mole fraction of the OM in the gas that is sent to the reactor, hyrides sources are delivered as high pressure gases, just as in the standard CVD. Due to toxicity of hydrides these lines are typically double contained, leak checked and continually sniffed for leaks. The vapour of these compounds can be recondense in the tubing a before it reaches chamber, it result in poor controlled deposition conditions unless tubing valve are heated too, the reactions related to MO are taken place in the chamber. At last the gas exhaust should allow the system to operate at controlled pressure. The scrubbing is made either in a high temperture furnace in which hydrides are oxidized after scrubbing the gases are flushes from the chamber.

Advantages of MOVPE

(*a*) Excellent surface / interface morphology and thickness control.
(*b*) Precisely controlled abrupt or graded heterojunctions.
(*c*) Complex heterostructures.
(*d*) High growth rates.
(*e*) Large area, multiple wafer scale-up.

Disadvantages of MOVPE

(*a*) Safely - large quantities of serious toxic gases (AsH_3).
(*b*) Moderately long transients for compositional and doping change.
(*c*) SiH_4 is highly explosive and toxic.
(*d*) H_2 is highly explosive.
(*e*) CH_2 is flammable.

4.4.4 Rapid Thermal Vapour Phase Epitaxy (RTVPE)

This method is also known as rapid thermal CVD processing, is an elaboration on conventional vapour phase epitaxy. It is a low cost method for developing advance devices structure. As we know that in the standard VPE the growth temperature is measured by thermocouples. In RTVPE growth temperature is measured by pyrometers sources gases (*e.g.*, hydrides, metalorganics, halides) react on low — thermal mass substrates heated by the radiation from external high intensity lamps (as shown in Fig. 4.10).

The thermal distribution should be proper because the edges of a wafers are cooler than the center of the wafer in simple RTP systems. The temperature uniformity can be optimized only for single process condition. In III–V semiconductors, first desorb substrate impurities at elevated temperature and then high quality epitaxial layer have been deposited. A very high quality lattice matched heteroepitaxial films can be grown by this CVD method.

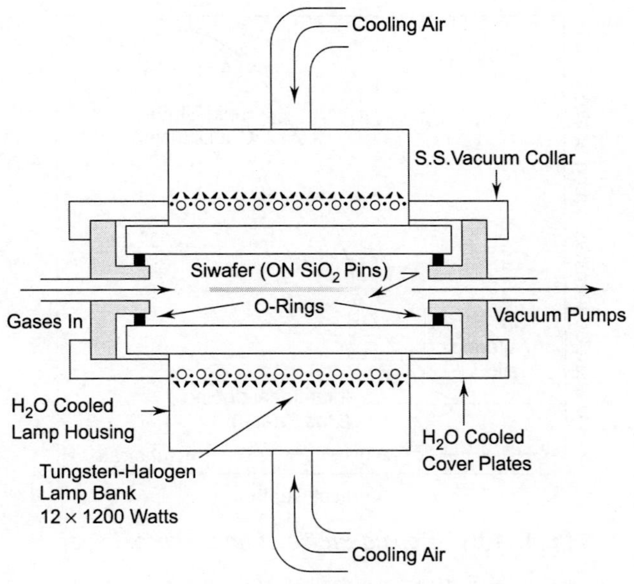

Fig. 4.10 *RTPVPE Reactor*

4.5 DOPING AND AUTODOPING

The consideration applied to epitaxial growth process are also applicable to doping. Typical hydrides of the impurity atoms are used as source of dopants. The reaction for arsenic dopant is as below:

$$2AsH_{3\,(gas)} \longrightarrow 2As_{(solid)} + 3H_{2\,(gas)}$$
$$\longrightarrow 2As_{(solid)} \longrightarrow 2As^{+}_{(solid)} + 2e$$

The hydride AsH_3 does not decompose spontaneously as it is stable because of the large volume of hydrogen present in the reaction. The interaction between the doping process and the growth process take place. In addition to intentional dopants incorporated into the layer, unintentional dopants are introduced from the substrate. This effect is known as Autodoping. Autodoping limits the minimum layer thickness that can be grown with controlled doping as well as the minimum doping level. There are variety of dopants is obtained as shown in Fig. 4.11(*a*). The dopant be released from the substrate from the suspector. The outgoing in the gaseous phase, when doping profile of epitaxial layer grown over a heavily doped substrate is shown 4.11(*b*).

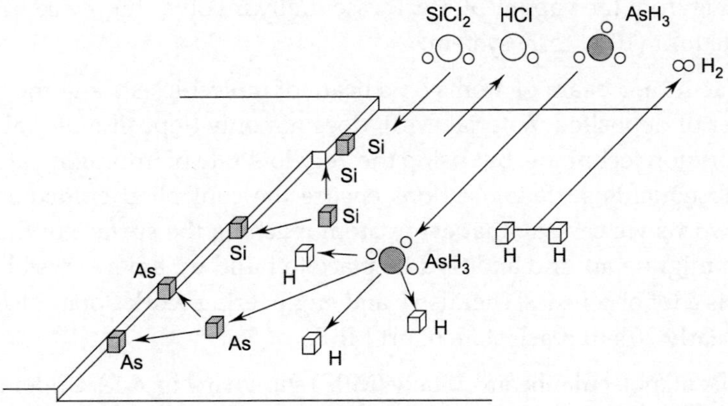

Fig. 4.11(a) *Schematic Representation of Arsenic Doping and the Growing Processes*

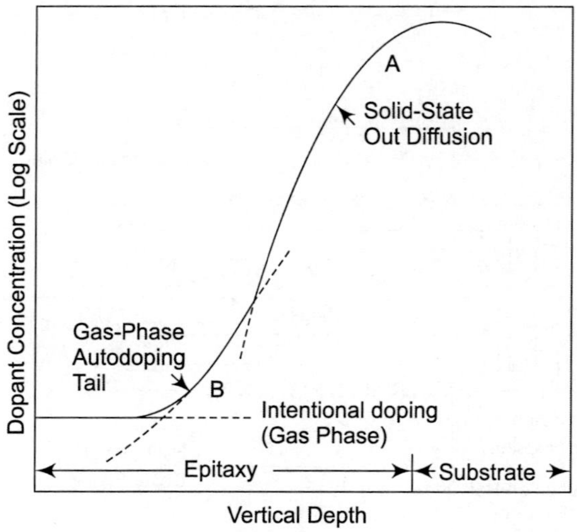

Fig. 4.11(b) *Doping Profile of an Epitaxial Layer*

The some points are observed from the profile plot are:

The first region close to the substrate interface out diffusion in solid phase from the substrate is predominant, resulting in a broad transition region. This effect is limited because the growth rate is normally much higher than the speed of solid state out diffusion. After this the dopant exceeds that intentionally introduced, an autodoping tail develops. This effect become is sufficient as the growth process proceeds, because of the coverage of a low doped layer over substrate and susceptor which slows down the supply of dopant of the surface. The doping tail is depends on dopants species of substrate and its concentration; reactor design, growth rate, growth pressure and temperature.

4.6 MOLECULAR BEAM EPITAXY (MBE)

4.6.1 Concept

Molecular beam epitaxy is conceptually a rather simple and versatile technique for growing thin epitaxial structures made of semiconductors, metals or insulators. The state of the art attainable in deposition processing from the vapour phase. It essentially involves highly controlled evaporation in an ultra high-vacuum (10^{-10} torr) system.

In a ultra-high vacuum, a beam of atoms or, a beam of molecules stick at the substrate's surface forming a new layer of deposited material. MBE does not only deposit material like. It is done by conventional evaporation technique, but using the very low rate of impinging atoms, migration on the surface and subsequents surface reactions ensure the controlled epitaxial growth of a new layer. In the other words we can say that every atom reaching the surface of the heated substrate has enough time to migrate around and find its place to build up a new crystal lattice. This looks simple but there was a lot of physics, chemistry and engineering work done since begining of MBE in the late 60's and early 70's to reach state of art MBE.

The basic principle of molecular beam epitaxy (MBE) shown in Fig. 4.12. Under ultra high vacuum (UHV) conditions the various constituents are coevaporated onto clean crystalline substrate surface which serves as seed crystal and is usually heated to provide high mobility of the impinging particle for crystalline growth. In conventional MBE especially group IV semiconductor (Si, Ge) group III/V (III = Al, Ga, In ; V = P, As, Sb) and group II/VI (II = Pb, Cd, Hg ; VI = S, Se, Te) compounds are widely used. These crystals exhibit covalent bonding in all three directions (3D) materials) and dangling bonds occurs at their surface which have to be matched at these 3D/3D heterojunctions (as shown in Fig. 4.12 since length and angle of covalent bonds at the interface can not easily be changed, the lattice mismatch Δa which defines as:

$$\Delta a = \frac{\left| a_f - a_s \right|}{a_s}$$

(a_f and a_s denote the in plane lattice parameter of film and substrate, respectively) is of crucial importance for the quality. For a lattice mismatch only 1%, the covalent interaction at the interface usually causes strong disturbances at interface like strain, structure imperfections and dangling bonds (edge dislocation) introducing a considerable defect density.

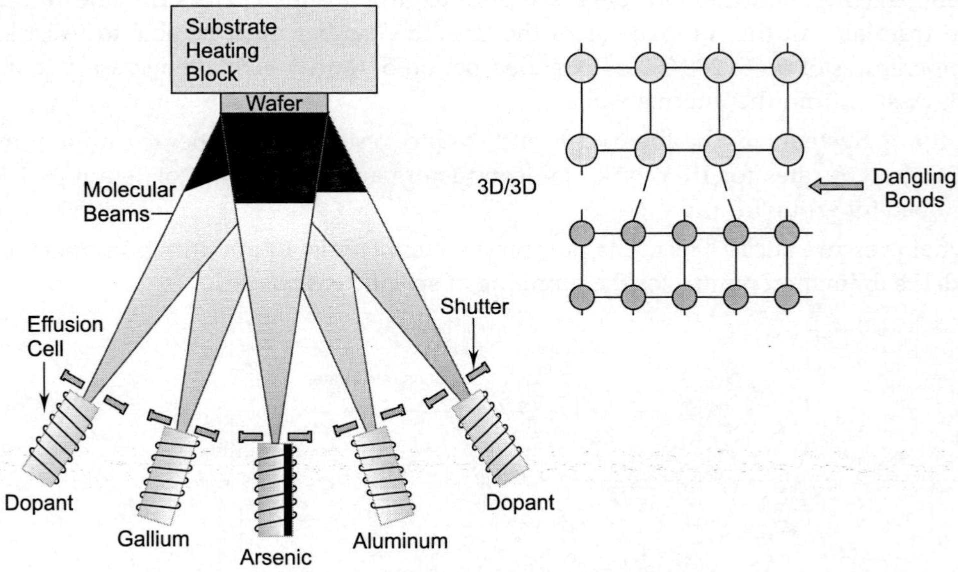

Fig. 4.12 *MBE Principle*

4.6.2 MBE Growth Equipment

MBE technique was developed with aim to obtain an integrate devices that require very percise controls over thickness and doping uniformity. It is performed at low temperature range *i.e.,* 500 to 900 °C. The silicon and other dopants are evaporated onto silicon substrate under UHV condition (The pressure in the chamber or equipment is approx between 10^{-9} to 10^{-11} Torr). The pressure is different at the different stages of the equipment as shown in Fig. 4.13. The wafers or substrate are loaded at 10^{-4} Torr and automatically transfer to the Growth Chamber where the transfer section has 10^{-6} torr pressure and chamber pressure is 10^{-9}.

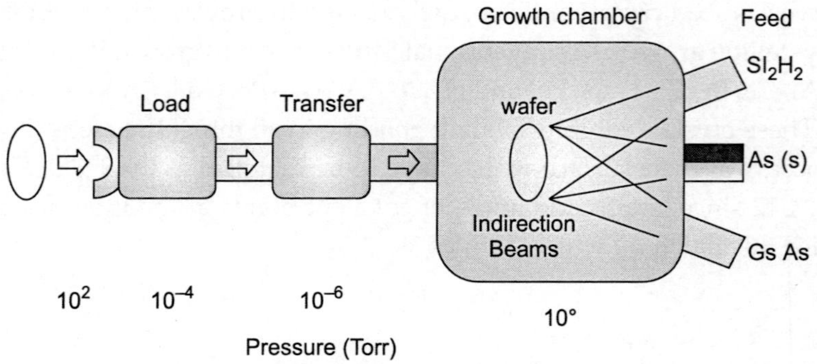

Fig. 4.13 *MBE Growth Equipment*

The MBE Growth Equipment (as shown in Fig. 4.14) is consist of the various components, the components and their functions are explained as:

1. **Vacuum System:** Consist in a strainless growth chamber, UHV connected to preparation chamber, where substrates are degassed prior to growth and load lock module for transfer to and from air. All the components of the growth chamber must be able to resist bake out temperatures of up to 200 °C for extended period of time, which are necessary to minimize outgrassing from the internal walls.
2. **Pumping System** must be able to efficiently reduce residual impurities to a minimum typical MBE growth rates for III-V type semiconductors are of the order of 1 m/n (~ 1 ML/sec) obtained for group III.

   Partial pressure aju 10^{-6} torr. The pumping system consist of ion pumps, cryogenic pumps and Ti sublimation pumps for the pumping of specific gas species.

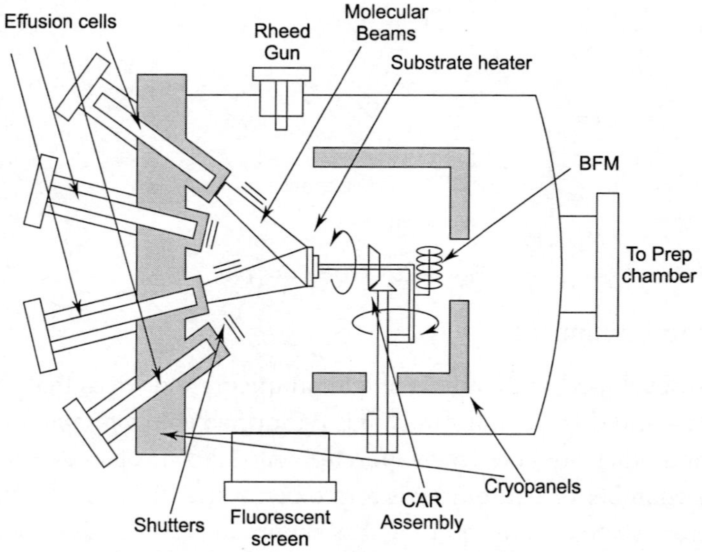

Fig. 4.14 *Molecular Bean Epitaxy Equipment*

3. **Effusion Cells** are the key components of an MBE system, because they most provide excellent flux stability. Uniformity and material purity. Being a part that must withstand the highest temperature (up to 1400°C) for the longest periods, they are often responsible for machine downtime, the effusion cell consist of cruable, filament, heating shield thermocouples etc. (as shown in Fig. 4.15)

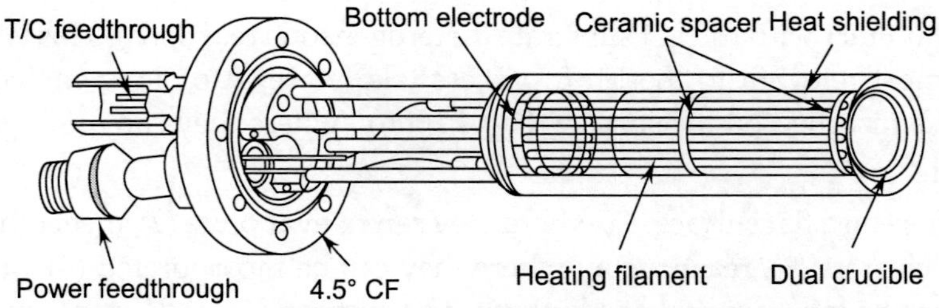

Fig. 4.15 *Effusion Cell*

The heating is provided by the filaments, while foils provide heating and radiation sheilding. The cruable is made of pyrolitic boron nitride, which be cylindrical or conical in shape with different tapering angles, depending on the material to be evaporated. It can stand temperature of up to ~ 1300°C without appreciable degrassing and Thermocouple (T/C) is used to measure the temperature of the material and precision PID regulators are used to provide temperature regulations. A mechanical or penumatic shutter, (usually made of T_a or M_o) is placed in front of the cell to trigger the flux.

4. **Substrate Manipulator** is capable of continuous Azimuthal Rotation (CAR) around its axis to improve uniformity across the wafer and wafer to multiwafer systems. The heater behind the sample is used to maximize temperature uniformity and minimize impurity out gassing and power consumption. Opposite to substrate holder, an ionization gauge is used (by rotating 180° the CAR assmebly) as a beam flux monitor (BFM).
5. **Liquid N_2 Cryopanels** surround internally, both the main chamber wall and source flange. MBE is a cold wall technique, cryopannels prevent re-evaporation from parts other than the hot cells. They provide thermal isolation among the different cells, as well as additional pumping of the residual gas.
6. **Analysis Tools** are available in MBE system, a number of diagnostic methods such as reflection high-energy electron diffraction (RHEED), Auger electron spectroscopy (AES), secondry ion mass spectroscopy (SIMS), X-ray induced, photoelectron spectroscopy (XPS), scanning tunneling Microscopy (STM) etc., can be performed for in-situ evaluation by incorporating necessary appratus in the system, together with a quadruple mass analyzer for measuring beam intensities and an ion gun for surface cleaning. The most valuable and peculiar one is Reflection High Energy electron Diffraction (RHEED). This technique employs a high energy (up to 20 KeV) electron beam, directed on the sample surface at grazing incidence, the diffraction pattern is imaged on symmetrically placed floroesent screen. These powerful facilities for control and analysis, eliminating much of the guess work, certainly give substantial advantages to MBE over other material preparation techniques.

4.6.3 Analysis of the MBE Growth Process

In general, three different phases can be identified in the MBE process.

The first is the crystalline phase constituted by the growing substrate, where short- and long-range order exists. On the other extreme, there is the disordered, gas phase of the molecular beams.

In MBE, the pressures of the residual gas and of the molecular beams are so low to make the mean free path for molecule collisions much larger than the source-to substrate distance (some tens of cm). Thus MBE fluxes are ballistic and no homogeneous reactions in the gas phase can occur. Between the two phases, a near-surface transition layer can be identified, where the impinging molecular beams interact with the hot substrate (heterogeneous reactions). This is the phase where the phenomena most relevant to the MBE process take place. Atomic or molecular species get physisorbed or chemisorbed on the surface [Fig. 4.16(*a*))], where they can undergo different processes [Fig. 4.16(*b*)].

Atoms can diffuse on a flat surface (1), where they can re-evaporate (2), meet other atoms to form two-dimensional clusters (3), reach a step where they can be incorporated (4), or further migrate along the step (5) to be incorporated at a kink (6). The energetic of each event, and thus their time and length scale depend on a number of factors.

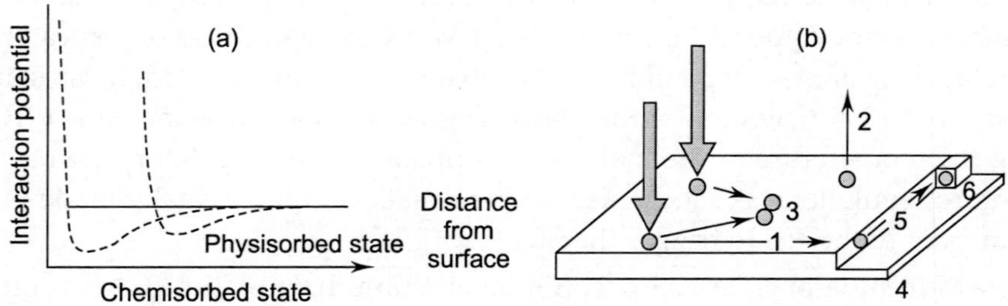

Fig. 4.16 (a) *Interaction Potential of an Arriving Atom or Molecule with the Sample Surface* (b) *Different Surface Elemental Processes in MBE*

Advantages of MBE

- Excellent surface / interface morphology and thickness control.
- Complex hetero structures.
- In-situ characterization and monitoring.
- Precisely controlled abrupt heterojunctions.
- High purity elemental solids readily available.
- No toxic gases, easily handled solid.
- Relatively simple chemistry.

Disadvantages of MBE

- Graded interface difficult.
- Flux transients.
- Reproducibility.
- Surface defects.
- Many sources needed, difficult to scale up.

4.7 APPLICATIONS OF EPITAXY

It has number of applications in nanotechnology and in semiconductor fabrication. Indeed, epitaxy is the only affordable method of high crystalline quality growth for many semiconductor materials, including technologically important materials as silicon-germanium, gallium nitride, gallium arsenide and indium phopshide.

Epitaxial silicon also used to grow layers of pre-doped silicon on polished side of silicon, wafers, before they are processed into semiconductor devices. This typical of power devices, such as those used in pacemaker, vending machine controller, automobile computer etc.

This is used in manufacturing of CMOS structures, layers of different doping *i.e., p*-layer on top of p+ substrate to avoid latch up problems. Epitaxial growth method also used to form the ultra-thin layers and Epitaxial silicon also had an important impact on the manufacturing yield of early bipolar integrated circuits by allowing electrical isolation of the individual circuit. Components to accomplished much more easily. It has also number of applications like BJT needed to produce. Baried layers, LED, Laser, Heterojunction BJT etc.

Comparison of Various Epitaxial Methods

Table 4.1: Comparison of Various Epitaxial Methods

	LPE	*MOCVD*	*MBE*
Pressure	760 torr	25 – 760 torr	10^{-10} torr
Reactant transport	Liquid phase	Vapour phase	Atomic and molecular beams
Temperature control	Source and substrate	Substrate	Source and substrate
Heterostructure formation	Sliding speed	Gas flow control	Shutter control
In-situ diagnostic	no	no	RHEED, reflectively Mass spectroscopy
Growth rate (μm/min)	0.1 ~ 10	0.005 ~ 1.5	0 ~ 0.05
Minimal thickness (A)	500	20	5
Uniformity	Good (small wafer)	good	good
Surface quality	bad	good	good
Interface sharpness	bad	good	good
Doping Range (cm10)	10^{14} ~ 10^{19}	10^{14} ~ 10^{19}	10^{14} ~ 10^{19}
Process yield	low	high	Very low

4.8 DEFECTS IN EPITAXIAL LAYERS

Defects in Epitaxial layers will degrade device properties deffects can reduce mobility and increase the leakage current, the defects in the Epitaxial layer can be categorized into five groups.

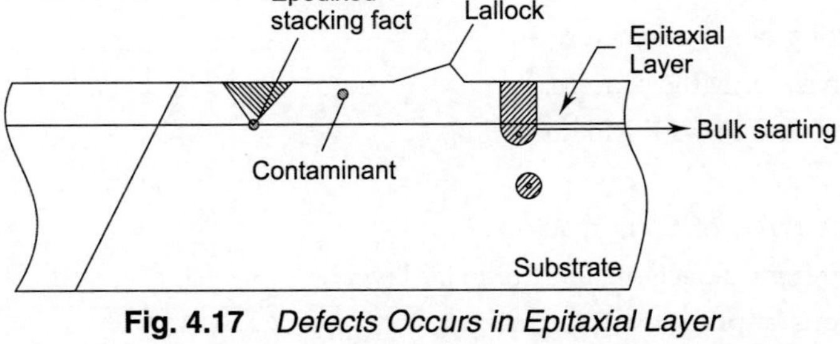

Fig. 4.17 *Defects Occurs in Epitaxial Layer*

1. **Edge dislocations:** These type of defects occurs in the heteroepitaxy of two lattice-mismatched semiconductors if both lattices are rigid, they will retain their spacing of lattice and the interface will contain rows of mis-bonded atoms described as misfit or edge dislocations. The edge dislocation can also form in a strained layer when the layer thickness becomes larger than the critical layer thickness.

2. **Impurity precipitates:** Impurity precipitates are one kind of surface defect that nucleate on an epitaxial stacking fault.
3. **Process related defects:** These defects include displacement of crystal planes past each other as the result of stress and impurity precipitates from the contamination.
4. **Hillocks:** The contaminations from the susceptor and the tweezers used in handling wafer also contaminate the surface of the wafer and form belock. The growth features can be related to the process or the surface finish of the wafer.
5. **Defects from the interface:** Any contamination at the interface of epitaxial layer and substrate may cause the formation of misoriented clusters or nuclear containing stacking faults.

SUMMARY

In the epitaxial growth process, the substrate wafer is the seed, high quality, single crystal films can be grown at the temperature 30% to 50% lower than the melting point, from the heteroepitaxial and homoepitaxial silicon structure, the homoepitaxial silicon structure will remain popular design choices. Cleaniness of both chamber and growth process is critical to the success of the process. The three methods of depositing epitaxial layers have their own advantages and disadvantages like LPE is less expensive and possible to work at low temperature whereas VPE is non-toxic, and inexpensive, but it require higher temperature for epitaxial growth. The MBE would be advantageous in fully ion-implanted VLSI circuits in which the total thermal cycle is minimized so that the doping capabilities of MBE can be exploited, the various types of defects are occur in the epitaxial growth which degrade the performance of the various means have been presented to minimize, or even eliminate. The epitaxial processes are well characterized and understood, the tend to thinner layers for bipolar and unipolar IC's will result in incremental process improvement and, continued study of flatness, autodoping effects and defect generation.

TUTORIAL SHEET

1. What is epitaxy? What are their types?
2. Describe the mechanism of epitaxial growth.
3. Explain and derive groove model of epitaxial growth.
4. Explain the principle and working of chemical vapour deposition method.
5. Explain the principle and working of vapour phase epitaxy equipment.
6. What is liquid phase epitaxy? How can we deposit a layer on the substrate with this method?
7. Explain the working of MBE Reactor.
8. Discuss the analysis of MBE process.
9. Compare MOVPE and RTVPE Methods.
10. What is autodoping?
11. Compare various types of CVD reactors.
12. What are the different defects in the Epitaxial layer?
13. Discuss the various applications of epitaxy.
14. What is effusion cell ? And where it is used?

5

THIN FILM DEPOSITION

5.1 INTRODUCTION

Thin film deposition is any technique for depositing a thin layer (few nano meters) of material on to a substrate or on to previously deposited layer. Deposited films are widely used in the fabrication of micro chips. These thin films provide: (*i*) Conducting regions within device (*ii*) electrical insulation between metals (*iii*) protection from the environment. There are three categories of these films: (*i*) Polysilicon (*ii*) Conductor (*iii*) Insulator (Dielectrics).

Polysilicon is polycrystalline silicon used as gate material, resistor material and for capacitor plates. Polysilicon is prepared by pyrolyzing selane at 575°C to 650°C. Polysilicon is mostly deposited without dopant because dopants like arsenic, phosphorus or boron reduces the resistivity of the polysilicon.

Conductors are usually made up of aluminium, although sometimes other metals such as gold are used siliudes also fall under this category. In the fabrication of IC's the conductor deposition layer are used in metallization of IC's.

Insulators refers to materials such as silicon dioxide, silicon nitride and P-glass (Phosphorous doped silicon dioxide) which serve as insulation between conducting layers, for diffusion and implantation masks for passivation protect devices from the environment.

Many methods are available for depositing thin films. Common methods are atmospheric pressure chemical vapour deposition (APCVD), Low pressure chemical vapour deposition (LPCVD) and Plasma Enhanced chemical vapour deposition (PECVD). These methods have their own advantages and disadvantages.

5.2 THIN FILM PARAMETERS

The thin film layers of the device must meet general and specific parameters. Some general parameters for the thin film deposition are:

(*i*) *Uniformity of thickness*: The thickness of the thin film is one of the most important parameter. Thickness of the film is required to meet both electrical and mechanical specifications. Deposited film must be free from contamination and pinhole voids etc., thickness of the layer also contribute to resistance. So, it should be uniform throughout the wafer and also uniform wafer to wafer if the layer is too thin at the edges, it may cause electrical shorts or unwanted induced charges in the device (as shown in Fig. 5.1).

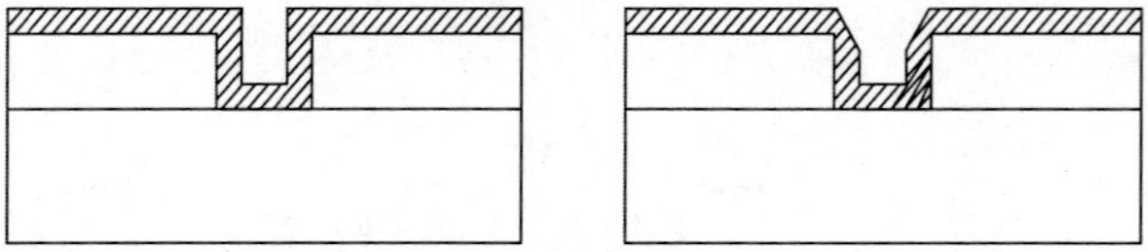

(a) (b)

Fig. 5.1 *Thinning of Deposited Layer (b) at Step*

(*ii*) *Stress free*: A film should be deposited without excess stress, because excessive stress may cause cracks. The cracks in the film causes surface roughness and it can allow contaminant to pass through wafer. Some metallic contaminants may cause electrical shortages of metallic lines.

(*iii*) *Surface flatness/roughness*: Surface roughness is one of the critical parameter. Excessive surface roughness can alter device performance and compromise the uniformity of layers deposited on top of the devices. The deposited layer method must minimize steps and cracks.

(*iv*) *Purity*: The thin film material must be free from unwanted chemical elements like O_2 in the Epitaxial film. O_2 is a type of contaminants which can change electrical properties of film.

5.3 DEPOSITION METHODS

There are two major classification of deposition techniques each having its own subsets of related techniques.

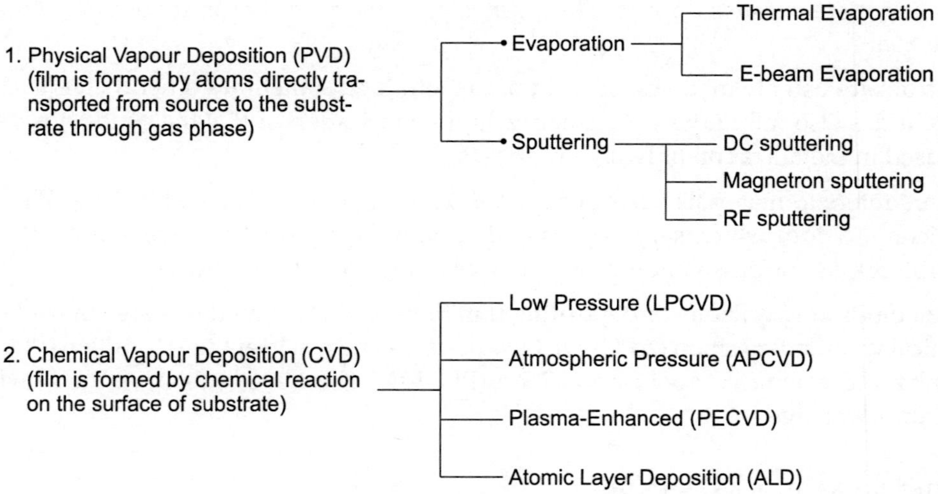

5.4 PHYSICAL VAPOUR DEPOSITION (PVD)

Physical deposition uses mechanical or thermal means to produce a thin film of solid phase. The physical deposition systems tend to require a low pressure vapour environment to function properly. The material to be deposited is placed in an energetic or entropic environment, so that particles of material can escape its surface. The whole system is kept in a vacuum deposition chamber, to allow the particles to travel as freely as possible. Since, particles tend to follow a straight path, films deposited by physical means are commonly directional, rather than conformal.

The principle of Physical Vapour Deposition (PVD) method is shown in Fig 5.2. In this method, firstly the solid or liquid material evaporates with the help of heater or *e*-beam.

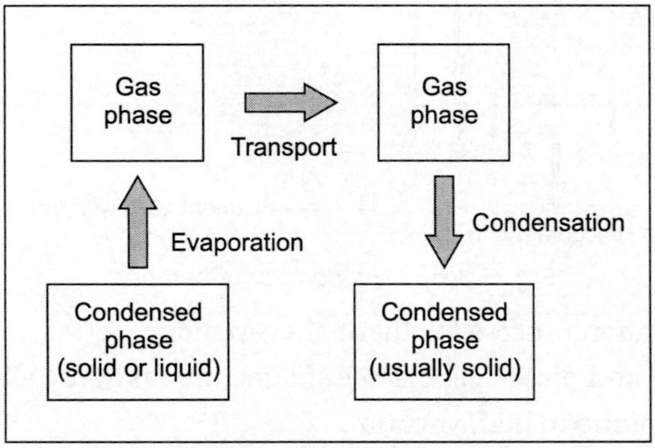

Fig. 5.2 *Principle of PVD Method*

Then in the vacuum, the material in the gas phase transport into the wafer or substrate, where material get condensed and deposit on the substrate and converted into solid phase.

5.4.1 Evaporation

The material to be deposited is heated until it evaporates, vapour then condenses on the sample surface. There are two types of evaporation. The only difference between the two is how the source material is evaporated.

(*a*) *Thermal evaporation*: Source material is heated with a resistive heater. The thermal evaporator uses an electric resistance heater to melt the material and raise its vapour pressure to a useful range. This is done in a high vacuum, both to allow the vapour to reach the substrate without reacting with or scattering against other gas phase atom in the chamber and reduce the incorporation of impurities from the residual gas in the vacuum chamber. The quartz crystal measures deposition rate. Deposition rate can be adjusted by changing the temperature of the filament. The rotating shutter masks sample until desired deposition rate is achieved as shown in Fig. 5.3 (*b*)

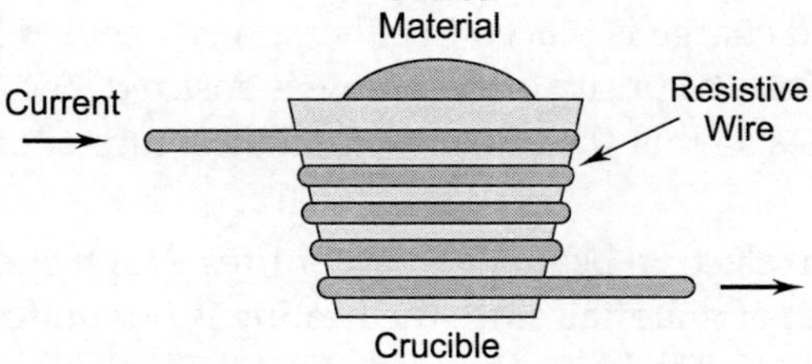

Fig. 5.3(a) *Principle of Thermal Evaporation*

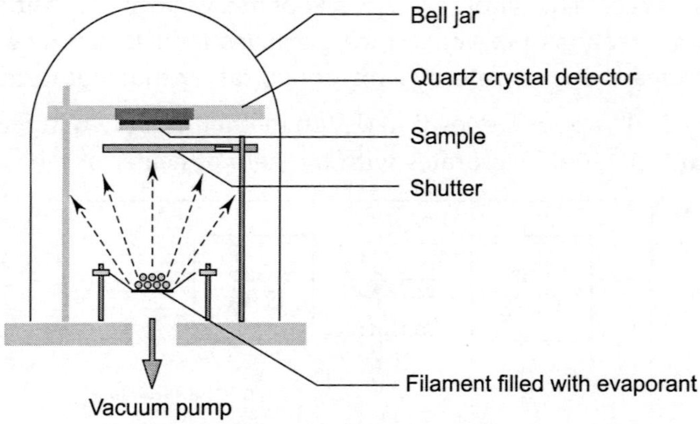

Fig. 5.3(b) *Evaporation Chamber*

The steps of evaporation process for thermal evaporator.

1. Mount sample and place evaporant in filament. Make sure shutter is closed.
2. Close bell jar, pump to high vacuum.
3. Turn on filament current to heat filament. Increase until desired deposition rate is achieved.
4. Open shutter, deposit desired thickness of material, then close shutter.
5. Vent the chamber, retrieve sample.

The resistively heated evaporation sources are available in the varity of forms. They can be divided into the following important categories:

1. Tungsten wire sources (Tungsten)	The furnaces are available in various forms like Helical or conical shapes.
2. Refractory metal sheet sources (Tungsten, tatalum, molybde-num) boat canoe, deep folded	The furnaces are available in dimpled strip.
3. Sublimation furnaces (Sulphides, solenoids, some oxide)	The furnaces are typically constructed of sheet tatalum, which is readily cut, beat and spot welded to form heaters.
4. Crucible source: (Pyrolytic BN, graphite, refractory metals)	These crucible is heated by external tungsten wire heating elements wound to fit snugly around them.

(*b*) *Election beam evaporation*: The *e*-beam evaporation method is very much similar to the thermal or filament evaporation method. Only the source of the chamber is changed with *e*-beam, the evaporated charge is placed in either a water cooled hearth or in the depression of a water cooled crucible the purity of evaporate is ensured, because only a small amount of charge melt or sublimes so that the effective crucible is unmelted skull material next to the cooled hearth.

As shown in Fig. 5.4 an electron beam evaporator fires a high energy beam from an electron gun to boil a small spot of material, since the heating is not uniform, lower vapour pressure materials can be deposited. The beam is usually bent through an angle 270° in order to ensure that the gun filament is not directly exposed to the evaporated flux. The *e*-beam is bent due to magnetic field arrangement. Typically deposition rates for electron beam evaporation range from 1 to 10 nanometers per second film contamination from the heated electron source is eliminated in this way. The comparison of thermal and e-beam deposition method is given in Table 5.1.

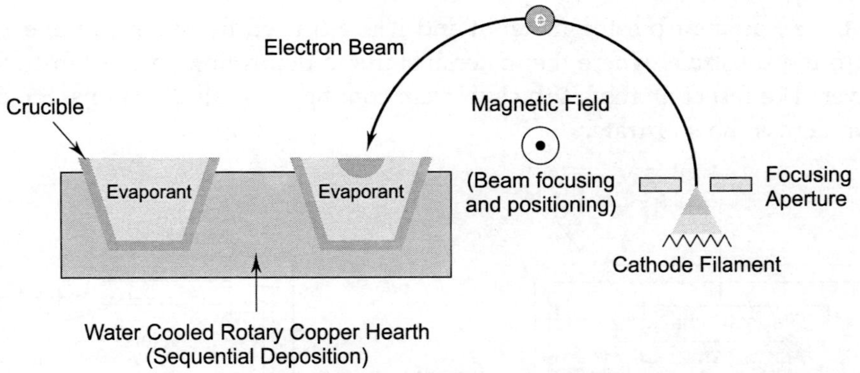

Fig. 5.4 *e-beam Heating Method*

Table 5.1: Comparison of Thermal and e-beam Evaporation

Deposition	*Material*	*Typical evaporant*	*Impurity*	*Deposition rate*	*Temperature range*	*Cost*
Thermal	Metal or low melt-point materials	Au, Ag, Al, Cr, Sn, Sb, Ge, In, Mg, Ga CdS, PbS, CdSe, NaCl, KCl, AgCl, MgF_2, CaF_2, $PbCl_2$	High	1 – 20 A/s	– 1800°C	Low
E-Beam	Both metal and dielectrics	Everything above, plus: Ni, Pt, Ir, Rh, Ti, V, Zr, W, Ta, Mo Al_2O_3, SiO, SiO_2, SnO_2, TiO_2, ZrO_2	Low	10 – 100 A/s	– 3000°C	High

5.4.2 Sputtering Process

A Tayet made of the material to be deposited is bombarded by energetic ions which will dislodge atom of the target *i.e.,* sputter them off. The dislodge atoms will have substantial kinetic energies, and some will fly to the substrate to be coated and stick there. It uses large area targets which gives uniform thickness over the wafer. It controls the thickness by depth time and parameters and also control film properties such as step coverage, grain structure. In sputtering the surface in vacuum cleaned prior to deposition.

The sputtering Methods are divided into three subsets categories:

(*a*) DC Sputtering
(*b*) RF Sputtering
(*c*) Magnetron

(*a*) **DC Sputtering:** In DC Sputtering, both target (source) and substrate are placed on two parallel electrodes (Diode). The wafer (substrate) is placed on the anode and target material is placed on cathode and chamber is filled with inert gas (Ar). (As shown in Fig. 5.5 (*a*) and 5.5 (*b*)). The DC voltage in (kV) is applied to the electrodes. Free electrons are accelerated by electric field in the chamber these free electrons collide with Ar atoms to form Ar^+ ions. After that Ar^+ will be attracted by target (cathode) electrode. The Ar^+ ions strike the target with energies which sputte the e^-/ atoms from the material and it is attracted by the anode and that material comes to the substrate, where the material starts to depositing on the substrate and forms thin layer. The target atoms hit the substrate coming from all directions. So, deposition is uniform across the substrate.

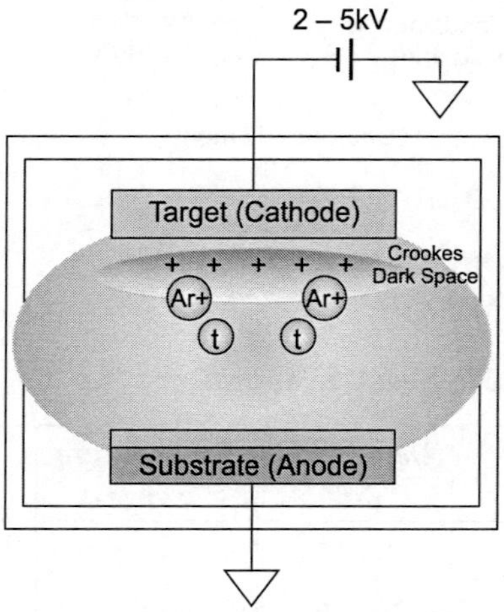

Fig. 5.5 (a) *Principle of Sputtering Process*

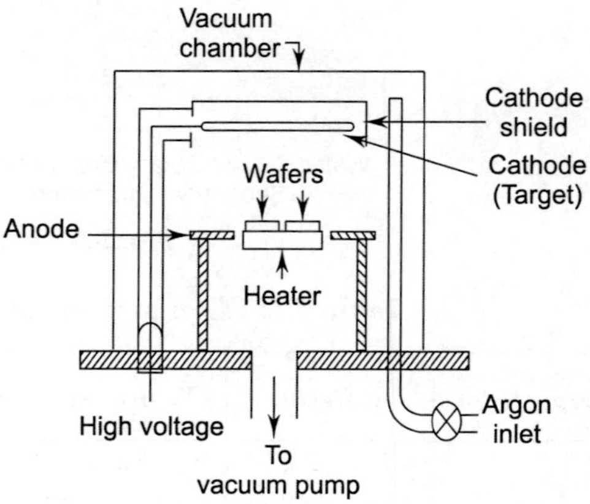

Fig. 5.5 (b) *DC Sputtering Chamber*

(*b*) **RF Sputtering:** The DC sputter deposition is not suitable for insulator (Dielectric) deposition, because the positive charge on the target surface rejects the ion flux and stop the sputtering process. The radio frequency (RF) voltages can be coupled capacitively through the insulating target to the plasma, so conducting electrodes are not necessary. The RF frequency is high enough to maintain the plasma discharge. The RF sputtering system shown in Fig. 5.6. The RF sputtering system consists of two electrodes same as in case of DC sputtering the working of this equipment is also same only the RF supply of 13.5 MHz is given instead of DC power supply. In this method, the wafer will be sputtered at the same rate as the target since the voltage drops due to the same at the both electrodes in symmetric system. Thus, it is very difficult to deposit any material in this way. By making the area of the target electrode smaller than the other electrode, the voltage drop at the target electrode will be much greater than at the other electrode. Therefore, almost all the sputtering will occur at the target electrode.

(*c*) **Magnetron Sputtering:** There is one more method of depositing thin film by magnetron sputtering. Magnetron sputtering is a powerful and flexible technique which can be used to deposit any thin film. With widerange of materials any solid metal or alloy and a variety of compounds. Prior to sputtering procedure a vacuum of less than one tenth millionth of an atmosphere must be achieved. From this point, a closely controlled flow of an inert gas such as argon is introduced. This raise the pressure to the minimum needed to operate the magnetrons, although, it is still only few ten thousandth of atmospheric pressure. Here, magnets are used to increase the percentage of electrons that takes parts in ionization events, increase probability of electron striking Ar, increase path length, so the ionization efficiency is increased significantly. During the sputter process a magnetic field can be used to trap secondary electrons close to target. The electrons follow helical paths around the magnetic field lines undergoing more ionizing collisions with neutral gases near the target than it would otherwise. This enhances the ionization of the plasma near target leading to a higher sputtering rate. It also means that the plasma can be sustained at a lower pressure the sputtered atoms are neutrally charged and so are unaffected by the magnetic trap.

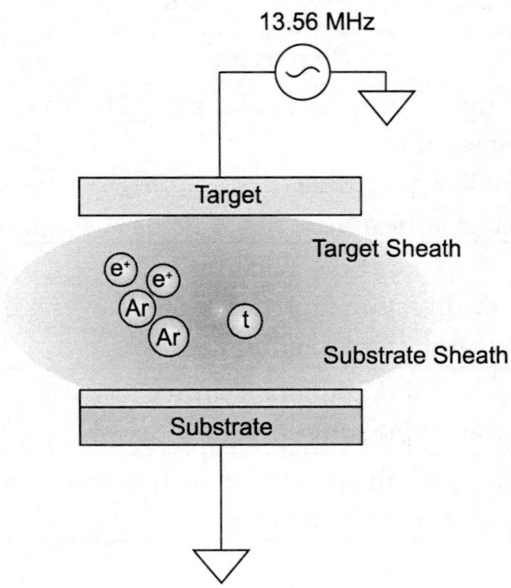

Fig. 5.6 (a) *Principle RF Sputtering*

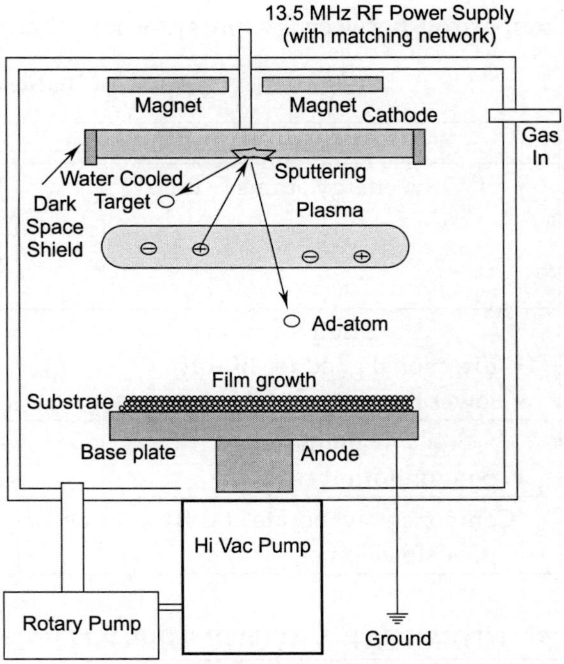

Fig. 5.6 (b) *RF Sputtering Equipment*

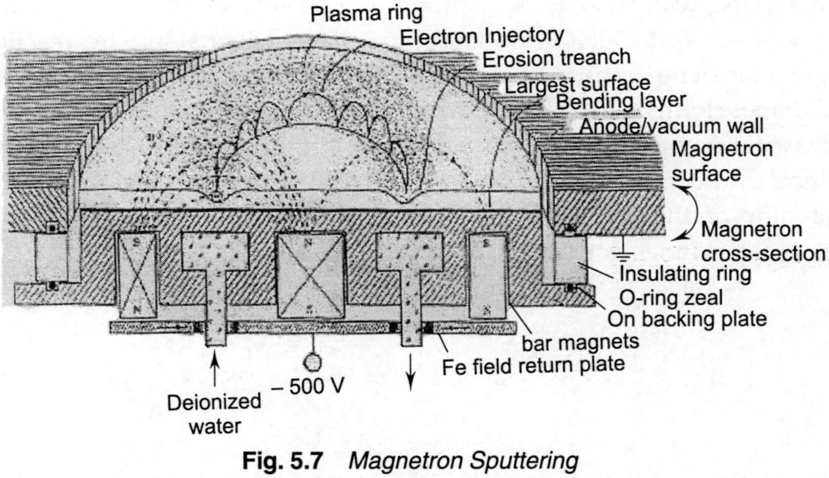

Fig. 5.7 *Magnetron Sputtering*

Applications of Sputtering

(*i*) Thin film deposition – Microelectrics, decorative coating, protective coating.

(*ii*) Etching target – Microelectrics patterning, depth profile microanalysis

(*iii*) Surface treatment – Hardening, corrosion treatment

Comparison between Evaporation and Sputtering

Table 5.2: Comparison between Evaporation and Sputtering

Evaporation	*Sputtering*
Low energy atoms (~ 0.1 eV)	High energy atoms/ions (1 – 10 eV) • denser film • smaller grain size • better adhesion
High Vacuum • directional good for lift-off • lower impurity	Low Vacuum • poor directionality, better step coverage • gas atom implanted in the film
Point source • poor uniformity	Parallel plate source • better uniformity
Component evaporate at different rate • poor stoichiometry	All component sputtering with similar rate • maintain stoichiometry

5.5 CHEMICAL VAPOUR DEPOSITION (CVD)

Chemical vapour deposition (CVD) is the process of chemically reacting volatile compound of a material to be deposited, with other gases, to produce a non volatile solid that deposit automatically on a suitably placed substrate. Firstly, the source gases are delivered into the reaction chamber at approximately ambient temperature. As they pass over or come in contact with a heated substrate, they react or decompose forming a solid phase which are deposited onto substrate (as shown in Fig. 5.8) then, after the reaction by-product are ejected out from the exhaust system. The substrate temperature is critical and can influence what reaction will take place. CVD methods are used to deposit the thin films on the devices which include epitaxial layer gate oxide, metallization layer etc. The film material includes polysilicon, dielectric SiO_2 and SiN.

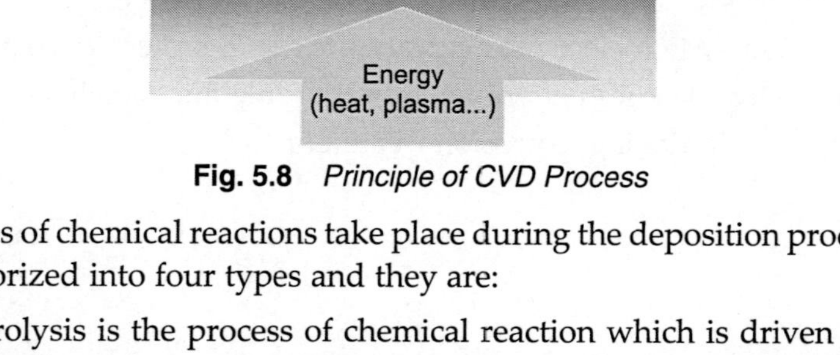

Fig. 5.8 *Principle of CVD Process*

The various types of chemical reactions take place during the deposition process, these chemical reactions are categorized into four types and they are:

1. **Pyrolysis:** Pyrolysis is the process of chemical reaction which is driven by heat alone. Hydrides, carbonyls and organ metallic compounds are thermically decomposed on the substrate.

$$AB_{(gas)} \rightarrow A_{(solid)} + B_{(gas)}$$

Example α - Si deposited at 580 – 650° C

$$SiH_{4(gas)} \rightarrow Si_{(solid)} + 2H_{2(gas)}$$

2. **Reduction:** Reduction is the process of chemical reaction in which hydrogen is used as a reducing a gent to effect the reduction of such gaseous species as hydrides, or other oxyhalides or other oxygen containing compounds

$$AB_{(gas)} + H_{2(gas,\ commonly\ used)} \leftrightarrow A_{(solid)} + HB_{(gas)}$$

Example W deposited at 300° C

$$WF_{6\,(gas)} + 3H_{2(gas)} \leftrightarrow W_{(solid)} + 6HF_{(gas)}$$

Reversible process, can be used for chamber cleaning

3. **Oxidation:** Oxidation is the process of chemical reaction in which an element combines with oxygen.

$$AB_{(gas\ or\ solid)} + O_{2(gas,\ commonly\ used)} \leftrightarrow AO_{(solid)} + [O]\,B_{(gas)}$$

Example 1 Low temperature SiO_2 deposited at 450° C

$$SiH_{4(gas)} + O_{2(gas)} \rightarrow SiO_{2(solid)} + 2H_{2(gas)}$$

Example 2 SiO_2 formed through dry oxidation at 900 – 1100° C

$$Si_{(solid)} + O_{2(gas)} \rightarrow SiO_{2(solid)}$$

4. **Nitration:** Nitration is the process of forming silicon nitride by exposing a silicon wafer to nitrogen at a high temperature.

$$AB_{(gas\ or\ solid)} + XY_{(gas\ or\ solid)} \leftrightarrow AX_{(solid)} + BY_{(gas)}$$

Example

$$3SiH_2\,Cl_{2(gas)} + 4NH_{3(gas)} \rightarrow Si_3N_{4(solid)} + 6\,H_2 + 6\,HCl$$

5.5.1 Grove Model of the CVD Process

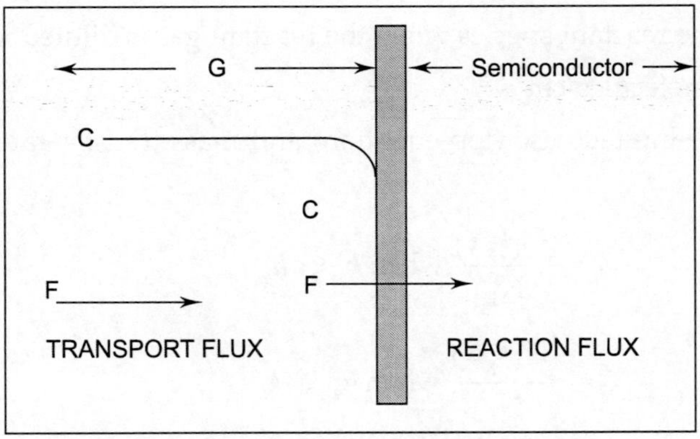

Fig. 5.9 *Grove Model of CVD Process*

At first, each of these steps can be described as independent flux equation, where the surface reaction can be written as

$$F_1 = h_g\,(C_g - C_s)$$
$$F_2 = k_s\,C_s$$

Where,

h_g: gas phase mass transfer coefficient

C_g: concentration of the reactant in the bulk of the gas
C_s: concentration of the reactant at the surface of the substrate
Flux: (F) Number of atoms or molecular crossing a unit area in a unit time, *i.e.*, atoms/cm^2/sec
Assuming equilibrium

$$F_1 = F_2$$

$$C_s = \frac{C_g}{\left(1+\frac{k_s}{h_g}\right)}$$

$$G = \frac{F}{N_1}$$

$$G = \frac{k_s h_g}{k_s + h_g}\left(\frac{C_g}{N_1}\right)$$

Where,
G: rate = F/N_1
N_1: number of atoms incorporated into a unit volume of the film.
K_s: surface reaction rate constant
When the gas is diluted in an inert carrier gas, the concentration of the reactant in the gas phase

$$C_g = C_T Y$$

$$G = \frac{k_s h_g}{k_s + h_g}\frac{C_T}{N_1}Y$$

$$G = \frac{(C_T k_s Y)}{N_1} \text{ when } k_s << h_g$$

Y: mole fraction of the reactant species when the reactant gas is diluted in an inert gas
C_T: total number of molecules/cm^3

Surface: reaction-rate-limited case (top equation) and mass-transfer-rate-limited case (bottom equation)

$$G = \frac{(C_T k_s Y)}{N_1} \text{ when } k_s << h_g$$

$$G = \frac{(C_T h_g Y)}{N_1} \text{ when } h_g << k_s$$

CVD Deposition Condition

Mass-Transport Limited deposition

- At high temperature the reaction rate exceeds the gas delivering rate (as shown in Fig. 5.10).
- Gas delivering controls film deposition rate.
- Film growth rate insensitive to temperature.
- Film uniformity depends on whether reactants can be uniformly delivered across a wafer and wafer to wafer.

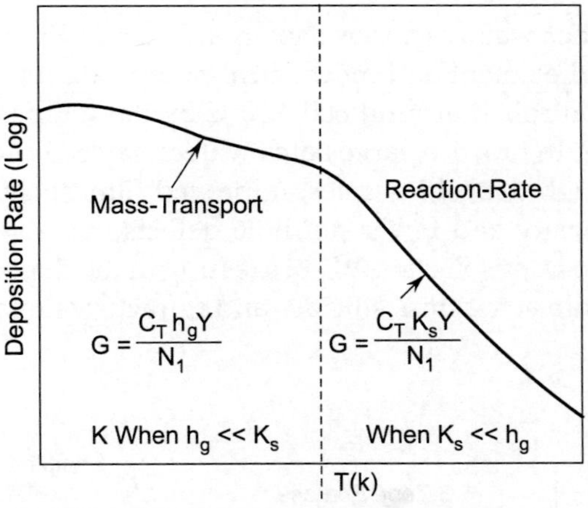

Fig. 5.10 *CVD Deposition Process*

Reaction rate limited deposition

- At low temperature or high vacuum such that the reaction rate is below gas arriving rate.
- Temperature controls film deposition rate.
- Film uniformity depends on temperature uniformity across a wafer and wafer to wafer.

5.6 CVD SYSTEMS

As we know that, CVD methods are used to deposit a variety of materials on the wafer substrate and hence, number of CVD systems are available in the market in variety of designs and options. Regardless of the process type; however the associated system must have the following components:

- *Gas delivery system*: for supply of gaseous source to the reactor chamber.
- *Reactor chamber*: chamber within which deposition take place.
- *Substrate loading mechanism:* a system for introducing and removing substrate.
- *Vacuum system*: a system which removes all the gases other than the required for reaction/ deposition.
- *Process control equipments*: gauges, control etc to monitor process parameters such as pressure, temperature and time alarm and safety devices.
- *Energy source*: provide the heat that is required to get source gaseous to react/decompose.
- *Exhaust system*: system for removal of all volatiles, by products from the reaction chamber.

5.7 TYPES OF CHEMICAL VAPOUR DEPOSITION (CVD) SYSTEMS

There are mainly five types of CVD systems available:

5.7.1 Low Pressure Chemical Vapour Deposition (LPCVD) System

LPCVD methods were first employed to deposit polysilicon films with greater control over stiocheometry and contamination problems. In this case, wafers are stacked upright without loosing uniformity as long as they have same temperature and deposition is performed at reduced pressure. The system works at vacuum (~ 0.1 – 1.0 torr), low pressure reduces gas-phase reaction which cause particle cluster that contaminates the wafer and system. The commercial LPCVD systems commonly used horizontal hot wall reactor as shown in Fig. 5.11. It consists of cylindrical quartz tubes heated by wire wound element and mechanical pumps are used to maintain the required pressure. Temperature is controlled around 600–900°C by flat temperature zone using multiple heaters as shown in Fig. 5.12. In practice, large batch wafers, say 100 and more can be processed at a time. This coupled with high deposition rates, improved film thickness uniformity better step coverage, lower particle density and fewer pin hole defects has given LPCVD important economic advantages relative to atmospheric CVD processing in the deposition of dielectric films. In addition to polysilicon and dielectric film, silicides and refractory metals have been deposited by LPCVD method.

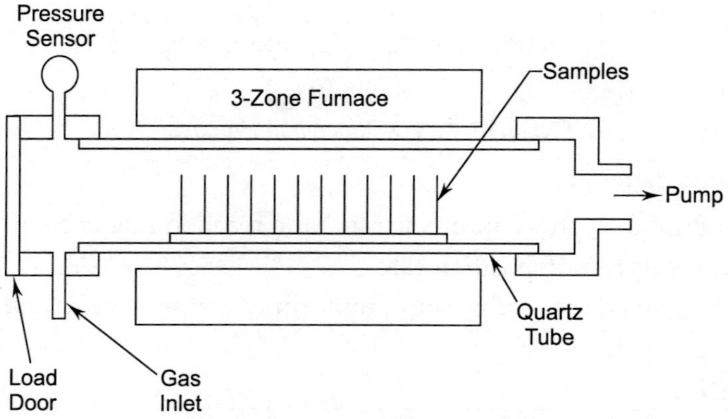

Fig. 5.11 *Low Pressure CVD*

These reactors typically have a "flat zone" in which the axial temperature is very nearly constant. The flat zone temperature can be controlled to within better than one degree C. The size of the flat zone determines the number of wafers then it can be loaded in a single run.

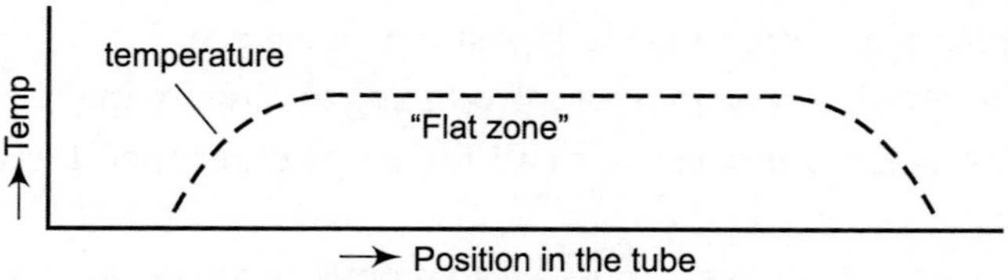

Fig. 5.12 *Flat Temperature Zone of LPCVD*

Tube furnaces are "hot wall" reactors: everything is at the same temperature and in general the film being deposited ends up on every exposed surface of the tube. The tube and wafer holding hardware must be cleaned periodically to avoid spalling and particle formation. Cleaning can be performed by removing the tube and quartzware and immersing them in appropriate wet chemical baths, although this procedure is complex, relatively dirty, time-consuming and expensive, since large quantities of toxic acids must be handled.

The list of LPCVD system advantages includes:

- Lower chemical reaction temperature.
- Good step coverage and uniformity.
- Vertical loading of wafers for increased productivity and lower exposure to particles.
- Less dependence on gas flow dynamics.
- Less time for gas phase reaction particles to form.
- Can be performed in standard tube furnaces.

5.7.2 Atmospheric Pressure Chemical Vapour Deposition (APCVD)

As the name implies, atmospheric CVD system reactions and deposition take place at atmospheric pressure. The design of these reactors are simple and allow high film deposition rates due to operation at atmospheric pressure. Since, APCVD is generally operated in the mass-transport limited region; the reactant flux to all parts of every substrate in the reactor must be precisely controlled. There are a number of system designs that falls under this category. There are two types of wall reactions *i.e.*, hot wall and cold wall.

Let us assume T_w = wall temperature

T_s = substrate temperature

When $T_w = T_s$ it is called hot wall CVD reactor and it has uniform reaction but the disadvantage is film forming reactions become pronounced on the chamber walls, causing flaking off.

When $T_w < T_s$

It is called cold wall CVD reactors. In this method, less material is deposited on the walls of chambers but it don't deposit the films in uniform manner.

Another difference is cold wall versus hot wall. The hot wall systems heat the wafers, the wafer holder and the chamber walls and the reaction occurs throughout the chamber, leaving products in the inside walls of the chamber. It requires frequent cleaning after the reaction to avoid contamination of the wafers. The cold wall systems directly heat the wafer holder or wafers with the help of induction or radiation heating, the walls of the chamber remains cold.

(*a*) **Horizontal tube induction heated APCVD (cold wall):** This type of CVD's are used for the deposition of silicon epitaxial films for bipolar devices. The furnace is placed horizontally with square cross section. Wafers are placed on a flat graphite slab. The tube is surrounded with copper coils that are connected to RF generator. The RF waves travelling in the coil pass through the quartz tube and the flowing gas in the tube without heating the walls of the chamber. This heating method is known as induction heating as shown in the Fig. 5.13(*a*). This method is contamination free, but one disadvantage in that the system is downstream depletion of the reactants in the laminar gas flow. Laminar flow is required to minimize turbulence. The deposited film is thinner if the wafers are placed flat in the chamber and wafers are depleted due to the closeness of the gas layer. So, wafers are tilted with the accurate angle in the tube to over come the problem as shown in Fig. 5.13(*b*).

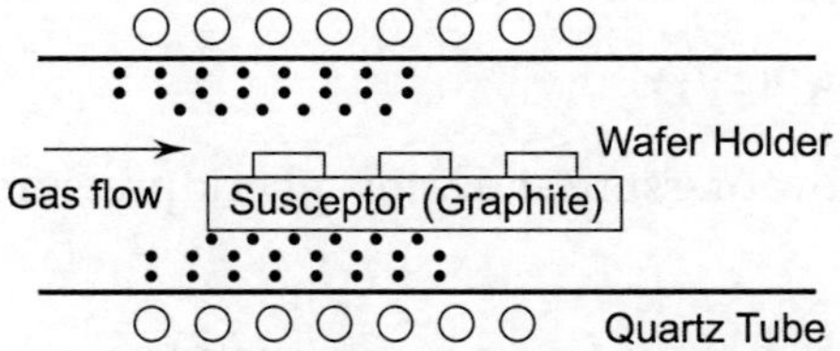

Fig. 5.13 (a) *Horizontal Tube Induction APCVD*

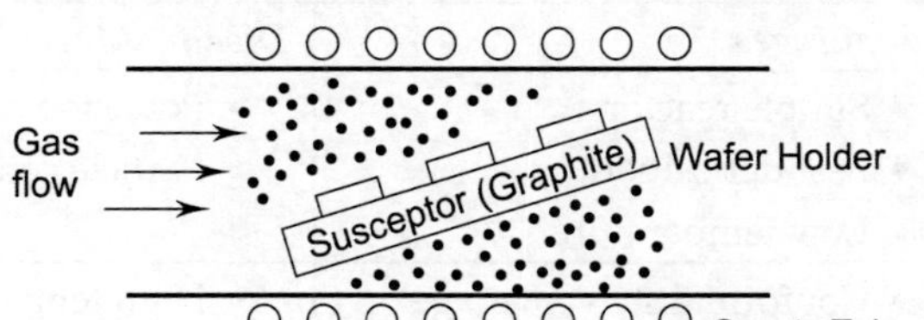

Fig. 5.13 (b) *Horizontal Tube Induction APCVD with Tilted Susceptor*

(*b*) **Horizontal Conduction heated APCVD (hot wall):** In this method, the gases are mixed outside the chamber and shower them onto the wafers. It consist of the belt under a plenum that dispenses the reacted gases and an injector which is used to inject the mixer of the gases into the chamber and resistive coils are used to provide heat. The walls of chamber also get heated due to the conduction heating as shown in Fig. 5.14.

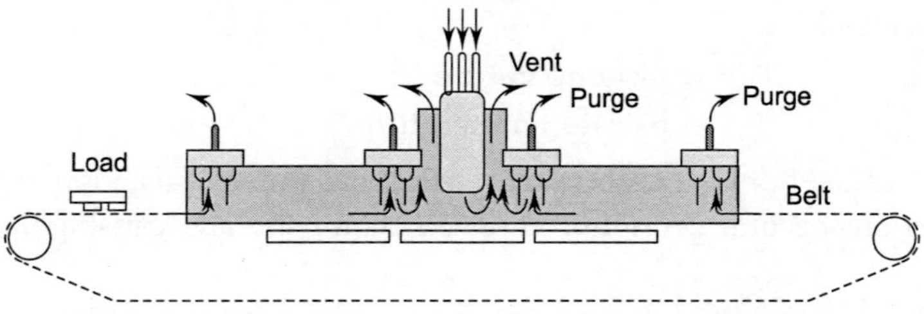

Fig. 5.14 (a) *Horizontal Conduction Heated APCVD*

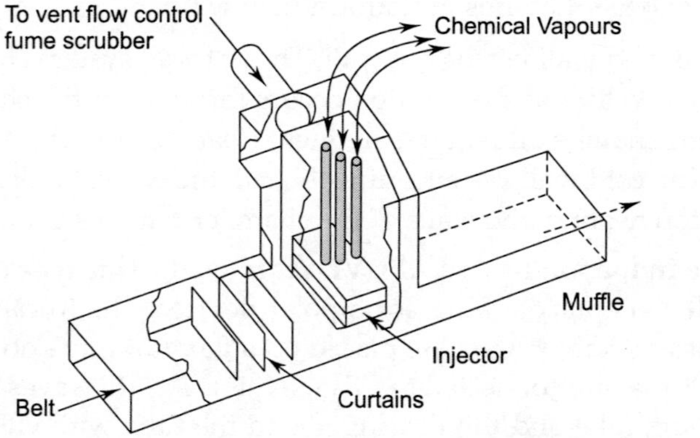

Fig. 5.14 (b) *Horizontal Conduction Heated APCVD Reactor*

The gases are injected from the cooled nitrogen shrouded nozzle and pass through narrow slots of the injector head. The gas mixing takes place millimeter above the wafer (substrate) surface and exhausted gases are vented to the scrubber by high nitrogen flow to avoid recirculation and gas phase reactions. The cleaning is required after the process to avoid contamination, because the reaction leaves the byproducts or chemicals deposition on the walls.

Comparison of LPCVD Vs APCVD

It is important to compare the low pressure and atmospheric pressure based CVD. The comparison is given in the Table 5.3:

Table 5.3: Comparison between APCVD and LPCVD

Process	*Advantages*	*Disadvantages*
APCVD	• Simple reactor • Fast deposition • Low temperature	• Poor step coverage • Particle contamination
LPCVD	• Conformal step coverage • Uniformity • Large wafer capacity	• High temperature • Low deposition rate

5.7.3 Plasma Enhanced CVD

In plasma enhanced CVD, glow discharged plasmas are sustained within chambers. PECVD uses a radio frequency power to generate glow discharge which transfers the energy into the reactant gases. Plasma deposited silicon nitride is commonly used as the final passivation layers for devices. Plasma deposited nitride and oxide can be used as insulators between metal layers since, the underlying metal have low melting point. PECVD's are similar to the plasma etch systems. They feature a parallel plate chamber operated at low pressure and low temperature, which provide good uniformity and throughput.

The system follows the design of a bottom heated pancake vertical flow CVD. (as shown in Fig. 5.15). The plasma is generated by the application of a RF field to a low pressure gas, which creates free electrons in the discharge region. Now, electrons start gaining energy from the electric field. When these electrons (energiesed) collide with gas molecules the reactant gases *e.g.*, silane and nitrogen or O_2 containing species dissociated and ionized. Then, it absorbed on the film surface (substrate). After absorption on the substrate, the radicals are subjected to ion and electron bombardment and rearrangement and react with other absorbed species to form new bonds and resulting film. The film deposition speed is generally increased, but it must be controlled to prevent film cracking or stress.

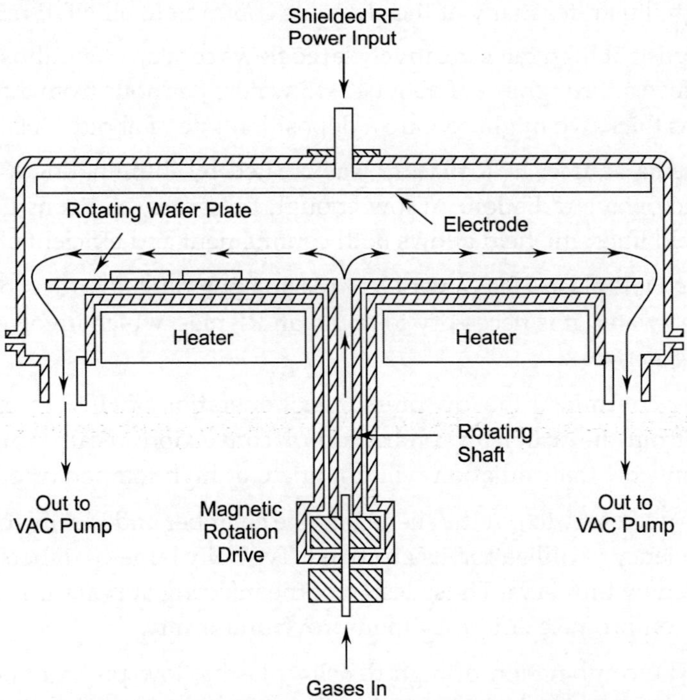

Fig. 5.15 *Plasma Enhanced CVD with Vertical Flow Pancake*

5.7.4 High Density Plasma CVD

Silicon dioxide films deposited by plasma-enhanced chemical vapour deposition (PECVD) under normal conditions in a showerhead reactor incorporates quite a bit of hydrogen, typically present mostly as silanol, Si-OH [see discussion of SiO_2 deposition under FILMS.

The first commercial HDP oxide deposition reactor designs were based on electron cyclotron resonance (ECR) plasmas.

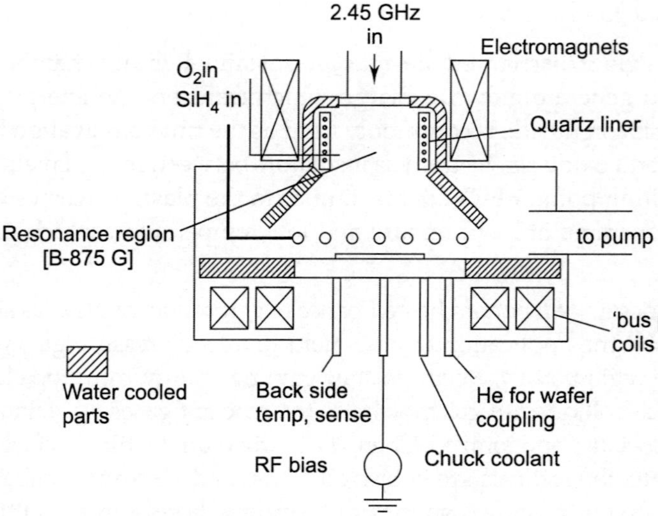

Fig. 5.16 *High Density Plasma CVD Reactor*

This reactor design illustrates many of the elements common to all HDP reactors:

Big pumps: to deposit at high rates, relatively large flows of silane (and thus oxygen and argon) are required. To achieve a throughput of about 12–15 wafers per hour from a module for a deposited film 0.75 microns thick, we might require a deposition rate of about 5000 A/minute.

High-density plasma source: ECR plasma was created by illuminating a magnetized region with high-power microwave radiation. At low enough pressures, electrons gain energy on each circuit around the field lines: the field allows both confinement and efficient electron heating.

Wafer RF bias: The ECR plasma has a very low plasma potential. In order to achieve significant sputtering rates at the wafer, it is necessary to apply an RF bias, which involves a separate power supply and matching network.

Wafer temperature control: At the low pressures of operation of HDP reactors, very little heat will be transported from the wafer by conduction or convection; recall from our discussion of showerhead heat transport that radiation will dominate at high temperature and low pressure.

Gas injection: It is necessary to put the silane into the chamber and ideally to dispense it close to the wafer so that efficiency of utilization is improved. Typically large chambers employed in HDP, transport is dominated by diffusion. Thus, design of the injecting apparatus is difficult and it may also be necessary to compromise efficiency to improve uniformity.

Chamber clean: The combination of high density plasma, low pressure and high sputtering rates at the wafer surface means that a lot of the silicon dioxide formed ends up on the chamber walls. It is also necessary to remove the deposit from the walls periodically, typically after 10–25 wafers.

Operation of HDP CVD

The method of turning on the process is of particular importance in the use of HDP reactors for intermetal dielectric deposition. Deposition and sputtering both occur in the process. If RF power is applied to the chuck before silane is present in the chamber, the underlying metallization will be sputter-etched. This can lead to undesirable faceting of the metal lines (reducing their cross-section, with as a consequence reduced reliability and conductivity) and to electrical leakage between lines due to the deposition of a thin metallic film on the lower insulator surface. On the other hand, if silane is added to the reactor before the RF bias is ramped up the initial deposited oxide will have relatively poor conformality, and may form a "lip" which prevents filling high-aspect-ratio features. Therefore, it is vital to carefully synchronize power and gas flows. Placing gas flow controls a long distance from the reactor is undesirable. One possible approach is to provide "run/vent" valves at the reactor, which can immediately switch the chamber inlet from argon to an argon/silane mixture with no change in flow and minimal time delays.

5.7.5 Atomic Layer Deposition (ALD)

Atomic Layer Deposition (ALD) is a technique that deposits ultra-thin films one atomic layer at a time. Reactants are introduced one by one with pump/purge cycles in between, resulting in a self-saturating surface reaction limited to a single layer on the exposed surface. The result is the deposition of a 100% conformal film, with sequential cycles of these reactions enabling precise control of film thickness. The majority of ALD reactions use two chemicals, typically called precursors. These precursors react with a surface one-at-a-time in a sequential manner. By exposing the precursors to the growth surface repeatedly, a thin film is deposited.

ALD is a self-limiting (the amount of film material deposited in each reaction cycle is constant), sequential surface chemistry that deposits thin-films of materials onto substrates of varying compositions. ALD is similar in chemistry to chemical vapour deposition (CVD), except that the ALD reaction breaks the CVD reaction into two half-reactions, keeping the precursor materials separate during the reaction. Due to the characteristics of self-limiting and surface reactions, ALD film growth makes atomic scale deposition control possible. By keeping the precursors separate throughout the coating process, atomic layer control of film growth can be obtained as fine as ~ 0.1 Å (10 pm) per monolayer. ALD can be used to deposit several types of thin films, including various oxides (*e.g.*, Al_2O_3, TiO_2, SnO_2, ZnO, HfO_2), metal nitrides (*e.g.*, TiN TaN, WN, NbN), metals (*e.g.*, Ru, Ir, Pt), and metal sulfides (*e.g.*, ZnS).

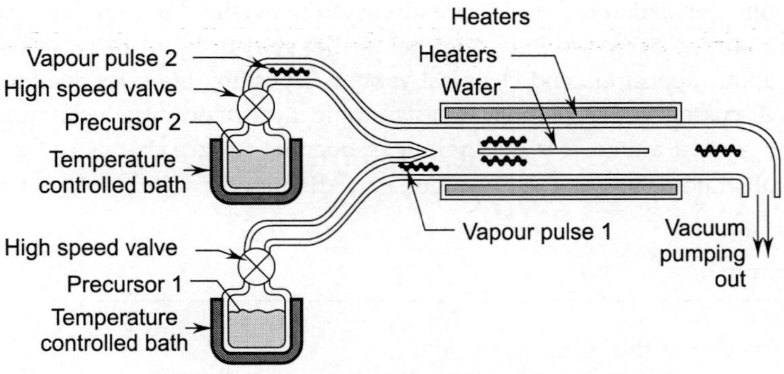

Fig. 5.17 *Atomic Layer Deposition Reactor*

The ALD system consist of two controlled temperature liquid baths provide vapour to the system as shown in Fig. 5.17. The wafer is held in the centre of a low volume reaction chamber for good throughput the chamber must purge quickly between each gas pulse. The reaction chamber is heated to initiate the surface reactions for film deposition and each of the liquid baths is isolated from the system by high actuation speed valves. The valves turn on and off the precursor flows to the system providing the precursor vapour pulses. A vacuum pump maintains flow through the system and facilitates rapid purging of the precursor gases between cycles. The system must be able to achieve rapid reactions during gas pulses and rapid purging in between gas pulses. ALD is a critical technology for future IC manufacturing process.

Advantages

Using ALD, Film thickness depends only on the number of reaction cycles, which makes the thickness control accurate and simple. Unlike CVD, there is less need of reactant flux homogeneity, which gives large area (large batch and easy scale-up) capability, excellent conformality and reproducibility and simplifies the use of solid precursors. Also, the growth of different multilayer structures is straight forward. These advantages make the ALD method attractive for VLSI manufacturing of future generation integrated circuits. Other advantages of ALD are widerange of film materials available and high density and low impurity level. Also, lower deposition temperature can be used in order not to affect sensitive substrates.

Disadvantages

The major limitation of ALD is its slowness; usually only a fraction of a monolayer is deposited in one cycle. Fortunately, the films needed for future generation ICs are very thin and thus the slowness of ALD is not such an important issue. Although the selection of film materials grown by ALD is wide, many technologically important materials (Si, Ge, Si_3N_4, several multi-component oxides, certain metals) cannot currently be deposited by ALD in a cost-effective way. ALD is a chemical technique and thus there is always a risk of residues being left from the precursors. The impurity content of the films depends on the completeness of the reactions. In typical oxide processes, where metal halides of alkyl compounds are used together with water as precursors, impurities found in the films are at the 0.1–1 atom % level.

SUMMARY

In this chapter, thin film deposition techniques are discussed in detail. Thin film deposition parameters like uniformity, thickness, stress, party are important before the start of fabrication processes. In deposition Methods, Physical vapour deposition and chemical vapour deposition are classified and also the comparison of each other is discussed in this chapter. In last Atomic layer deposition technique that deposit ultra-thin films one atomic layer at a time. It is self limiting, sequential surface chemistry that deposits this films of materials on to substrate of varying compositions are discussed.

TUTORIAL SHEET

1. Describe parameters of thin film deposition.
2. Explain physical vapour deposition. Describe methods of PVD.
3. Differentiate between evaporation and sputtering.
4. Explain Groves model of CVD process.
5. Explain chemical vapour deposition (CVD)
6. Describe types of CVD system. Also give advantage and disadvantages of each.

6

OXIDATION

6.1 INTRODUCTION

A Passivation layer of dioxide is formed over the silicon through the oxidation of silicon. It is necessary throughout the modern integrated circuit fabrication process. To have a high-quality IC, understanding of basic oxidation mechanism not enough, but it requires the ability to form a high-quality oxide in a controlled and repeatable manner.

Whenever, a silicon surface is exposed to oxygen, it is converted to silicon dioxide. Silicon dioxide is composed of one silicon atom and two oxygen atoms (SiO_2). Silicon dioxide layers are formed on bare silicon surfaces at elevated temperatures in the presence of an oxidant. The process is called thermal oxidation.

Silicon dioxide has following uses in fabrication of IC.

(*a*) To serve as a mask against implant or diffusion of dopant into silicon.
(*b*) To have surface Passivation that is, to protect semiconductor devices from contamination.
(*c*) To isolate one device from another.
(*d*) To act as component in MOS (Metal Oxide Surface) structures.
(*e*) To provide electrical isolation of multilevel metallization systems.

Several techniques like thermal oxidation, wet oxidation, vapour phase oxidation and plasma oxidation of silicon are develop to obtain high quality oxide layer. First, we will discuss the kinetics behind the oxide growth.

6.2 OXIDE GROWTH

When the silicon wafer is exposed to oxygen, the silicon surface easily reacts with oxygen to form a oxide layer.

The following reactions take place:

$$Si + O_2 \rightarrow SiO_2 \text{ in oxygen}$$

$$Si + 2H_2O \rightarrow SiO_2 + 2H_2 \text{ in water vapour}$$

SiO_2 formation involves sharing of valance electrons between silicon and oxygen that is covalent bond formation. As shown in Fig. 6.1, SiO_2 grows in both the directions upwards as well as downwards with some thickness approximate.

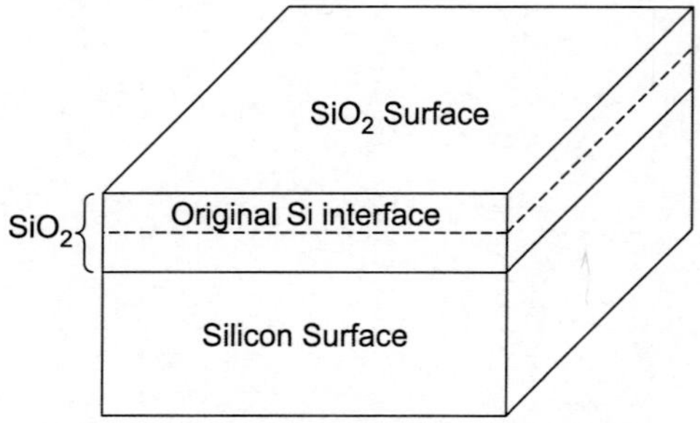

Fig. 6.1 *Grown of Silicon Dioxide*

The structure of thermally grown dioxide is a silicon ion surrounded tetrahedrally by four oxygen atoms as shown in Fig. 6.2. Silicon oxygen internuclear distance is 1.6 A° and oxygen-oxygen internuclear distance is 2.27 A°.

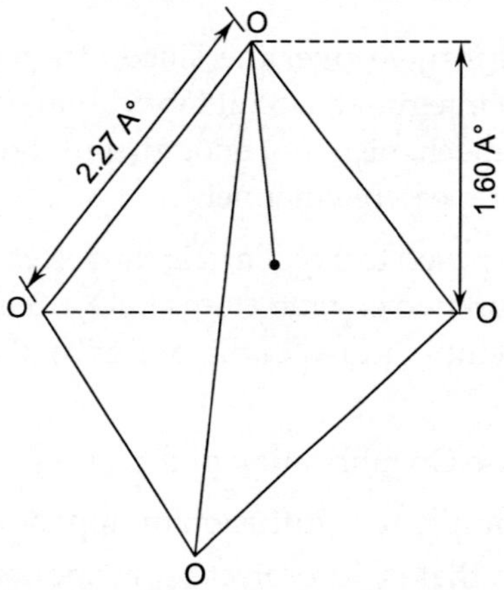

Fig. 6.2 *Silicon Dioxide Structure*

6.3 OXIDE FORMATION

As we know oxidation is carried out by subjecting the silicon wafer to dry oxygen or water vapour at an elevated temperature. When, we use water vapours for oxidation that is called wet oxidation, in which an inert gas is bubbled through water, which is held at 95°C.

In order to have continuous and regular layer of oxide, the oxidation spices must have to move through the growing oxide layer to reach silicon surface. The reactions that occur are mentioned below.

(*a*) Water vapour reacts with the bridging oxygen ions in the silica structure to form non-bridging hydroxyl groups.

$$H_2O + Si - O - Si \rightarrow Si - OH + OH - Si$$

(*b*) At the interface of oxide silicon, the hydroxyl groups react with the silicon lattice to form silicon Polyhedral and hydrogen.

$$\begin{matrix} Si - OH & & Si - O - Si \\ & + Si - Si \longrightarrow & + H_2 \\ Si - OH & & Si - O - Si \end{matrix}$$

(*c*) Hydrogen leaves the oxide layer by rapid diffusion. Some of hydrogen reacts with bridging oxygen ions in the silica to form hydroxyl groups $\frac{1}{2}H_2 + O - Si \rightarrow OH - Si$ which weakens silicon structure.

6.4 OXIDE THICKNESS MEASUREMENT

The thickness of grown oxide is measured by the deal and grove's model that describes the kinetics of silicon oxidation. To apply this model, conditions are:

(*a*) Temperature should be between 700°C and 1300°C

(*b*) Partial pressures between 0.2 and 1.0 atm and

(*c*) Oxide thickness between 300 and 20,000 A° for oxygen and water ambients.

Now, consider the basic model for the thermal oxidation of silicon as shown below.

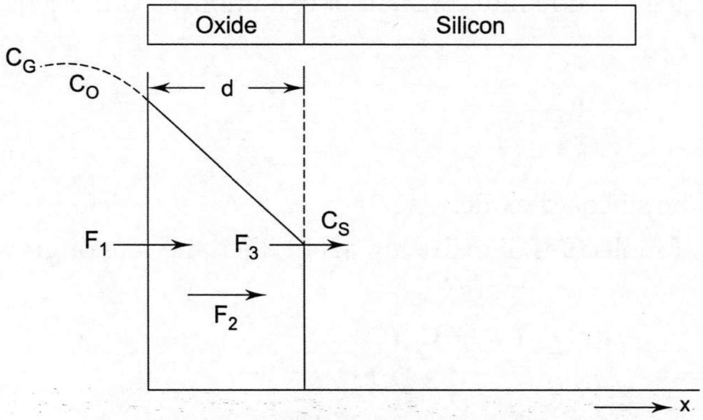

Fig. 6.3 *Basic Model for Thermal Oxidation*

Figure 6.3 shows the silicon substrate covered by an oxide layer that is in contact with the gas phase. In this figure, it is assumed that an oxide of some thickness '*d*' is already present on silicon wafer.

Oxidizing species are transported to silicon through three phases.

(*i*) First of all oxidizing species are transferred from bulk of the gas phase to the gas-oxide interface that is at C_G to C_O (C_G gas phase concentration) and (C_O is oxide phase concentration).

(*ii*) Then across the existing oxide towards the silicon.

(*iii*) At the last react at the $Si - SiO_2$ interface with silicon at silicon surface C_S. (silicon concentration).

Due all these three phases three fluxes exist that are:

F_1 at cross and oxide interface = $h_G (C_G - C_S)$, where h_G is the gas-phase mass-transfer coefficient.

F_2 through the oxide and

F_3 at the silicon surface.

The oxide grows by diffusion of the oxidizing species to the oxide-silicon surface or interface, where a simple chemical reaction such as:

$$Si + O_2 \rightarrow SiO_2$$
$$Si + 2H_2O \rightarrow SiO_2 + H_2$$

As, it is earlier mentioned that three fluxes say F_1, F_2, F_3 are exist at their respective places, at steady state,

$$F_1 = F_2 = F_3 \quad ...(1)$$

The oxidizing agent diffuses through the silicon dioxide layer resulting in a concentration C_S at the surface of silicon at oxide-silicon interface. The flux F_1 is

$$F_2 = D\frac{dc}{dx} \approx \frac{D(C_O - C_S)}{x} \quad ...(2)$$

Where 'D' is the diffusion coefficient of the oxidizing agent and 'x' is the thickness of the oxide layer already present.

Assuming that reaction rate is proportional to the concentration of the oxidizing species at the silicon surface, flux F_2 is written as

$$F_3 = kC_S. \quad ...(3)$$

Where 'k' is the surface reaction rate constant of oxidation, so combining equations (2) and (3) we can write

$$F = \frac{DC_O}{\left(x + \frac{D}{k}\right)} \quad ...(4)$$

This reaction forms the silicon dioxide.

If C_1 is the number of molecules of oxidizing agent, then the rate of growth of the oxide layer thickness is given by:

$$\frac{dx}{dt} = \frac{F}{C_1} = \frac{DC_O/C_1}{(x + D/k)} \quad ...(5)$$

Now, by applying set of condition on 'x' for time t,

At $t = 0, x(0) = d_O$

d_O is the initial value of oxide thickness, by applying this condition in equation (5), we have:

$$x^2 + 2\frac{D}{k}x = \frac{2DC_O}{C_1}(t + \tau) \quad ...(6)$$

Where τ is a constant given by

$$\tau = \frac{(d_O^2 2Dd_O/k)\,C_1}{2DC_O} \quad ...(7)$$

τ represents a shift in time coordinate. After time 't' the oxidizing thickness is given by

$$x = \frac{D}{k}\left[\sqrt{1 + \left[\frac{2C_O k^2 (t + \tau)}{DC_1}\right]} - 1\right] \quad ...(8)$$

equation (8) can be simplified to

$$x = \begin{cases} \dfrac{C_O k}{C_1}(t+\tau) \text{ for small } t \\ \sqrt{\dfrac{2DC_O}{C_1}(t+\tau)} \text{ for large } b \end{cases} \qquad ...(9)$$

From this it is clear that, initially in oxidation the surface reaction is rate limiting factor, the thickness of oxide is directly proportional to time. After sometime when thicker oxide layer is formed, the oxidizing agent have to diffuse through the oxide layer to react at $Si - SiO_2$ interface, at this step thickness of oxide proportional to the square root of the oxidizing concentration.

So, equation (5) can be written as:

$$d_O^2 + Ad_O = B(t+\tau) \qquad ...(10)$$

where, $$A = \frac{2D}{k} \quad \text{and} \quad B = \frac{2DC_O}{C_1}$$

and here $\tau = \dfrac{d_i^2 + Ad_i}{B}$, where d_i is initial oxide layer

Now, solving equation (10) for d_O as a function of time

$$\frac{d_O}{A/2} = \left[1 + \frac{t+\tau}{A^2/4B}\right]^{1/2} - 1 \qquad ...(11)$$

Case 1:

When $t \gg \tau$ and $t \gg A^2/4B$.

$$d_O^2 = Bt \qquad ...(12)$$

The above equation is parabolic function, where B is the parabolic rate constant.

Case 2: For short oxidation time.

where, $(t+\tau) \ll A^2/4B$. So, we have

$$d_O = \frac{B}{A}(t+\tau) \qquad ...(13)$$

This is linear law, where B/A is the linear rate constant. Equations (12) and (13) are the diffusion limited and reaction controlled cases, respectively.

6.5 PREOXIDATION CLEANING

For a successful oxidation process, the removal of surface contamination and unwanted native oxides is essential. If contamination diffuse into the wafer, then structural integrity problems and electrical problems can occur in the device.

Preoxidation processes start by removing the particulate matter either by mechanical or ultrasonic scrubbing. For chemical cleaning methods, immersion processing techniques were preferred until the development of centrifugal spray methods, which eliminate buildup of contaminants as cleaning progresses.

Mechanical scrubber, followed by a wet cleaning sequence to remove organic and inorganic contamination. Then, HF or diluted HF etch is applied to rid the surface of native or chemically grown oxides.

Clean wafers are essential at all stages of the fabrication process, but are especially necessary before any of the operations performed at high temperature.

6.6 DIFFERENT STEPS OF OXIDATION PROCESS

1. Transport of reactants to the deposition region.

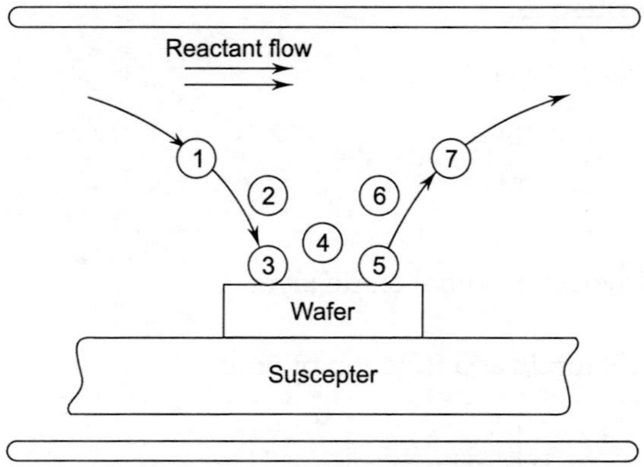

Fig. 6.4 *Different Steps for Oxidation*

2. Transport of reactants from the main gas stream through the boundary layer to the wafer surface.
3. Adsorption of reactants on the wafer surface.
4. Surface reactions including: chemical deposition or reaction, surface migration to attachment sites, emission and redeposition.
5. Deposition of by-products.
6. Transport of by-products from the boundary layer.
7. Transport of by-product away from the deposition.

6.7 THERMAL OXIDATION METHOD

In thermal oxidation the reaction takes place is

$$Si(soild) + O_2(gas) \xrightarrow[\Delta]{} SiO_2(soild)$$

This reaction takes place at room temperature. In order to have quality oxides for practical use in circuits and devices, an elevated temperature is required, mostly between 900°C and 1200°C.

To understand growth mechanism consider the wafer placed in a heated chamber and exposed to oxygen gas as shown in Fig. 6.5(*a*).

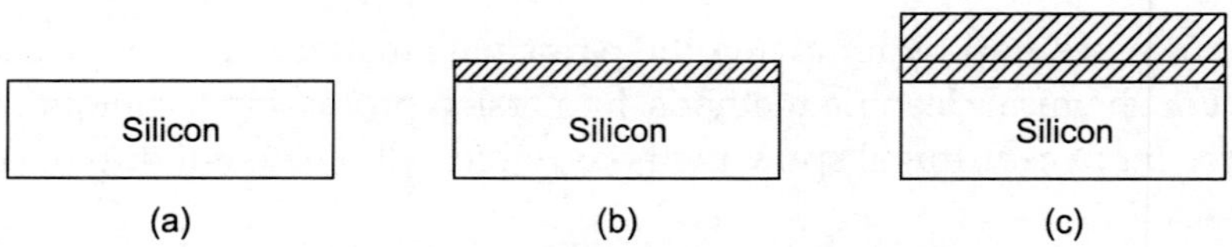

Fig. 6.5 *(a) Initial (b) Linear(c) Parabolic*

The oxygen atoms combine readily with the silicon atoms. This stage is called **linear**, because the oxide grows in equal amounts for each unit of time Fig. 6.5(*b*). After 500 A° of oxide is grown a limit is imposed on the linear growth rate. For oxide layer to keep growing, the oxygen and silicon atoms must come in contact, but initially grown layer of dioxide separates the oxygen in chamber from the silicon atoms of the wafer. To have a continue growth, either silicon in wafer must migrate through the already grown oxide layer to oxygen in the vapour, or oxygen must be able to migrate to the wafer surface. In the thermal growth of silicon dioxide, the oxygen migrates through the existing oxide layer to the silicon wafer surface. Thus, the layer of silicon dioxide consumes silicon atoms from the wafer surface. The oxide layer grows into the silicon surface. This stage of oxidation is called parabolic stage Fig. 6.5(*c*). The linear and parabolic stages of growth are illustrate as:

$$X = B/A\, t \qquad\qquad X = \sqrt{Bt}$$

Linear oxidation of silicon — Parabolic oxidation of silicon

where, X = oxide thickness

B = Parabolic rate constant

B/A = linear rate constant.

From Parabolic rate it is clear that thicker oxide require much more time to grow than thinner oxides. This longer time presents a problem for semiconductor processing. When dry oxygen is used as the oxidizing gas, the growth of thick oxide layers requires even longer oxidation time, especially at lower temperature.

One way to achieve faster oxidations is to use water vapour (H_2O) instead of oxygen as the oxidizing agent. The reaction that takes place with water vapour as oxidizing agent for the growth of silicon dioxide is:

$$\text{Si(solid)} + 2\text{H}_2\text{O} \xrightarrow[\Delta]{} \text{SiO}_2\ \text{(solid)} + 2\text{H}_2\ \text{(gas)}\uparrow$$

In vapour state, the water is in the form H–OH^-. OH^- is called **hydroxyl ion.** This ion diffuses through the oxide layers already on the wafer, faster than oxygen. Doping concentration effects the oxidation rate, vary with the dopant type and concentration level. In general, higher doped regions oxidize faster than more lightly doped regions. Highly doped phosphorus regions can oxidize 2 to 5 times the undoped oxidation rate. Certain impurity, particularly chlorine forms hydrochloric acid (HCl), are included in the oxidizing atmosphere for inclusion in the growing oxide, in case of HCl, the growth rate can increase from 1 to 5%.

6.7.1 Types of Thermal Oxidation Methods

Silicon dioxide layers are grown either at atmospheric pressure or at high pressure. An atmospheric pressure takes place in a system without intentional pressure control. The pressure is simply that of the atmospheric for the location. There are two atmospheric techniques

1. Tube furnaces
2. Rapid thermal system.

Tube furnace is further of two types.

(*a*) Horizontal tube furnace

(*b*) Vertical tube furnace.

6.7.1.1 Horizontal Tube Furnace

This technique is in use in industry since the early 1960s for oxidation, diffusion, heat treating and various deposition processes.

A cross-section of a basic single horizontal three-zone tube furnace as shown in Fig. 6.6

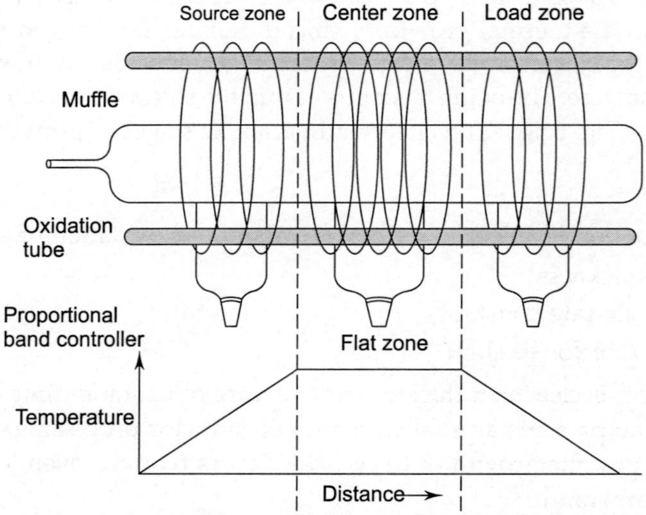

Fig. 6.6 *Cross-Section of Single Horizontal with Three Heating Zones*

This cross-section consists of a long ceramic tube made of mullitie, with coils of copper tubing on the inside surface whole section is divided into three sections or zones and each section is connected to a separate power supply which is operated by a proportional band controller. A furnace can have up to seven zones. Reaction chamber is made of quartz in which oxidation takes place. The reaction tube may itself be inside a ceramic linear called a muffle. Muffle acts as a heat sink for distribution of heat along the quartz tube. Thermocouples send the temperature information to the proportional band controllers.

To heat the reaction tube, coils get the power from proportional controllers. The place, where the coils touch the tube, conduction takes place. For the oxidation, wafer are placed on a holder and positioned in the flat zone. The place, where oxidation takes place in the tube, oxidant gas is passed.

A production tube furnace is an integrated system of various sections as shown below in Fig. 6.7.

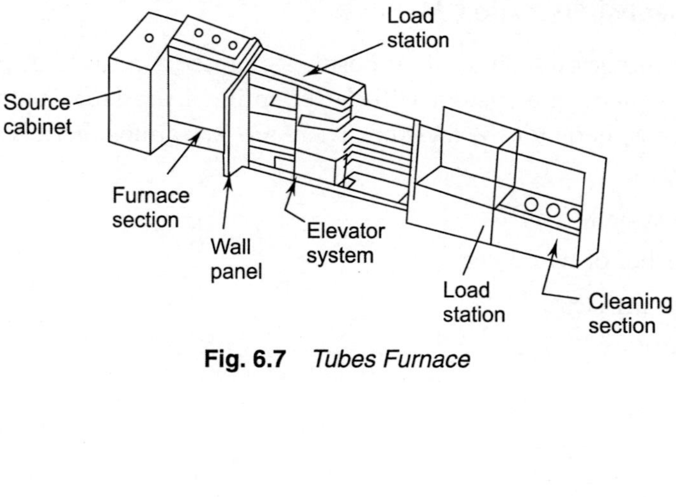

Fig. 6.7 *Tubes Furnace*

Various sections of tube furnace are:

1. Reaction chamber
2. Temperature control system
3. Furnace section
4. Source cabinet
5. Wafer cleaning section
6. Wafer load station
7. Process automation

1. **Reaction chamber:** The traditional reaction chamber material is high-purity quartz. Drawback of quartz are its fragility and the presence of some metallic ions. Sodium ions from the heating coils to pass into the reaction chamber. At temperature above 1200°C quartz break up and sag. This is called **diversification.**

   Quartz tubes require periodic cleaning. One method is in a tank of HF(Hydrofluoric acid). The cleaning process take place out of fabrication area. The cooling and heating hasten the diversification process. The HF cleans by removing a thin layer of the quartz.

   In place of quartz silicon carbide can be used. It is structurally strong and does not break down with repeated heating and cooling.

2. **Temperature control system:** It connects thermocouples touching the reaction tube to proportional band controllers that feed the power to the heating coils. Proportional band controllers maintain even temperatures in the tube. The closer the tube is to the set point temperature, the smaller the amount of power that is fed to the coils.

   In this system, one process called ramping exists, during which the furnace at a temperature several hundred degrees below the process temperature is maintained.

   The wafers are slowly inserted into the furnace, at this lower temperature, after a short stabilization period, the controllers automatically take up the furnace at up to the process temperature.

   During the ramping process, the controllers must maintain the temperature control in the flat zone.

3. **Furnace section:** Tube furnace will contain three or four tubes and separate temperature control system for each of the tubes. The tubes open into an exhaust chamber that draws away the spent and heated gases as they exit the tube. This section of furnace is called scavenger.

4. **Source cabinet:** Number of gases are required to accomplish the desired chemical reaction. Almost, every tube process has nitrogen-flow capability. Each of the tube process requires that the gases be delivered to the tube should be in a specified sequence, at a specified pressure and temperature.

   The equipment used to regulate the gases is located in cabinet attached to the furnace section of the system and is known source cabinet. There are separate units called, a gas control panel, gas flow controller connected to each tube. Flow amount into the tube is controlled by a flow meter.

5. **Wafer cleaning station:** After reaching at tube furnace, the wafers are sent to the wafer-cleaning station. Contamination left on the wafer surfaces can hit the surface, diffuse into the surface.

   The cleaning station may be a Vertical Leminar Flow (VLF) "wet" bench that has built in tanks for cleaning chemicals, units for rinsing the wafers in deionized water, and drying units.

6. **Wafer loading station:** When cleaning is over, the wafers are passed to the in-line load station. Here, wafers are inspected for cleanliness and loaded into holders for insertion into the tube. Wafer loading can be automatic or manual.

   The challenge to any wafer boat-loading system is the correct placement of the test wafers within the load of device wafers as well as dopant source wafers one more challenge is to load two wafers back to back in the same slot.
7. **Process automation:** Modern production tube furnace is automated from wafer load to wafer unload in a one-button operation. A goal is to automate the process from wafer clean to unload. The operator places the cassettes of cleaned wafers on the load mechanism and pushes the start button. The computer takes over and directs the tube furnace through the predetermined process.

6.7.1.2 *Vertical Tube Furnaces (VTF)*

To have greater contamination control and move to ever increasing wafer diameters as well as the need for more productive processing has forced us to move on VTF. As the larger wafer diameter require larger tube diameters, which puts additional pressure on the maintenance of extended temperature flat zones. Larger diameter tubes causes uneven reactions within the tube.

To solve all these problems, VTF is used in which loading takes place from the top or bottom. Tube materials and heating systems are the same as for the horizontal system. VTF is shown below in Fig. 6.8.

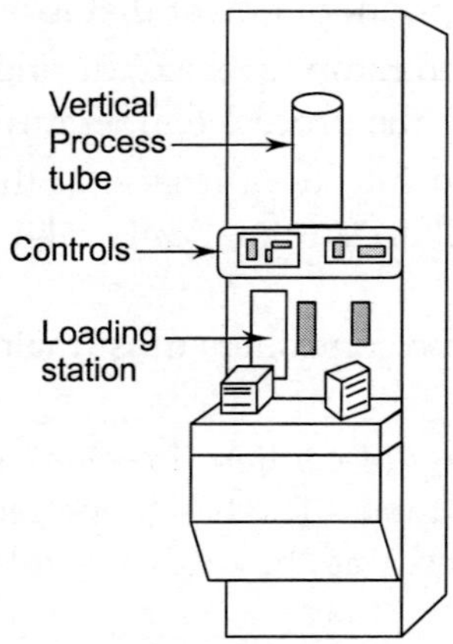

Fig. 6.8 *Vertical Tube Furnace*

The wafers are loaded in standard cassettes and lowered or raised into the flat zone. In this configuration, the wafers are in the most dense loading for a tube furnace. By rotating the wafers in the tube a more uniform temperature across a wafer can be obtained.

In vertical system, the gasses move parallel to gravity minimizing the gas separation problem and the boat rotation minimizes gas turbulence. VTF are capable of producing process variations 60% of those in horizontal furnaces.

Vertical system offer the possibility of locating the furnaces outside the clean room. In this arrangement, the clean-room footprint of the furnace is practically zero and maintenance can take place from the service chose. The system is smaller than a conventional-four-stack system.

6.8 RAPID THERMAL PROCESSING (RTP)

RTP is based on the principle of radiation heating. The wafer is automatically placed in a chamber fitted with gas inlet and exhaust outlets.

Inside chamber a heat source above (and possibly below), the wafer provides a rapid heating. Heat source include graphic heaters, microwave, plasma arc and tungsten halogen lamps. Tungsten halogen lamps are most popular.

The radiation from the heat source couples into the wafer surface and brings it up to process temperature at rates of 75 to 125°C. In this cooling takes place in seconds. Use of RTP reduces the thermal budget required for a process.

RTP technology is a natural choice for the growth of thin oxide used in MOS gate. Smaller feature sizes on the wafer surface has brought along with it a decrease in the thickness of layers added to the wafer. Oxides of gate less than 100 A° thick are required in advanced production. Oxides with this thin dimensions are sometimes hard to control in conventional tube furnace due to problem of quick supplying and removing the oxygen from the system. RTP systems used for oxidation called rapid thermal oxidation (RTO) systems, are similar to the annealing systems, but have an oxygen atmosphere instead of inert gas.

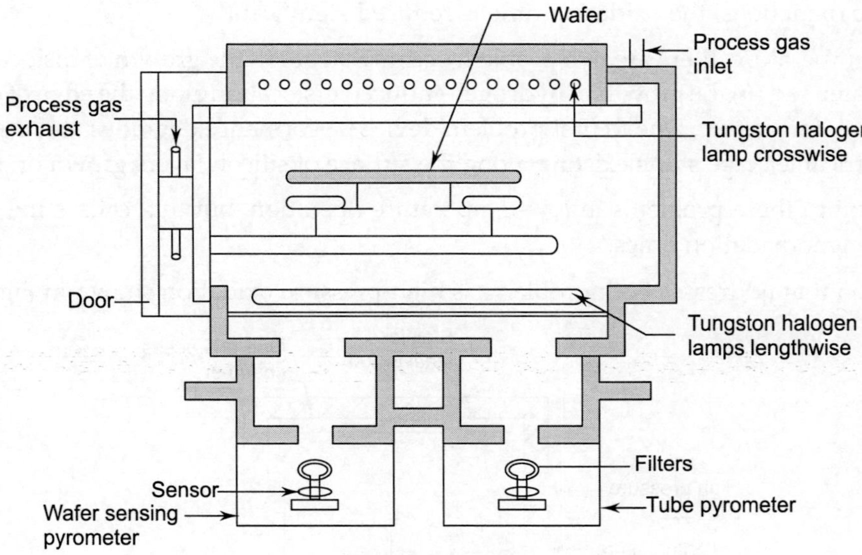

Fig. 6.9 *RTP Design*

Disadvantages of RTP

1. In an RTP system, the wafer never comes to thermal stability. This problem is particularly occur at the wafer edges.
2. Another problem comes from the number and different layer already on an in-process wafer. These layers absorb the heating radiation in different way resulting in temperature difference across the wafer. This phenomenon is called emissivity and is a property of the particular material and the wavelength of the heating radiation. Solution to the problem include lamp placement and control of individual lamps in the system along with top and bottom lamps.

In RTP process, temperature is measured by thermocouples, but they require back contact with the heated wafer, which is impractical in a single-wafer system, and thermocouple have a response time that is longer than some RTP heating cycles.

Optical pyrometers are some time are used for measuring temperature by characterizing the energies given off by the heated object. However, in this also some chances of occurrence of error, especially on wafers with a number of layers. The difficulty is relating the emission given off by the wafer to the actual temperature on the surface.

Solution to this problem includes a backside seal layer of silicon nitride to minimize backside emissivity variations and open loop control. Open loop control is based on converting lamp control to direct current to get away from voltage variations to the lamps.

6.9 HIGH PRESSURE OXIDATION

Oxidation in high pressure steam produces a substantial acceleration in the growth rate. High-pressure oxidation of silicon is particularly attractive, because thermal oxide layers can grow at relatively low temperature in run times as compared to typical high-temperature. One advantage of high pressure oxidation is that oxidation induced defects are suppressed. For high-temperature, high-pressure oxidations, the oxidation time is reduced significantly.

For high-temperature, there are two problem occurs that are (*i*) the growth of dislocations in the bulk of the wafer and (*ii*) the growth of hydrogen-induced dislocation along the edge of openings in layers on the wafer surface. Due to first problem device performance degrades and due to second problem electrical leakage starts existing along the surface of silicon layers grown on the wafer.

The solution to these problems in low-temperature operation, but that causes the production problem of longer oxidation times.

The solution that addresses both problems is high-pressure oxidation shown in Fig. 6.10.

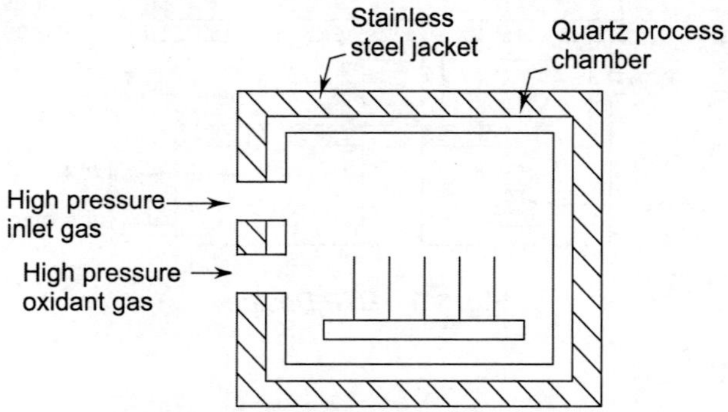

Fig. 6.10 *High Pressure Oxidation*

These systems are configured like horizontal tube furnace, with some difference that is, the tube is sealed and the oxidant is pumped into the tube at pressure 10 to 25 atm. At this pressure, oxidation proceeds at a faster rate. The main rule followed is that a 1 atm increase in pressure allows a 30°C drop in temperature. So, in high-pressure systems drops of 300 to 750°C in temperature.

6.9.1 Wet Oxidation

One way to active faster oxidation is to use water vapour (H_2O) instead of oxygen as oxidizing gas. Water vapour is composed of one atom of hydrogen and one molecule of hydroxyl ion (OH^-) ($H - OH^-$). This ion diffuses through the oxide layer already grown on wafer surface at faster rate than oxygen.

6.9.2 Dry Oxidation

Dry oxidation or HCl dry oxidation is straight forward, using microprocessor-controlled equipment. The desired insertion and withdrawl rates, ramp rates, gas flows, and temperatures are all programmable care must be taken in handling HCl, especially with the exhaust, because HCl corrodes metal parts.

6.10 THIN AND THICK OXIDES

With the advances in micro electronics technology the thickness of the oxide layer is also shrinking. So, it is required that the thin oxides should also shows good electrical characteristics and can be used for long. Dry oxidation is usually used to grow thin oxides. Thin oxides are used in MOSFET technology to form the gate oxide. In dynamic RAM's which utilize the storage property of the capacitors also use thin oxides.

To have accurate and uniform oxide layers the growth of thin oxides should be very slow. For this dry oxidation is used with HCl, using different temperatures. For wet oxidation low pressure techniques are used. The rate of oxidation also depends on the temperature. At lower temperature growth is lower.

Hot nitric acid, boiling water and air at room temperature are used to produce thin oxides. Other techniques use rapid thermal oxidation in a controlled oxygen ambient heating by tungsten halogen lamp to produce oxides with this technique oxides of 40–130 A° thickness with breakdown field can be obtained.

Oxidation performed at reduced pressure provides an attractive way to grow oxides in a controlled manner. Oxides so obtained etch at the same rate as dry oxides. Oxidation in dry $O_2 - HCl$ ambient followed by a heat treatment in N_2, O_2 and HCl at 1150°C produces oxides which are uniform, reproducible and with small defect density. High pressure, low temperature steam oxidation of silicon produces thicker (≈ 300 A°) oxides. Oxides of various thickness can be grown by time, pressure and temperature considerably changing.

6.11 OXIDATION INDUCED DEFECTS

1. **Stacking Faults:** These faults can be generated during the thermal oxidation of silicon laying on <III> at high temperatures. These faults are structural defects in the silicon lattice that are extrinsic in nature and are bounded by partial dislocations. They are caused by incomplete oxidation at $Si - SiO_2$ interface which results in the formation of interstitial silicon in the bulk or at the surface.

   Both wet and dry oxidations results in the formation of oxide induced stacking faults. These faults can be 40 – 50 mm long at their point of penetration into the silicon surface. They can be shrunk by heat treatment at high temperatures in a non-oxidizing atmosphere like argon.

The cause of shrinkage is related to extra plane of atoms diffusing outwards vacancies in silicon. Fault formation is strongly affected by steam oxidation than by dry oxidation and also dependent on crystal orientation, conductivity type and defect nuclei present.

2. **Oxide Isolation Defects:** During selective isolation, stress along the edge of the oxidized area may produce damage in the silicon. Such defects cause increased leakage in nearby devices. So, the stress generated by growing oxide must be retrieved without damaging the silicon. To do this the oxidation temperature must be sufficiently high to allow the stress in the oxide be received by viscous flow. High temperature will present stress induced defect formation in a recessed structure. Also, other stress relieving mechanisms should be used to reduce the stress.

SUMMARY

During fabrication, to protect the underlying silicon oxidation is necessary. During which dioxide layer is formed over silicon, when ever a silicon is came in contact with oxygen. That layer serve as a mask during the etching process. The thickness of that grown droved is measured by using 'Grove's model which operates between. High-pressure oxidation allows thick oxides to be grown at low to moderate temperature while supressing oxidation-induced stocking faults. For growing thin oxide layer rapid thermal processing is very promising technique, in which surface of the silicon is subjected to a high temperature *i.e.*, 1150°C for a very short period of time. For a successful oxidation process, the removal of surface contamination and unwanted native oxides is essential, far that pre oxidation cleaning is done, with HF or diluted HF. Thermal oxidation process takes place atmosphere pressure without intential pressure control, with horizontal tube furnace or vertical one. VTF is preferred to have greater contamination control. High pressure oxidation is one which accelerate the oxidation growth rate.

TUTORIAL SHEET

1. What do you mean by oxidation and what are the uses of oxidation in fabrication of an IC?
2. What are the various steps followed during the oxidation?
3. Write a short note on the oxide growth during oxidation and also explain the silicon dioxide structure.
4. Explain the difference between wet oxidation and dry oxidation.
5. Write the chemical reactions occur during the wet oxidation and write the benefits of hydroxyl ions.
6. Why wet oxidation is better than dry oxidation?
7. Write a note on Grove's law and how it can be used for oxide thickness measurement?
8. What is need of preoxidation cleaning?
9. What is thermal oxidation and what are the different methods of thermal oxidation?
10. Write a short on horizontal tube furnace. Explain with the help of diagram.
11. Explain various section of horizontal tube furnace.
12. Draw and explain vertical tube furnace. Why it is better than horizontal tube furnace?
13. Write a short on rapid thermal processing and explain with the help of diagram. What are the disadvantage of RTP?
14. Write a note on oxidation induced defects.

7

DIFFUSION

7.1 INTRODUCTION

In previous chapters, we have discussed about properties of silicon, crystal structure, crystal growth oxidation etc. In the process of VLSI fabrication, we can start from the basic raw silicon dioxide (SiO_2) which is converted into metallic grade silicon which in turn converted into electronic grade silicon and the crystal growth is done which follows the wafer preparation. After that different fabrication processes will be carried out such as oxidation, diffusion, annealing, etching, lithography, etc. Diffusion is done to implant impurity atoms in semiconductor.

7.2 DEFINITION

It is a spontaneous and automatic process. Broadly, it can be defined as a process of movement of small molecules from an area, where they are highly concentrated to an area, where they are less concentrated.

With respect to VLSI fabrication it is a process by which controlled amount of impurity is introduced into the semiconductor, by subjecting the semiconductor at high temperature in an ambient containing the dopant impurity.

The process at atomic level

It occurs when there is a concentration gradient of impurity atoms.

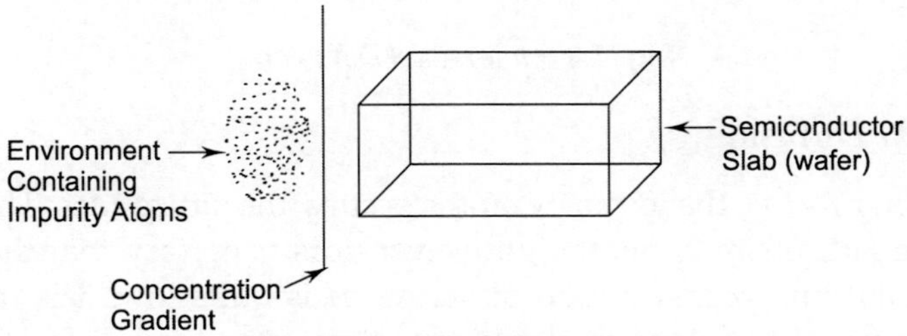

Fig. 7.1 *The Process of Diffusion*

The impurity atoms will move inside the semiconductor in above case. Diffusion is governed by the following factors:

1. Concentration gradient
2. Temperature
3. Nature of impurity atoms
4. Crystal structure of farget material
5. Defects in crystal
   (*a*) Vacancy
   (*b*) Interstitial
6. Time of diffusion the total time for which semiconductor is placed under diffusing environments.

On the basis of above discussion, we can say that diffusion can be viewed as the atomic movement of diffusing impurity atoms in the lattice through the vacancies or through the interstitial sites or the combination of both or by some other mechanisms.

1. Interstitial Diffusion
2. Substitutional Diffusion
3. Interstitial-Substitutional Diffusion
4. Inter Latency Diffusion
5. Interchange Diffusion
6. Gain Boundary Diffusion

7.2.1 Interstitial Diffusion

The process in which an impurity atom moves through the crystal lattice by jumping from one interstitial site to the next. Since, the large number of vacant sites are available in a semiconductor, the process is relatively fast. The diffusion of sodium and lithium in semiconductor involves such type of mechanism.

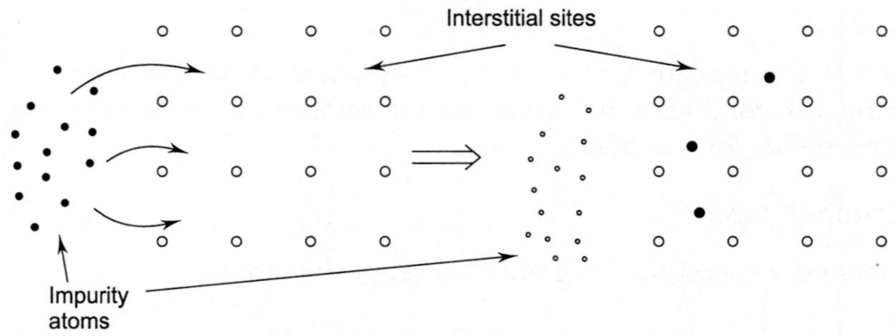

Fig. 7.2 *Interstitial Diffusion*

7.2.2 Substitutional Diffusion

In this type of diffusion process, the impurity atoms occupy the lattice sites. The process continues by jumping from one lattice site to another, however, it is necessary that these adjacent site be vacant. Since the equilibrium concentration of vacancies is quite low, it is reasonable to expect substitutional diffusion to occur at much slower rate than interstitial diffusion.

Many dopants, used in silicon microcircuit fabrication are substitutional diffusers.

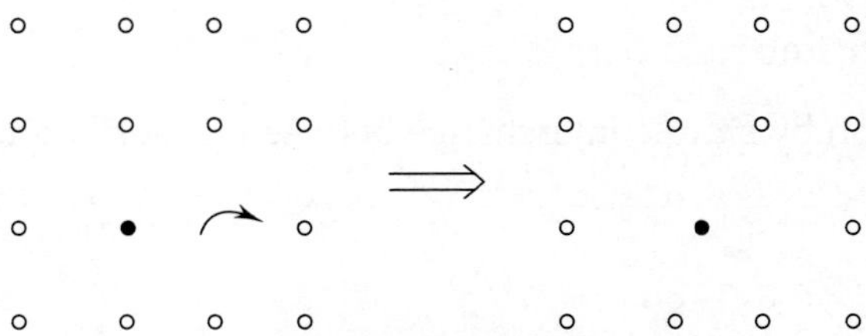

Fig. 7.3 *Substitutional Diffusion*

7.2.3 Interstitial-Substitutional Diffusion

When impurity atoms occupy both, interstitial as well as substitutional sites the diffusion is so called interstitial-substitutional diffusion. This occurs by two mechanisms. One is the dissociative mechanism in which a substitutional impurity atom can become an interstitial, leaving behind a vacancy, which can be a controlling factor for the process. Copper and nickel in silicon and zinc, cadmium and copper in GaAs diffuses by this mechanism.

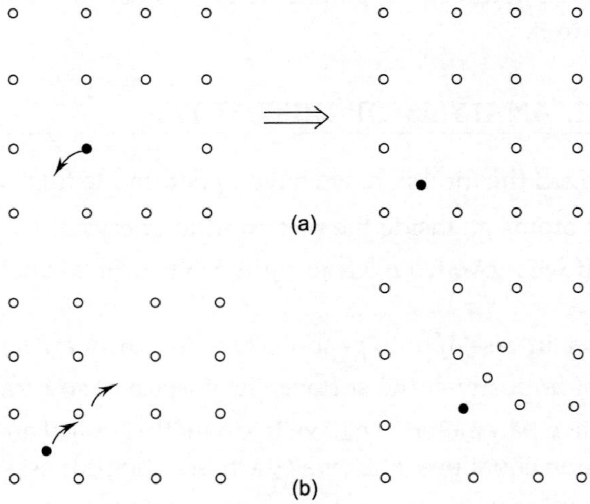

Fig. 7.4 *Interstitial-Substitutional Diffusion by (a) Dissociative (b) Kick-out Mechanism*

Second is the kick-out mechanism in which the impurity atoms at interstitial sites kick-out the lattice site base crystal atoms and occupies the position as describe in Fig. 7.4(*b*). Gold and platinum diffuses in this way.

7.2.4 Inter Latency Diffusion

Here, interstitial host atoms can be annihilated by pushing substitutionally located impurity atoms into interstitial sites. These impurity can diffuse to adjacent substitutional sites and create new self interstitials. Thus, the interstitial position of the diffusing impurity atom is purely a transition state, in moving from one substitutional site to another. Boron and phosphate diffuse in silicon in this manner also.

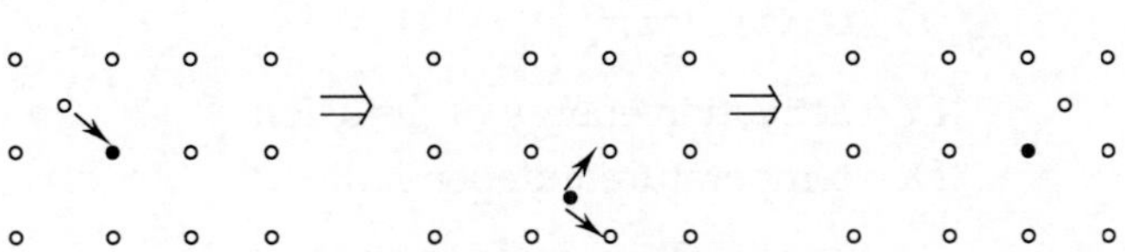

Fig. 7.5 *Inter Latency Diffusion*

7.2.5 Interchange Diffusion

It is the process of diffusion by simple interchange between impurity and host atoms.

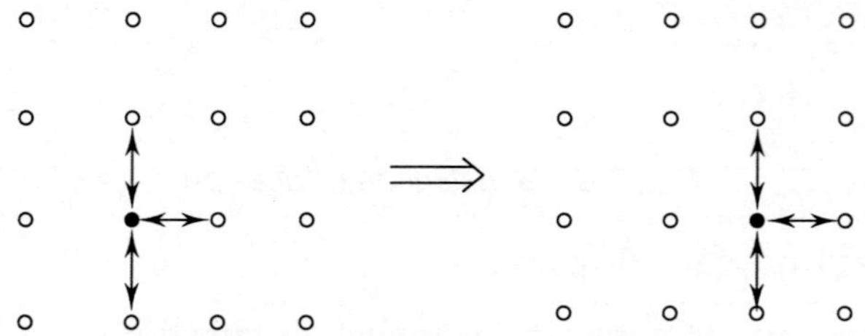

Fig. 7.6 *Interchange Diffusion*

7.2.6 Gain Boundary Diffusion

Diffusion can also occur by movement along dislocations and gain boundaries. This process is anisotropic being much faster in directions parallel to the dislocation core or to the gain boundary edges then at right angles to it.

7.3 MATHEMATICAL ANALYSIS OF DIFFUSION

In order to fully characterized the diffusion, we have to determine following parameters:

(*a*) How far the dopant atoms go inside the semiconductor crystal.

(*b*) Doping profile *i.e.*, if we move from left to right or vice versa what is the change in doping concentration.

(*c*) Where is the junction, in case, if both *p*- and *n*-type impurity are being dopped.

(*d*) The concentration of impurity at the surface, the surface concentration.

To determine all the above parameters we have to do mathematical analysis by modeling some equations known as diffusion equations. Flick was the first, who proposed the diffusion equations.

7.3.1 Flick's Laws

First Law: The flux density is directly proportional to the concentration gradient. Flux density is defined as the rate of transfer of dopant per unit a area

$$j \propto \frac{\partial N}{\partial x} \Rightarrow \boxed{j = D\frac{\partial N}{\partial x}} \quad ...(1)$$

where, j = Flux

D = Diffusion coefficient

N = Concentration

k = Distance in the direction of measurement

D can also be defined as,

$$D = D_o \exp \frac{-E_A}{kT} \quad ...(2)$$

E_A = activation energy of diffusion

D_o = temperature independent.

From the equation (2) we can say that with the increase in temperature the diffusivity increases. A plot between will give the value of E_A. E_A signifies the ease in diffusion. For example, E_A for substitutional diffusion is around 3 to 4 ev (electron volt) and for interstitial diffusion it is around 6 to 12 ev.

$$E_{A,\text{substitutional}} > E_{A,\text{interstitital}}$$

Hence, we can say that movement through interstitial space is much faster than substitutional diffusion.

Second Law: Consider a slab of semiconductor in which the dopant atoms are moving from left to right as shown in Fig. 7.7.

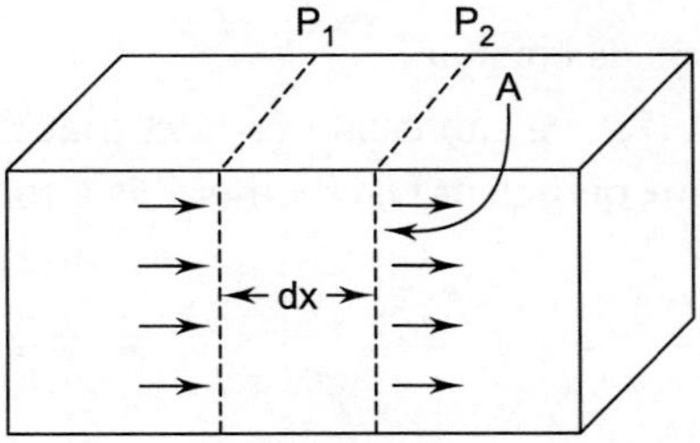

Fig. 7.7

Consider a plane P_1 normal to the direction of diffusion and another plane P_2 parallel to P_1 at a distance dx from P_1.

Considering the accumulation of dopant atoms between the planes P_1 and P_2.

if $\frac{\partial N}{\partial t}$ is the rate of change in concentration and A dx is the incremental volume, volume between two planes. Then, $\frac{\partial N}{\partial t}$. A dx will be the rate of accumulation of dopant atoms in between planes, P_1 and P_2.

Again, difference in flux at P_1 and P_2 multiplied by the area is also the accumulated atoms.

$$\therefore \quad \frac{\partial N}{\partial t} A\,dx = A\,[j(x) - j(x+dx)] \qquad ...(3)$$

$$\text{or,} \quad \frac{\partial N}{\partial t}\,dx = [j(x) - j(x+dx)]$$

$$\text{or,} \quad \frac{\partial N}{\partial t} = -\frac{j(x+dx) - j(x)}{dx} \qquad ...(4)$$

Putting the limit dx tending to zero, on both sides

$$\lim_{dx \to 0} \frac{\partial N}{\partial t} = \lim_{dx \to 0}\left[-\frac{j(x+dx) - j(x)}{dx}\right] \qquad ...(5)$$

$$\text{or,} \quad \frac{\partial N}{\partial t} = -\frac{\partial j}{\partial x} \qquad ...(6)$$

Putting the value of j from equation (1) in equation (6) we get,

$$\boxed{\frac{\partial N}{\partial t} = \frac{\partial}{\partial K}\left(\frac{D\,\partial N}{\partial K}\right)} \qquad ...(7)$$

Equation (7) is known as flick's second Law. Both equations (1) and (7) are one dimensional diffusion equations. Flick's first law gives the statement of the steady state flux and Flick's second law gives the statement of the dynamics of the redistribution of dopant during diffusion. Flick's Laws are based on the following assumptions.

(*a*) Validity of Maxwell—Bolt—Mann statistics.

(*b*) Quasi–equilibrium description of the particle is adequate.

(*c*) Isothermal conditions are prevalent.

(*d*) Electric field inside the Lost lattice can be neglected.

(*e*) Relaxation time for collisions is constant.

From the analysis of equation (7), we can easily predict that the differential is completely dependent on the nature of diffusion coefficient D, whether it is a constant or a varies with *k*. We will consider both cases one by one.

Case 1: D is constant.

We can rewrite equation (7) as

$$\frac{\partial N}{\partial t} = D\frac{\partial^2 N}{\partial x} \quad ...(8)$$

Equation (8) is known as Flick's simple law of diffusion. The doping profile will again be of various types as

(*a*) **Infinite or Constant Source:** It means that the surface concentration is always constant and is given by its maximum value at surface. It can be achieved by setting a constant supply of dopant in the ambient. Maximum value is the solid solubility limit. Boundary condition will be,

$$N(0, t) = N_0$$

$$N(\infty, t) = 0$$

Where, $N(0, t)$ = concentration at $x = 0$ and at time +

$N(\infty, t)$ = concentration at $x = \infty$ and at time +

N_0 = surface concentration which is constant in this case.

Also, at the instant when diffusion starts, *i.e.*, $t = 0$, the doping concentration is 0 independent of x,

$\therefore \quad N(x, 0) = 0$

which is the initial condition,

on solving equation (8) subjected the mentioned boundary and initial conditions, we get the doping concentration in the crystal as

$$N(x, t) = N_0 \operatorname{erfc}\frac{x}{2\sqrt{Dt}} \quad ...(9)$$

$$Q = 2N_0\sqrt{\frac{Dt}{\Pi}} \quad ...(10)$$

Graphically,

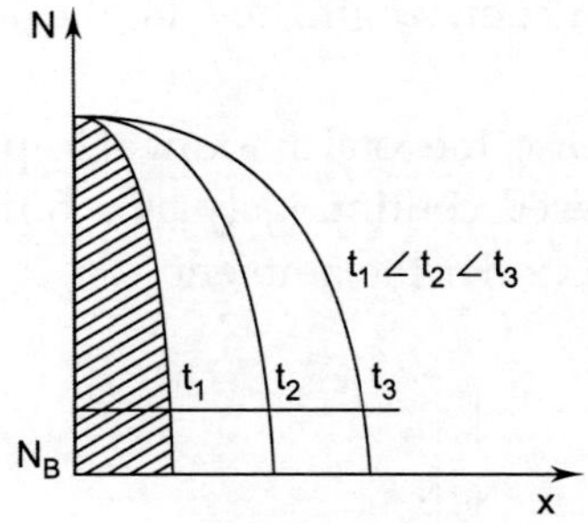

Fig. 7.8

In equations (9) and (10),

Q = Total amount of impurity

Erfc is the error function compliment and is given by

$$\text{erfc}\left(\frac{x}{2\sqrt{Dt}}\right) = 1 - \text{erf}\left(\frac{x}{2\sqrt{Dt}}\right) \qquad ...(11)$$

and
$$\text{erf}\,\frac{x}{2\sqrt{Dt}} = \frac{2}{\sqrt{\Pi}}\int_0^{x/2\sqrt{Dt}} \exp(-z^2)\,dz \qquad ...(12)$$

To calculate total amount of impurity which has been introduced in the semiconductor during diffusion, we can calculate the area under the curve shown by hashed lines for $t = t_1$ and can be computed by simple definite integral laws as

$$Q = \int_0^\infty N(x,t)\,dx$$

$$Q = \int_0^\infty N_0 \text{ exfc } \frac{x}{2\sqrt{Dt}}\,dx$$

$$Q = N_0\, 2\sqrt{Dt}\int_0^\infty \text{exfc } z\,dz \quad \left(\text{by putting } \frac{x}{2\sqrt{Dt}} = z\right)$$

$\therefore$ $$dx = 2\sqrt{Dt}\,dz.$$

The value of the integral is $1/\sqrt{\Pi}$

$\therefore$ $$Q = 2N_0\sqrt{\frac{Dt}{\Pi}}$$

Which is the equation (10)

From the above discussion, we have the following points to be noted:

(*a*) The doping profile follows the complementary error function.

(*b*) The surface concentration is constant and is given by putting $x = 0$ in equation (9), which is,

$$N(0, t) = N_0 \text{ erfc } \frac{0}{2\sqrt{Dt}}$$

$$N(0, t) = N_0$$

which is the solid solubility limit.

(*c*) Total amount of impurity is a function of time in constant source diffusion.

(*d*) $\sqrt{Dt}$ is in turn the major of junction length, how far the dopant atoms are penetrating inside the semiconductor.

Constant Total Impurity: In this case, the total amount of impurity atoms are constant and hence the surface concentration is decreased continuously as a function of time, because it is being absorbed by the semiconductor put under the ambient.

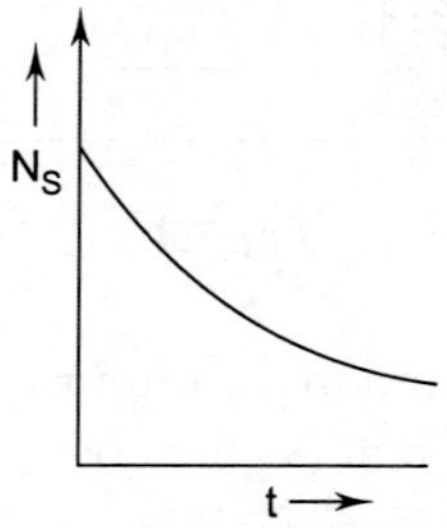

Fig. 7.9 *Change in Surface Concentration Vs Time*

To determine the doping concentration profile we have to determine some boundary conditions which are,

$$\int_0^\infty N\,dx = Q \quad \text{total dopanent concentration is constant.}$$

$$N(x, t) = 0 \quad \text{concentration at } x = \infty \text{ is 0.}$$

The initial condition will remain same,

$$N(x, 0) = 0$$

Solving the equation (8), Flick's simple law of diffusion subjected to the above boundary and initial conditions we have,

$$\boxed{N(x, t) = \frac{Q}{\sqrt{\Pi Dt}} \exp\left(\frac{-x^2}{4Dt}\right)} \qquad \text{...(13)}$$

Where,
Q = Total impurity
D = Diffusion coefficient
t = time
x = distance

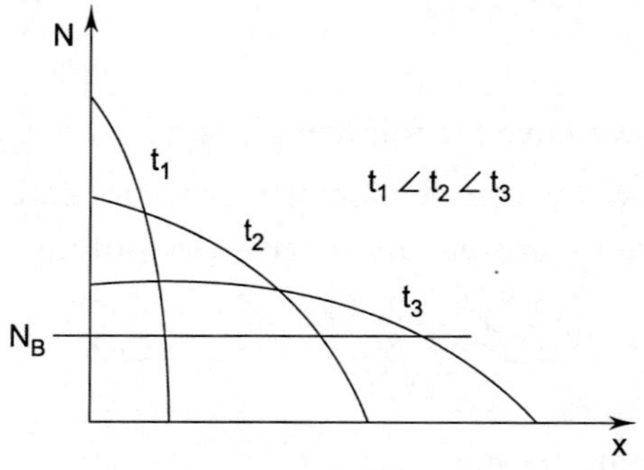

Fig. 7.10

the equation (13) is of the form of diffusion profile. Surface concentration can be calculated by simply putting $x = 0$,

$$N_s = N(0, t) = \frac{Q}{\sqrt{\Pi Dt}} \quad \text{...(14)}$$

and the diffusion depth can be calculated as,

$$x_j = \sqrt{4\,Dt \ln \frac{N_S}{N_B}} \quad \text{...(15)}$$

Where x_j = junction depth

N_B = background doaping.

From above equation (15), we can say that junction depth will depend upon the following two factors,

- Product of diffusion constant and time (Dt).
- ratio N_S and N_B.

The factor Dt is coming in almost all equations, Form equation (9) to (15) and is a very important factor. Since, the unit of D is cm^2/s and unit of t is sec. hence Dt will have a unit of cm^2 and $\sqrt{Dt}$ will have a unit cm. It signifies that how much the dopants move inside the semiconductor.

We have discussed the case of constant D under infinite source and constant total source. But what happens practically? It is a question which one wants to know, and the answer is, practically both the type of processes are carried out the second after the first. Initially the semiconductor is subjected under constant supply of dopant for specified temperature and given time period. So that sum dopant atoms diffuses in the semiconductor, the diffusion profile follows the complementary error function and hence the depth of diffusion is very less. Hence, the step is also known as **prediffusion** step. Then, the source is shut off and the sample with thin sheet of dopant atoms is already incorporated very close to the surface is subjected farther to high temperature process. As the source is shut off there is no possibility at any new impurity coming in to the semiconductor and the case becomes the constant total impurity and the step is also referred to as **drive in step.**

Case 2: D is a function of N

In this case, the diffusion coefficient is itself a function of concentration *i.e.*,

$$D \propto N^r$$

$$D = kN^r \quad \text{...(16)}$$

Where, k is the proportionality constant and $r \in 1, 2$ and 3

The equation (7) is written below again,

$$\frac{\partial N}{\partial t} = \frac{\partial}{\partial x}\left(D\frac{\partial N}{\partial x}\right)$$

Since, N is the function of x hence D is in turn is a function of x hence we can't take it out from the differential.

In case of surface conditions equation (16) becomes,

$$D_S = kN_s^r \quad \text{...(17)}$$

From equations (16) and (17) we can write,

$$D = \left(\frac{N}{N_s}\right)^r D_s \qquad \text{...(18)}$$

Equation (18) is the normalized equation.

For different values of r we have,

$$D = \frac{N}{N_s} D_s \qquad \text{...(19)}$$

$$D = \left(\frac{N}{N_s}\right)^2 D_s \qquad \text{...(20)}$$

$$D = \left(\frac{N}{N_s}\right)^3 D_s \qquad \text{...(21)}$$

The solution of these equations are obtained by computer simulation tables given the values for $r = 1, 2$ and 3. The graph can be plotted between N/Ns and $x/2\sqrt{D_s t}$ which is,

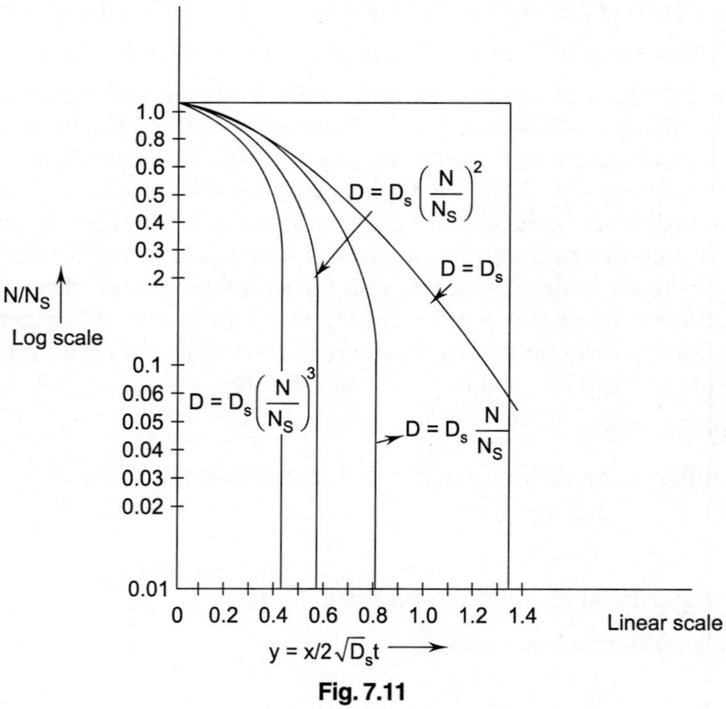

Fig. 7.11

From Fig. 7.11 it is seen that, for $r = 1, 2$ and 3, doping profiles are extremely steep at low concentrations. Consequently, highly abrupt junctions are formed when diffusions are made into a background of opposite impurity type. Furthermore, the steepness of the doping profile results in a junction depth which is given by the value for which $N \simeq 0$. That is, it is almost independent of the background concentration.

For different cases, junction depth can be calculated as,

$$x_j = \begin{bmatrix} 1.616\sqrt{(D_S t)} & \text{for} & r=1 \\ 1.019\sqrt{D_S t} & & r=2 \\ 0.872\sqrt{D_S t} & & r=3 \end{bmatrix} \qquad \text{...(22)}$$

A number of computer aided process simulation programs can be used to provide numerical solutions of these equations. The programs provided two and three dimensional simulations, and can include complex effects such as impurity consumption during the oxidation phase.

Successive Diffusion: The process of diffusing the impurity atoms in semiconductor is known as successive diffusion. Suppose, a semiconductor wafer is doped by *n*-type impurity in the fabrication of BJT for base region and then it is doped by *p*-type impurity for the formation of emitter and collector region then during the drive in step of *p*-type impurity the diffused *n*-type impurity will again starts diffusing due to high temperature of the wafer and hence re-distribution of the previously doped impurity atoms occur which is known as successive diffusion and is given by,

$$(Dt)_{eff} = D_1 t_1 + D_2 t_2 + \ldots\ldots\ldots \qquad \text{...(23)}$$

Where, $(Dt)_{eff}$ is the product of effective diffusion coefficient and time.

D_1, D_2 ….. are the diffusion coefficients for process1, process2 ….. respectively.

t_1, t_2 …. are the total time of process1, process2 ….. respectively.

The important aspect is that any high temperature process will modify both the

- diffusion profile
- diffusion length

and it is the region why one goes for as low as possible temperature process while designing a process. So, as not to effect the already obtained diffusion profiles.

Diffusion is sensitive to various parameters such as:

(*a*) **Dependence on concentration:** When the concentration of dopanent atoms is very less as compared to intrinsic concentration then the D is independent from the concentration on the other hand if intrinsic concentration is very less as compared to dopanent concentration, then it is a function of N.

(*b*) **Dependence on temperature:** Diffusivity D is a strong function of temperature is rewritten as

$$D = D_0 \exp\left(-\frac{E_a}{kT}\right)$$

D_0 is a constant known as frequency factor and E_a is the activating energy. In atomic diffusion models D_0 is related to the atomic jumping frequency or the lattice vibration frequency and a jumping distance of impurity a defect or defect impurity pair. Activation energy E_a is related to the energies of motion and the energies of formation of defect impurity compliances. Diffusivity of some common dopants in silicon is given in Table 7.1 .

Table 7.1

Element	*D (cm²/sec)*	*Eₐ (ev)*	*T (°C) range*
P	10.5	3.69	950 – 1235
As	0.32	3.56	1100 – 1360
Sb	5.6	3.95	1100 – 1360
B	10.5	2.69	950 – 1275
Al	8.0	3.7	1100 – 1375
In	16.5	3.9	1100 – 1360
Ga	3.6	3.51	1100 – 1360

(*c*) **Dependence of multiple vacancy types:** Diffusivity due to single vacancy type is considered so far. To calculate accurately, the diffusion process is described by multiple vacancy diffusion model. If there are M different vacancy diffusion mechanisms present, then the effective diffusivity at a given temperature is obtained by summing the contribution due to each of them.

7.4 STUDY OF DIFFUSIVITIES OF SOME COMMON DOPANTS

To form *p-n* junction in silicon some common dopants are boron, phosphorus, arsenic and antimony and hence the study of their diffusing properties is important. The explanation under various conditions are:

1. **Effect of low impurity concentration:** The intrinsic diffusivities of boron, phosphorus, arsenic and antimony are shown in Table 7.2, where Di stands for intrinsic diffusivity, and Fig. 7.12 shows the variation of intrinsic diffusivity expression represented by equation (2).

Table 7.2

Parameter	*Boron*	*Phosphorus*	*Arsenic*			*Antimony*
			CS	PD	IS	
	$(Di)_B$	$(Di)_P$		$(Di)_{AS}$		$(Di)_{Sb}$
Do(cm²/s)	0.76	2.85	24	22.9	6.0	0.24
E(ev)	3.46	3.66	4.06	4.1	4.2	3.65

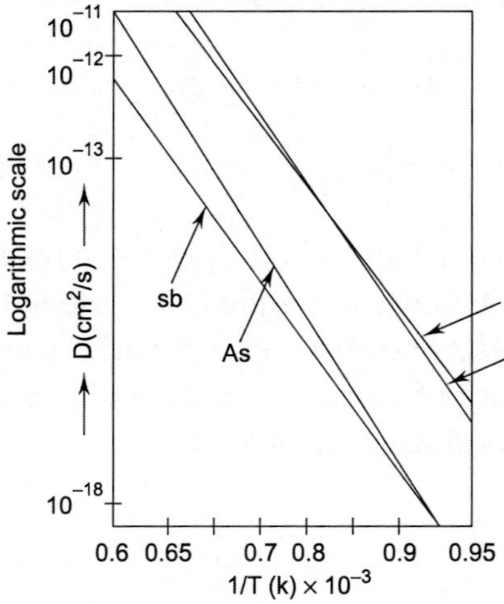

Fig. 7.12 *D Vs 1/T for Different Materials*

2. **Effect of electric field:** For ionized impurity atoms local electric field exists between the ionized impurity atoms and the electrons/holes. Also the concentration gradient of these ionized impurity atoms produces an internal electric field that increases their diffusivity. This internal field is expressed as,

$$Ex = -\frac{\partial}{\partial x}\phi(x, t) \quad ...(24)$$

where, $\phi(x, t)$ is the electric potential energy given by,

$$\phi(x, t) = \frac{E_C - E_F}{q} \quad ...(25)$$

with, E_C as conduction band energy and E_F as Fermi level energy.

3. **Effect of band gap:** The effect of band gap will affect the concentrations of the electrically active point defects because the concentrations of charged point defects is related to the Fermi level and the relative location of the Fermi level w.r.t. the conduction/valance band edge is related to the energy band gap. At high impurity doping concentrations band gap variation is more prominent. Lattice stress and strain also causes band gap narrowing at highest concentration of dopant atoms. Band gap narrowing effect causes the effective intrinsic electron density to reduce from the intrinsic electron density. For phosphorus diffusion, the band gap energy changes and it can be expressed as:

$$\Delta Eg = -1.3 \times 10^{-22}(Cs - Cz) \quad ...(26)$$

where, Cs = surface concentration

Cz = Empirical concentration

The energy band gap narrowing in phosphorus from a $POCl_3$ source and a P ion implanted diffused layer is given in Table 7.3.

Table 7.3

	Temperature	*Time (min)*	*Cs($10^{20}cm^{-3}$)*	*C_2($10^{23}gap^{-3}$)*	*ΔEg(ev)*
	900	30	7	3	0.052
$POCl_3$	1000	60	9	2.85	0.08
	1100	25	5.6	2.6	0.039
	1200	15	4	1.5	0.033
P implant	1040	55	–	–	0.03

4. **Effect of high concentration:** When the surface concentration of the dopant atoms is greater than several phenomena like clustering a reduction in electric activity have been observed which is discussed below for different dopants.

   (*a*) **Arsenic:** For concentration as greater than $10^{20}/cm^3$, the measured profile has a nearly flat portion near the surface region. Arsenic atoms form clusters at this concentration. A curve between C_{max} and temperature is drawn for different values, shown in Fig 7.13.

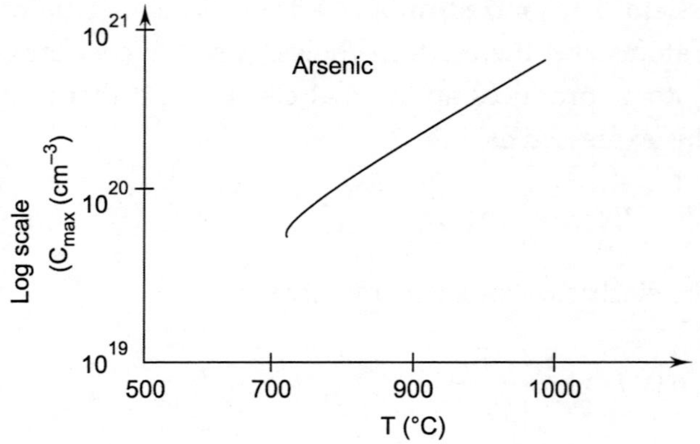

Fig. 7.13

(*b*) **Boron:** The diffusivity of boron in high concentration region is reduced to almost zero. The maximum carrier concentration of boron versus temperature at highest concentration is shown in Fig. 7.14.

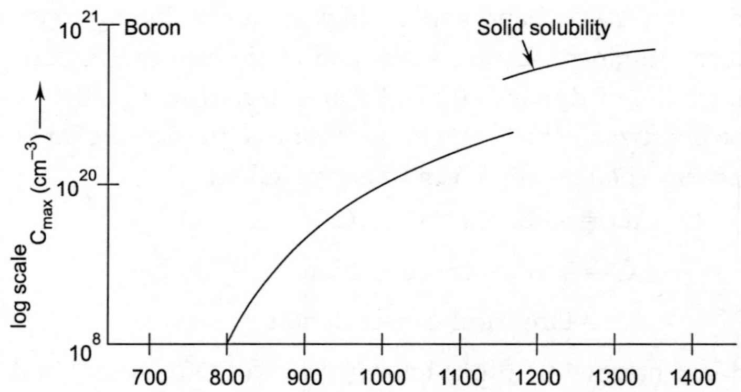

Fig. 7.14

(*c*) **Phosphorus:** It is mainly used in small, shallow junctions and low temperature processing. It diffuses mainly via a vacancy mechanism. In high concentration regions, a fraction of phosphorus ion (P^+) concentration pairs with v^{2-} vacancies to form $(Pu)^-$. The concentration $(Pv)^-$ is proportional to the surface electron concentration in npn narrow base transistor with phosphorus diffused emitter and boron diffused base, the base region under the emitter region is deeper than outside emitter region by 0.2 μm to 0.6 μm. This phenomenon is known as emitter push effect. The curve will be,

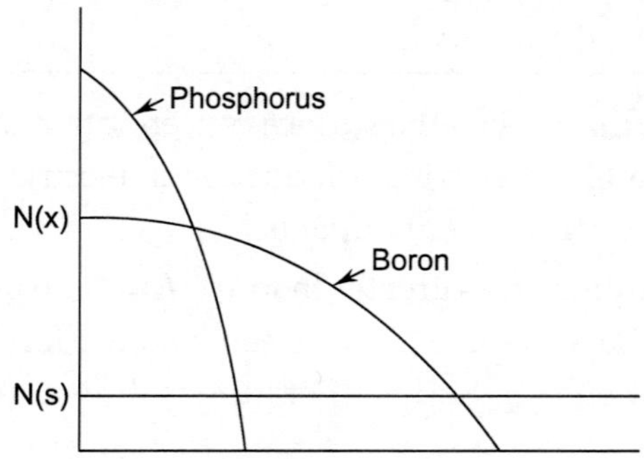

Fig. 7.15 *Emitter Push Effect*

7.5 DIFFUSION SYSTEMS

Diffusion systems can be categorized as:

1. Predisposition, where the dopant is to be transported in to the wafer.
2. Redistribution or drive in where the impurity is redistributed with the wafer as an increased temperature effect.

The main requirement of any diffusion system is that means which is provided for bringing the diffusion impurity in contact with a suitably prepared wafer. The surface concentration should be controllable over a wide range of values. Diffusion process should not damage the wafer surface. After diffusion, the material lying on the wafer surface should be easily removable. The diffusion system should give reproducible results from one diffusion to the next and from slice to slice within a single run. The system should be capable of batch process.

There are two types of diffusion systems used in the industries, these are:

1. **Sealed tube type system:** The dopant and wafers are enclosed in a clean, evacuated quartz tube prior to heat treatment. After the diffusion process the slices are removed by breaking the tube. This is a contamination free diffusion sealed tube system operate by thermal evaporation of the dopant source, transport in the gas phase, adsorption on the surface of the semiconductor and the tube walls and finally the diffusion of the dopant into the wafer/slices surface concentrations obtained in sealed tube diffusion can be made to approach the solid solubility limit for the impurity at the diffusion temperature.
2. **Closed tube system:** Usually, the open tube method is used, where the wafer is kept in a highly temperature controlled furnace. These furnaces, have a ceramic core on which (allow of Al, Cr and Ni) wire is wound as an heating element. The heating coil has three separately temperature controlled zone. For temperature control, a Pt – Pt Rh thermocouple is used to sense the temperature in each of the zones. The central zone thermocouple compares, with a reference set temperature and the outer zone thermocouple compare with additional thermocouples placed in the central zone. The quartz furnace tube is open at both ends. The wafers are loaded in a quartz boat.

Diffusion systems: The dopant can be a gaseous liquid or a solid source. And, hence the dopant diffusion system can be of following three types:

(*a*) **Solid source diffusion system:** In solid source diffusion system, a solid compound of the dopant is taken. A solid source diffusion system is shown in Fig. 7.16.

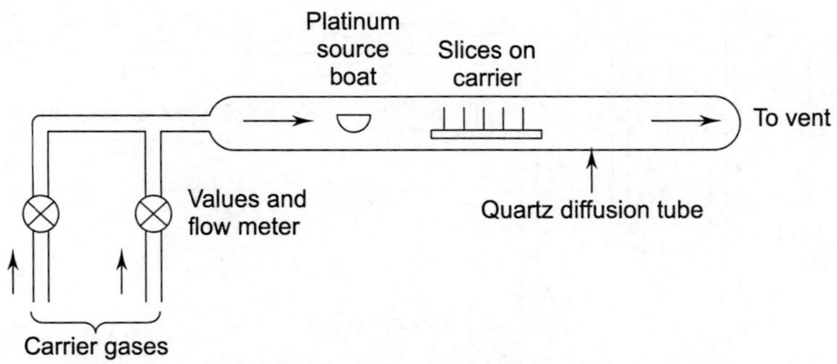

Fig. 7.16 *Solid Source Diffusion System*

Here, a platinum boat is used to hold a solid source of dopant species upstream from the carrier with the semiconductor wafers. In operation, the carrier gas transports vapours from this source and deposits them on the semiconductor slices. Source shutoff is usually accomplished by moving the dopant source to a colder region of the furnace.

(*b*) **Liquid source diffusion system:** Here, a carrier gas is bubbled through the liquid which is transported in vapour form to the surface of the slices. It is common practice to saturate the gas with the vapour so that the concentration is relatively independent of gas flow. The surface concentration is thus set by the temperature of the bubbler and of the diffusion system. Additional lines are also provided for other gases in which the diffusion is performed and for flushing out the system after use.

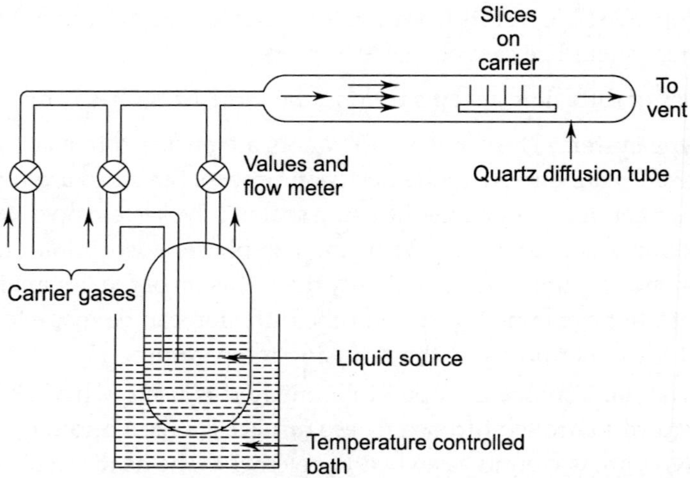

Fig. 7.17 *Liquid Source Diffusion System*

(*c*) **Gaseous source diffusion system:** It is more convenient than the liquid source diffusion system. Here, provision is made for an ambient carrier gas in which the diffusion takes place. A chemical trap is often incorporated to dispose of unrelated dopant gases, which are generally highly dangerous. All vapour transport methods rely on the fact that the surface concentration of the incorporated dopant is solid-solubility limited, so that it is relatively insensitive to the vapour pressure of the reactant species.

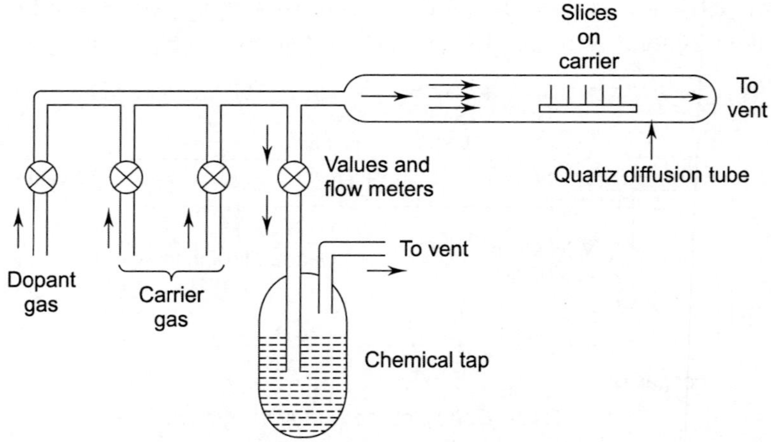

Fig. 7.18 *Gaseous Source Diffusion System*

Spin-on technique: This is used for obtaining a uniform source coverage over the slice. Here, a dilute mixture of a dopant compound in an organosilane is used as a source chemical. The semiconductor slice is kept in a vacuum chunk and rotated at high speed (2500–5000 rpm). A drop of the source chemical is next applied to form a thin layer (≈ 5000A°) across the slice by means of centrifugal force.

SUMMARY

In this chapter first the process of diffusion has been discussed with the different types of diffusion and their comperative merits and demerits. As diffusion is the main part of fabrication, the quantitative analysis is also necessary to achieve an enact diffusion depth and profile during VLSI Fabrication. The mathematical analysis reveals some laws (Flick's laws) and equations. There arises a coefficient called diffusion coefficient (D) on which diffusion depend. Further mathematical analysis has been done for constant D and D as a functions of N.

A focused study of some common dopants like arsenic, boron, phosphorus is also carry out to develop a practical understanding of readers. After that the different types of diffusion system used in VLSI industries are discussed.

TUTORIAL SHEET

1. What is diffusion? Explain Briefly.
2. What are the different mechanisms (at atomic level) through which diffusion can occur?
3. What are the different parameters/factors on which diffusion depends?
4. What are Flick's laws? Explains Flick's first and second laws.
5. What is diffusion coefficient (D)? What is its significance in diffusion? On what factors does it depend?
6. What are the assumptions on which the Flick's laws are based?
7. What is infinite source diffusion? Explain the concentration profile in that case.
8. What is constant total source diffusion? Explain the concentration profile in that case.
9. Explain the concentration profile when diffusion coefficient is a function of concentration.
10. What is successive diffusion?
11. Explain the diffusivity of some common doponants.
12. What is solid source diffusion system?
13. What is liquid source diffusion system?
14. What is gaseous source diffusion system?

8

LITHOGRAPHY

8.1 INTRODUCTION

By the process of lithography, windows can be opened in layers of SiO_2 and fine-line patterns can be defined in polycrystalline silicon and aluminium layers.

Starting with a circuit diagram, it is necessary to lay out the art work for a set of individual photo masks, corresponding, for example, to buried layers, isolation regions, transistor bases and resistors, emitters, contact windows, metalization patterns, and so on. Typically, a mask is required for each structurally defined step in the fabrication of an integrated circuit. The mask layout process results in a set of scaled drawings, typically 200 to 1000 times octal size, on Mylar-base graph paper.

Microfabrication or micromachining in the narrow sense comprises the use of a set of manufacturing tools based on batch and thin film fabrication techniques commonly used in the electronics industry. In the broader sense, microfabrication describes one of many precise engineering disciplines which takes advantages of serial direct write technologies as well as more traditional precision machining methods, enhanced or modified for creating small three dimensional (3D) structures with dimensions ranging from subcentimeters to submicrometers, involving sensors, actuators, or other microcomponents and micro systems.

Both micromachining and micro electronics fabrication start with lithography, the technique used to transfer copies of a matter pattern onto the surface of a solid material. Such as a silicon wafer. Using a computer operated digitizer, the mask pattern information is recorded and stored in computer memory. The layout information can then be plotted or displayed on an oscilloscope. Interactive graphic display are often used so that the designer can make layout changes with a keyboard or a light pen.

Completed mask layout can then be stored digitally on magnetic tapes. Libraries of standard circuits or cells are developed in this manner. Standard cell information from the "library" can be used for new IC layouts. Interactive graphics can be used to modify input and output circuits and interconnection patterns. The usefulness of mask libraries has been limited, however, by the rapid evolution of IC technology.

8.2 PHOTOLITHOGRAPHY

The essence of *photolithography* is to print temporary circuit structure on a wafer that will later be used to assist etch and ion implant processes. The structures are first created as patterns on a quartz template that is referred to as a mask. Ultraviolet light passes through the mask to transfer the pattern to a light-sensitive film on the wafer surface. As with common photography work, the photographic film is developed and the image appear on the wafer. Later a chemical etching is used to image the film pattern on the underlying wafer surface, or the wafer is sent to the ion implant area where selective doping can be performed through the patterns on the wafer. The various patterns transferred to the wafer define the numerous device feature, vias, and wiring interconnects necessary for the device layers as well as determine area of the silicon to be doped.

Photolithography is closely related to the cost performance of a microchip. The processing cost of a wafer is largely independent of the number of chips on a wafer. That is, a wafer with few chips costs almost as much as the same wafer with many chips since the number of processing steps, amount of materials, wafer handling and so on, are nearly identical for the two wafers. More chips at the same cost means that the overall cost of each chip is lower, simply by putting more chips on the wafer.

More than any other single fabrication technology, photolithography has contributed to the revolutionary advancement of chip performance, affirming Moore's law over the fifty plus years of semiconductor manufacturing.

Photolithography produces a three-dimensional pattern on the surface of the wafer using a light-sensitive photoresist material and controlled exposure to light. Other terms process are *photo, lithography, masking,* and *pattering.*

8.2.1 Concepts of Photolithography

Photolithography is at the center of the wafer fabrication process (see Fig. 8.1), as evidenced by how wafer repeatedly flow into and out of photolithography operation in the fabrication process.

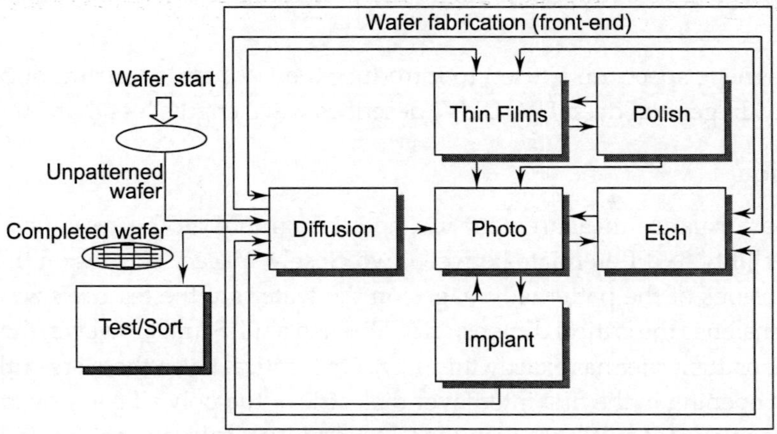

Fig. 8.1 *Wafer Fabrication Process Flow*

Photolithography is often considered the most critical step in the IC process, requiring superior performance to achieve high yield in other process areas. It is estimated that lithography accounts for nearly one third of the total wafer fabrication cost.

The shape of the photolithography pattern transferred to the wafer surface depends entirely on the wafer layer being constructed. Possible pattern are semiconductor devices in the silicon wafer, isolation trenches, constants, the metal interconnects, and vias to interconnect metal layers. The patterns are transferred into a light-sensitive photo resist material to prepare the substrate for etch or ion implantation. The patterned photoresist image are three-dimensional because there are width, length, and height to the patterned in the photo resist. There may be hundreds of identical chips on a wafer, each needing the appropriate pattern to be transferred to each die.

Photolithography technology employs a light-sensitive photoresist, or resist, that is applied on the substrate surface as a polymer solution. The resist is then baked, which drives out the solvent. Next, the resist is exposed to a controlled light. The light passes through a reticle that defines the desired pattern. The photo resist is a temporary material placed on the wafer surface solely to transfer the necessary pattern and is removed once the pattern is etched or implanted onto the wafer surface material.

8.2.1.1 Critical Dimension Generation

Microlithography has traditionally been the technology in semiconductor manufacturing that defines the critical dimension (CD) on a wafer (Example : the polysilicon gate length on a microprocessor). Since CD is the most difficult dimension to control during fabrication, it becomes the required dimension for other process areas to achieve. The critical dimension in photolithography is often used to describe device technology nodes or generations. Some sub-0.25 μm technology nodes are 0.18 mm, 0.15 μm and 0.1 μm. The ability to reduce the critical dimension permits die to be placed on a single wafer, thus lowing fabrication costs significantly and improving profitability.

8.2.2 Light Spectrum

Energy is required to activate the photoresist and transfer the pattern from the reticle. The source of energy is in the form of radiation, typically an ultraviolet (UV) light source. Light-sensitive photoresists are designed to chemically respond to a certain UV wavelength. UV light has been the most common energy source for photolithography patterning and this will probably continue for the near future.

The electromagnetic spectrum is used to introduce the UV light spectrum of most interest for photolithography. In general deep UV (DUV) describes wavelengths less than 300 nm.

8.2.3 Resolution

An important performance measurement in photolithography is the *resolution* of each image. *Resolution* is the ability to differentiate between two closely spaced features on the wafer surface. The actual dimensions of the patterned images on the wafer are the feature sizes. The minimum feature size dimension is the critical dimension (CD), such as 0.18 μm. However, this does not mean that every feature on the wafer has exactly this (CD). On a critical layer, there are critical dimensions. Such as a contact opening in the first inter layer dielectric or the poly silicon gate length, and other features that are non-critical and large than the CD. Resolution is important for critical dimensions.

8.2.4 Overlay Accuracy

Photolithography requires alignment between the pattern on the mask and the existing features on the wafer surface. This quality measure is known as *overlay accuracy*. Alignment is critical because the mask pattern must be precisely transferred to the wafer from layer to layer. Since multiple masks are used during pattering, any overlay misalignment contributes to the total placement tolerances between the different features on the wafer surface. This condition is known as the *overlay budget*. A large overlay budget essentially reduces the circuit density, which limits device feature sizes and therefore, IC performance.

8.2.5 Process Latitude

There are many process variables involved in photolithography. Examples of process variables are equipment settings, types of materials, human performance, machine alignments, and the stability of materials over time. *Process latitude* represents the capability of the photolithography process to consistently produce products that meet the specified requirements.

8.3 PROCESSES OF PHOTOLITHOGRAPHY

Photolithography consists of two basic types of processes these are negative and positive lithography. *Negative lithography* prints a pattern on the wafer surface that is the opposite of the pattern in the mask. *Positive lithography* prints a pattern on the wafer that is the same pattern as that on the mask. The major difference between these two basic processes is the type of photo resist used. How the photo resist material reacts when exposed to light depends on whether it is negative or a positive resist material.

8.3.1 Negative Lithography

The primary feature of negative lithography is that the photo resist will become insoluble and harden by cross linking when exposed to light. Once hardened, the cross linked resist cannot be washed away in solvents. Resists of this type are known as *negative resists* because the image formed in the resist is negative of the pattern found on the reticle (See Fig. 8.2). Negative resists were the earliest types of photoresist used in semiconductor photolithography.

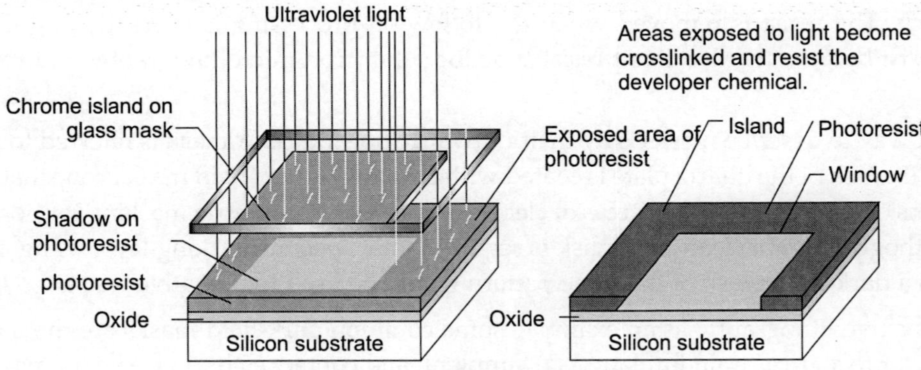

Fig. 8.2 *Negative Lithography*

It can be seen in the diagram in Fig. 8.2 that the mask for the negative is a transparent plate of quartz. The dark portion of the mask is a thin film of chromium deposited and patterned to form the desired mask artwork. The chrome is opaque and does not permit the passage of the ultraviolet light. For a negative resist, the regions underneath the opaque chrome of the mask are not exposed to light and therefore remain unchanged. The resist remains soft and dissolves when exposed to the developer chemical. UV light passes through the clear areas of the mask and hardens the resist so that it does not dissolve in the subsequent developer chemical. In this manner, the negative resist has a pattern that is the opposite of the pattern on the mask.

8.3.2 Positive Lithography

In positive lithography, the pattern printed on the wafer surface has the same image as the pattern on the mask. The areas of the resist exposed to the UV light undergoes a photochemical reaction and become soluble and soften in the developer. In this manner, regions of the positive resist exposed to light are removed in a developer solution, where as resist not exposed to light underneath the opaque mask pattern remains on the wafer (See Fig. 8.3).

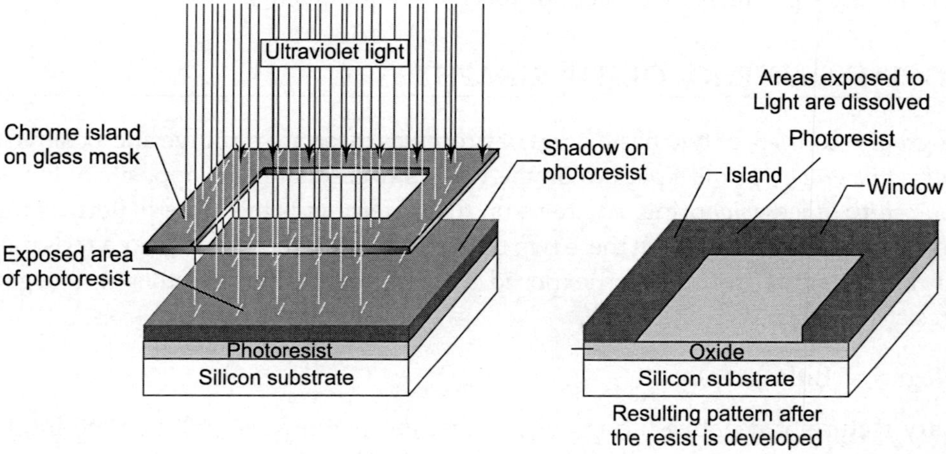

Fig. 8.3 *Positive Lithography*

This type of resist is referred to as a positive resist because the image formed in the resist is the same pattern formed on the reticle. The resist that remains was already hardened before light exposure and remains on the wafer surface to protect the top layer during subsequent processing, such as etch. The resist is removed when the follow-on processing is done. During the 1970's, positive resist become the dominant resist type for submicron microlithography and remains so today.

Another way to describe masks is by their outward appearance. A mask is referred to as a dark field mask if much of the quartz plate is coated with chrome. A clear-field mask is one that has very thin patterns of chrome with large areas of clear quartz. When comparing masking for positive and negative lithography, if a clear-field mask is required for a specific masking layer using a positive resist, then a dark-field mask of the same pattern would be used for negative resist.

Using positive lithography as an example, some common dark-field masks are the masks that are used prior to source/drain implant, LDD implant, and contact etch. A clear-field mask is used for application such as preceding gate etch and metal interconnect etch.

8.4 BASIC STEPS OF PHOTOLITHOGRAPHY

Photolithography is a complex process with many variables contributing to its process latitude, such as reduced feature size, alignment tolerance, number of masking layers, and cleanliness of the wafer surface. For convenience, we can divide the patterning process of photolithography into an eight-step procedure. Figure 8.4 provides a pictorial overview of the eight process steps.

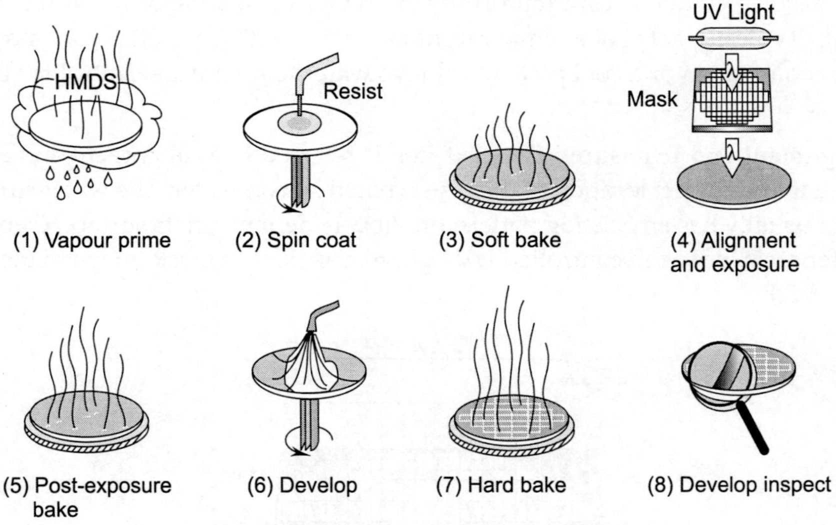

Fig. 8.4 *Eight Steps of Photolithography*

In the wafer fab these steps are often referred to as operations. Breaking the large photo process into these eight steps simplifies the various aspects of microlithography.

The photolithography process has undergone significant equipment integration. Automated equipment, referred to as a water/developer track system or tracks, employs robots, automated material handling and computers to perform all eight steps without human intervention. Integrated tracks offer many benefits over the previous stand-alone manual tooling for photolithography. This integration improved process control by controlling delays between process steps, processing wafers efficiently, increasing flexibility, reducing contamination due to environmental control and minimal operator handling, and increasing safety due to reduced operator exposure to chemicals. Eight steps of photolithography are as given below:

Step 1. Vapour Prime : The first step in photolithography is to clean, dehydrate and prime the surface of the wafer. The purpose of these steps is to promote good adhension between the photoresist and the wafer surface.

Wafer cleaning may involve a wet clean and DI water rinse to remove contaminates. Most wafer cleaning is performed before entering the photolithography work by. A dehydration dry bake is done in a closed chamber to drive off most of the absorbed water on the surface of the wafer. The wafer surface must be clean and dry.

Step 2. Spin Coat : After priming, the wafer is coated with the liquid photoresist material by a spin coating method. The wafer is mounted on a vacuum chuck, which is a flat metal or teflon disc that has small vacuum holes on its surface to hold wafer. A precise amount of liquid photoresist is applied to the wafer and then the wafer is spun to obtain in uniform coating of resist on the wafer.

Different resists require different spin coating conditions, such as an initial slow spin (*e.g.*, 500 rpm), followed by a ramp up to a maximum rotational speed of at 3,000 rpm or higher. Some important quality measures for photoresist application are time, speed, thickness, uniformity, particulate contamination, and resist defects such as pinholes.

Step 3. Soft Bake : After the resist has been applied to the wafer surface, it must undergo a soft bake. The purpose of this bake is to drive off most of the solvent in the resist. The soft bake process improves adhension, promotes resist uniformity on the wafer, and yields better line width control during etching. Typically soft bake temperature are 90° to 100°C for 30 seconds on a hot plate, followed by a cooling step or a cool plate to achieve wafer temperature control for uniform resist characteristics.

Step 4. Alignment and Exposure : The next step is referred to as alignment and exposure. The mask is aligned to the correct location of the resist coated silicon wafer. The wafer surface could be bare silicon but usually has an existing pattern previously defined on its surface. Once aligned, the mask and wafer are exposed to controlled UV light to transfer the mask image to the resist-coated wafer. (See Fig. 8.5.)

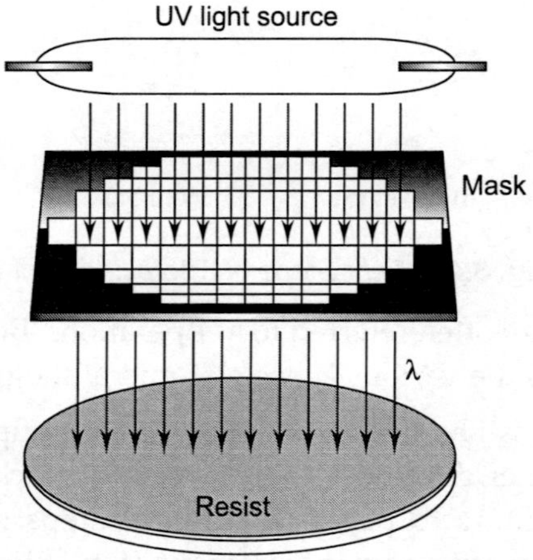

Fig. 8.5 *Alignment and Exposure*

The light energy activates the photosensitive components of the photoresist. Important quality measures for alignment and exposure are line width resolution, overlay accuracy and particles and defects.

Step 5. Post-Exposure Bake (PEB) : There is a post exposure bake on a hot plate at 100° to 110°C that is necessary for the deep UV (DUV) resist. This bake follows immediately after the photoresist exposure. A few year ago an optional step for non DUV conventional resists, but it has become a virtual standard even for conventional resists.

Step 6. Develop : Develop is the critical step for creating the pattern in the photoresist on the wafer surface. The soluble areas of the photoresist are dissolved by liquid developer chemicals, leaving visible patterns of islands and windows on the wafer surface. The most common methods for development are spin, spray, and puddle.

Step 7. Hard Bake : A post-development thermal bake, referred to as hard bake, is required to evaporate the remaining photoresist solvent and improve the adhension of the resist to the wafer surface. This is the critical step for stabilizing the resist for the following etch or implant processing. The hard bake temperature for positive resists is about 120° to 140°C. This is a higher temperature than soft bake, but it cannot be too high or else the resist will flow and deform the pattern.

Step 8. Develop Inspect : Once the resist is patterned on the wafer, an inspection is done to verify the quality of the resist pattern. The inspection system is nearly always automated for pattering on highly integrated critical layers. The purpose of this inspection is two fold : to identify wafers that have quality problem with the resist and to characterize the performance of the photoresist process to meet specifications. If the resist is determined to be defective, it can be removed through resist stripping and the wafer can be reprocessed.

The goal of the photolithography process, as in any manufacturing process, is a defect-free product. However, it would be catastrophic to not inspect and leave defects in the resist. Develop inspect is one of the few areas in the wafer fabrication process where errors can be detected and corrected. Once defective wafers are moved to next patterning step, usually etch, then there is no further change to correct mistakes. If a wafer is etched incorrectly, it has a total defect, is considered scrap, and is of no further value to the company. This is why inspection data to characterize and improve the photoresist process is so important.

8.5 SPIN COAT

Once the wafer surface has been cleaned, dehydrated, baked, and primed, it is ready for the application of photoresist. Photoresist is applied by spin coating where the wafer is spun to form a thin film of resist on the wafer surface. Before considering the photoresist spin process, we will review the different chemistries of photoresist.

8.5.1 Photoresist

Photoresist is an organic compound that experiences a change in solubility in a developer solution when exposed to ultraviolet (UV) light. Photoresists used in wafer fabrication are applied to the wafer surface as a liquid but dried into a film. The purpose of photoresist in wafer fabrication is :

(*i*) To transfer the mask pattern to the resist on the top layer of the wafer surface.

(*ii*) To protect the underlying material during subsequent processing.

Types of Photoresist

Optical photoresists are negative resist and positive resist. This classification is based on how the resist material responds to UV light. For a negative resist, the UV-exposed regions become cross linked and hardened. This makes the exposed photoresist less soluble in the developer solution ; the photoresist is not removed in the developer liquid. A negative mask image is patterned in the resist. For a positive resist, UV-exposed regions of the resist become more soluble and a positive mask image appears in the resist. A positive resist breaks down during exposure to light and the exposed areas are easily washed away in the development solution.

Negative Versus Positive Resist : The negative photoresist hardens when exposed to light and becomes insoluble when developed, the image is the opposite of the original mask pattern. Positive photoresist soften when exposed to light and becomes soluble. When developed, the image is the same as the original mask pattern. The earliest resists in the semiconductor microlithography were primarily negative resists until the mid - 1970's. Negative resists of this era exhibited good adhension to the wafer substrate and good resistance to etching. However, due to the swelling and distortion during develop, negative resists typically have a resolution limit of about 2 μm. As along as the circuit line width CD remained above this dimension, then negative resists were acceptable. With the introduction of VLSI and UlSI technology and their associated micron and submicron circuit image sizes, the negative resist type was replaced by positive resist. Positive resist had been around for many years, but it had poorer adhension and its capability for improved resolution was not needed.

Physical Properties of Photoresist : The photoresist must perform well under all the different process conditions, including coating, spinning, baking, develops ion implantation and etch. There are many different types of photoresists used in wafer applications, each with its own physical properties that are directly related to the photolithography process requirements. A particular resist is selected based on the following properties :

(*i*) Sensitivity	(*ii*) Viscosity
(*iii*) Resolution	(*iv*) Contrast
(*v*) Adhension	(*vi*) Etch resistance
(*vii*) Surface tension	(*viii*) Storage and handling.
(*ix*) Contamination and particles	

(*i*) **Sensitivity :** Sensitivity is the maximum amount of light energy at a certain wavelength that is needed to produce a good pattern in the photoresist on the wafer surface. The amount of light energy supplied to a resist is commonly called the exposure dose. Sensitivity is important for new resist because of low energy levels emitted at the shorter UV wavelength.

(*ii*) **Viscosity :** A quantitative measure of flow characteristics for the liquid resist represents the resist viscosity. Viscosity is time-dependent in that it increases as the solvent in the resist evaporates during usage. Viscosity is important because it affects resists resist thickness and the uniformity of the resist over the wafer surface topography. Such as steps and small gaps. As viscosity increases, the resist has less tendency to flow and its thickness on the wafer increases.

(*iii*) **Resolution :** Resolution is the ability to differentiate between two or more closely spaced patterned features on the wafer surface. A practical way to interpret resolution is by the smallest feature that can be printed on the wafer and meets specified quality requirements. The smaller the critical dimensions produced the better the resolution capability of the resist and the photolithography.

(*iv*) **Contrast :** The contrast represents the sharpness of the transition from exposure to non-exposure in photoresist. Contrast represents the ability of the resist to become exposed only in those areas defined by the clear region of the photo mask. It is desirable to have high-contrast to produce vertical resist side walls.

(*v*) **Adhension :** *Adhension* describes how strongly the resist sticks to the substrate. The resist must adhere to many different types of surfaces, including silicon, polysilicon, silicon dioxide, silicon nitride and different metals. Lack of resist adhension leads to distorted patterns on wafer surface.

(*vi*) **Etch Resistance :** The resist film must maintain its adhension and protect the substrate surface from the subsequent wet and dry etch processes. This property is known as etch resistance. Some dry etch processes are done at higher temperature which requires a resist that has thermal stability to maintain its shape.

(*vii*) **Surface Tension :** Surface tension refers to the molecular attraction forces in a liquid that tends to pull surface molecules toward the body of the liquid. An example of high surface tension is water beading up on a waxed automobile surface. Resist has molecular forces that create a relatively high surface tension so that the resist molecules hold together during the various process steps.

(*viii*) **Storage and Handling :** Photoresist chemicals are activated in the presence of energy, either as light or heat, which requires carefully controlled storage and usage conditions. Resists are specified with a shelf life and storage temperature environment. Crosslinking of negative resists and sensitizer-delay of positive resists can occur with extended storage time or elevated temperature. If the solvent on the resist was allowed to evaporate due to an unsealed container, the viscosity would change rapidly and solids could precipitate from the liquid.

8.6 OPTICAL LITHOGRAPHY

Optical lithography has traditionally been the major limiting factor to the continual reduction in microchip feature sizes since the beginning of wafer fabrication. It is now believed that optical lithography is capable of resolutions down to a CD of 0.1 μm and beyond. Lithography is often considered the engine that is driving the performance improvements of Moore's law.

Photolithography for wafer fabrication has been based largely an optical lithography. The longevity of optical lithography is attributed the basic improvements that been made to the equipment and process.

8.6.1 Light

A light source is needed in optical lithography to project the reticle pattern on the photoresist and cause a photochemical reaction. A practical description of light is an electromagnetic wave that is visible to the eye. Light is also radiant energy. These two descriptions reflect the dual nature of light as a wave and a particle. Light travels in waves similar to sound waves. Since light is a wave, it can be described by wavelength (l) and frequency (f). The relationship of the two is given by the formulation shown in Fig. 8.6.

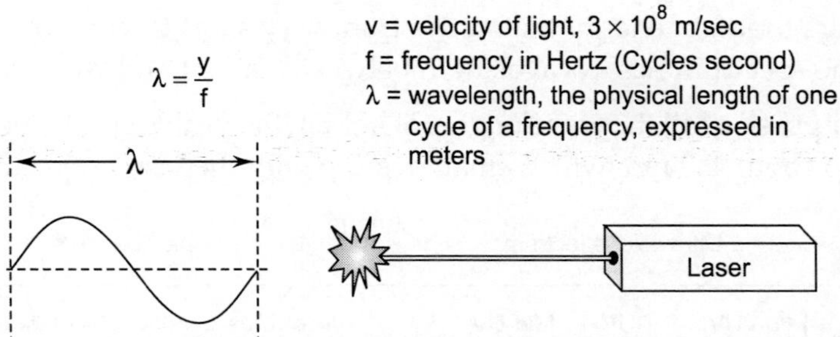

Fig. 8.6 *Light Wavelength and Frequency*

Interference of Light Waves

Waves are sinusoidal in nature. Sinusoidal waves of any type (Example : light, electrical, or sound) that have the same frequency (monochromatic) can have interference between the individual waves. For example, wave interference may be two water ripples interfacing with each other and partially cancelling each other out. There are two types of interference based on whether the waves are in phase or out of phase.

(*i*) Constructive Interference : Two waves in phase that add to the each other.

(*ii*) Destructive Interference : Two waves out of phase that subtract from each other.

Optical Filters

Optical filters use light interference to block unwanted incident light through either reflection or interference to obtain a light with a particular wavelength (See Fig. 8.7).

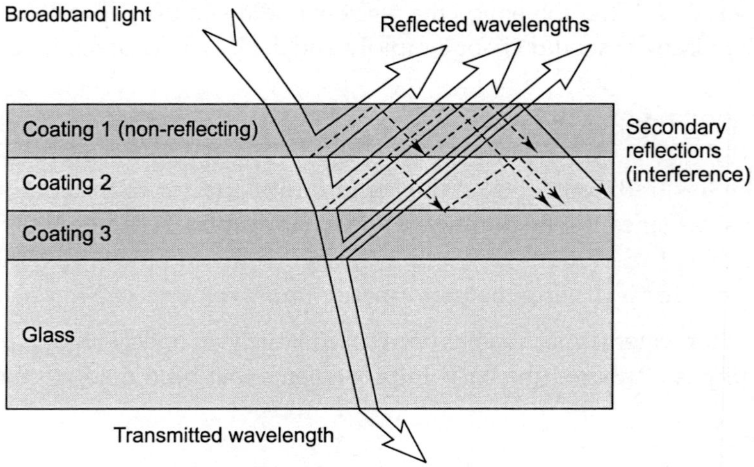

Fig. 8.7 *Optical Filtration*

Optical filters will typically be made of glass and will have one or more layers of thin film coatings. The type of coating and its thickness will determine which wavelength are blocked through destructive interference or passed through the glass.

Electromagnetic Spectrum

The entire group of visible and non-visible electromagnetic waves is called the electromagnetic spectrum, consisting of radiant energy that varies from very short to very long wavelengths. The visible portion of the spectrum has a wavelength between 390 nm and 780 nm.

White light is composed of all the wavelengths of light in the visible spectrum. The UV spectrum ranges from 4 nm to about 450 nm, with a small overlap with the visible spectrum (See Fig. 8.8).

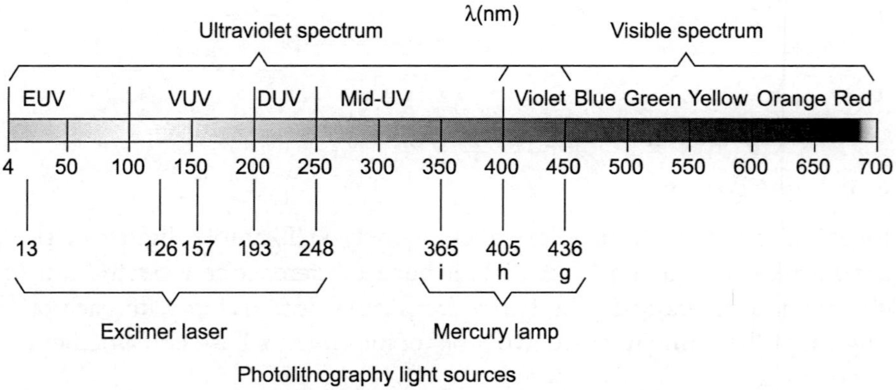

Fig. 8.8 *Ultraviolet Spectrum*

The deep ultraviolet (DUV) wavelengths are presently used for patterning critical layers in production. The vacuum ultraviolet (VUV) wavelength of 157 nm is most likely UV successor after the DUV wavelength of 248 nm and 193 nm. The short wavelength and high photon energy of VUV are strongly absorbed by oxygen absorption bands, therefore, operation is necessary in a vacuum or inert gas ambient (thus the name VUV). Research and development (R & D) is under way in the semiconductor industry into the future technology of using 13 nm UV wave-length, referred to as extreme UV (EUV).

Since, the UV spectrum overlaps a portion of the visible spectrum, white light also contains a portion of UV. Yellow light is commonly used in the photolithography area of the wafer fab because it is in the portion of the visible spectrum with little UV and, therefore, does not affect photoresists.

Exposure Source

During exposure of the resist, photochemical transformations occur in the resist material to transfer the pattern from the mask. This is a critical step in photolithography. It must occur in the shortest amount of time and be repeatable for high volume production of wafers.

Ultraviolet (UV) light is used to expose resist because the resist material reacts to light at this particular wavelength. Wavelength is also important because light with a smaller wavelength allows for the resolution of smaller features on the photoresist. The two types of UV exposure sources most often used today for optical lithography are :

(*i*) Mercury arc lamp.

(*ii*) Excimer laser.

Mercury Arc Lamp

The high-pressure mercury arc lamp is used as the UV illumination source in all conventional i-line steppers. In this type of lamp, electrical current is passed through a tube of mercury-error gas to create a discharge arc. This arc emits a characteristic light spectrum, with the useful UV wavelength emissions for a mercury arc lamp between about 240 nm and 500 nm in length (See Fig. 8.9).

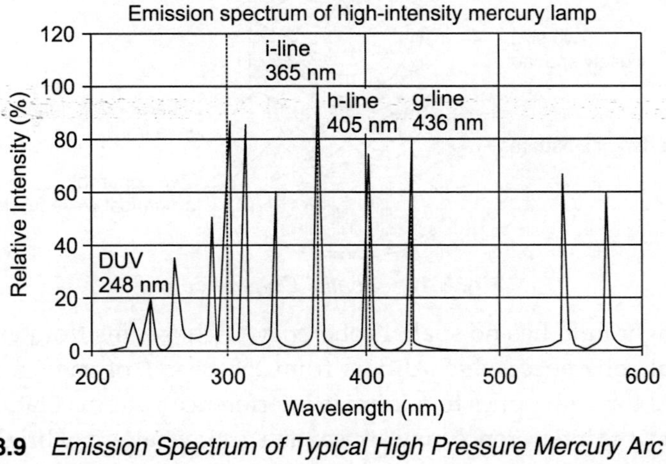

Fig. 8.9 *Emission Spectrum of Typical High Pressure Mercury Arc Lamp*

A conventional photoresist will have certain spectral response that corresponds to a particular UV wavelength during exposure. For instance, a DNQ non-violet i-line resist, which is used for CD feature size down to 0.35 μm, responds to the UV i-Line wavelength of 365 nm. To expose a resist to the proper UV wavelength, a set of filters is used to block the unwanted wavelengths and any infrared wavelengths. The exposure wavelength is chosen to match the critical feature size on the wafer.

Excimer Laser

Laser light sources for photolithography have been available since the mid-1980's, but reliability and performance concerns delayed their implementation in wafer production until the mid 1990's. Their main benefit has been to provide more light intensity at the DUV wavelengths of 248 and below, since the mercury arc lamp is an inefficient emitter at these wavelengths.

The only laser light source used thus for photolithography exposure is the excimer laser. An excimer is an exotic molecule formed from an atom of a noble gas and halogen, such as argon fluoride (ArF), where the molecule exists only in a quasi-stable, excited state. The word excimer actually comes from the expression excited dimer, which is a molecule formed from two identical atoms. It is also used to represent the noble gas and halogen molecules.

Most excimer lasers today contain a high pressure mixture of two or more elements that are bound in the excited state. Laser light emission occurs when the excited state decays and the exotic molecule falls apart into its two constituent atoms. The laser maintains a condition of more molecules in the excited state than in the ground state by using a high voltage pulse discharge across two flat-plate electrode to excite a high pressure mixture of the noble gas and halogen.

Spatial Coherence

Recall that light is an electromagnetic wave. A light beam has spatial coherence when points on different waves remain in phase to one another as the waves propagate. The light waves move in unison (See Fig. 8.10).

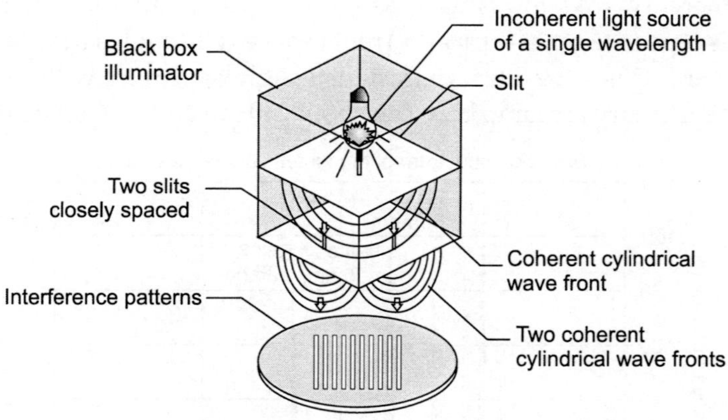

Fig. 8.10 *Spatial Coherence*

A standard room light bulb has no spatial coherency. Light beams from excimer lasers have a low amount of spatial coherence, which differs from the beams of conventional lasers. Spatial coherence is controlled through optics to minimize interference patterns that can be formed in the image. If not controlled, the interference may appear as a grainy pattern of bright and dark spots on the resist, referred to as speckle.

Exposure Control

A uniform dose of UV light is critical for resist exposure. The exposure dose must be repeatable from exposure to exposure and from wafer to wafer. The exposure latitude for a DUV resist is around 1% in dose variability. The variation in photolithography processes, equipment, and materials requires stringent exposure control.

Exposure control in photolithography is achieved by measuring the intensity of the UV light source at the surface of the wafer with a *dose monitor*. The exposure dose is measured at different locations in the exposure field, permitting the calculation of the percent uniformity of the dose. A photodetector and an electronics circuit are used on automated steppers and step- and scan systems to monitor and control the exposure dose through use of a shutter or by varying the speed of the scan.

8.6.2 Optics

Optics is the study of the physical properties and composition of light. Optics is important because lithography is based on an optical imaging process used to transfer the reticle pattern to the photoresist. The quality of the resist pattern can be limited by poor-quality optics. All photolithography equipment currently used for high-volume wafer fabrication is based on optical lithography.

Reflection of Light

The *law of reflection* describes the relationship between an incident light ray and the corresponding reflected ray, stating that the angle of incidence is equal to the angle of reflection (See Fig. 8.11). This law is true whether the reflecting surface is rough or smooth.

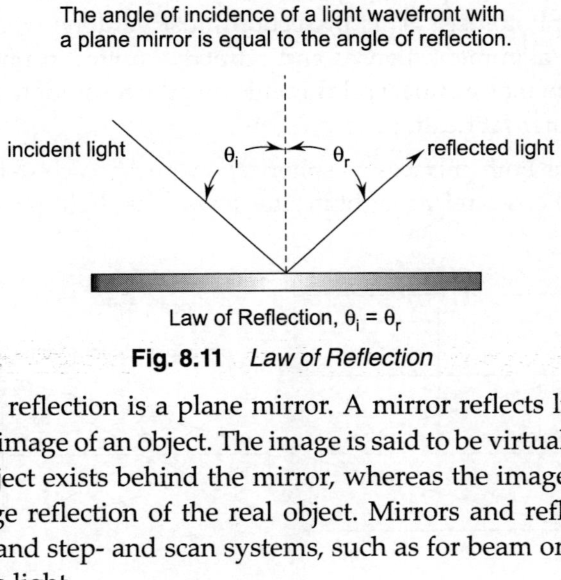

Fig. 8.11 *Law of Reflection*

An example of law of reflection is a plane mirror. A mirror reflects light from its surface and creates a virtual referred image of an object. The image is said to be virtual because it appears to the viewer as though the object exists behind the mirror, whereas the image on the mirror surface is actually a reversed image reflection of the real object. Mirrors and reflective optics have many applications in steppers and step- and scan systems, such as for beam orientation or for illuminators that shape and focus light.

Refraction of Light

When light passes from one transparent medium into another. Such as from air through a glass window, the light changes direction. The change in direction of a light ray when passing from one transparent medium into another is called refraction.

Refraction of light is caused by the different in the speed of light in two different media. Light travelling through a new medium, such as glass, its speed actually decreases. This decrease occurs because the glass is optically a more dense material. The *relative index of refraction, n,* represent how much the light ray bends as it passes through the interface of the two media based on its change in velocity (See Fig. 8.12).

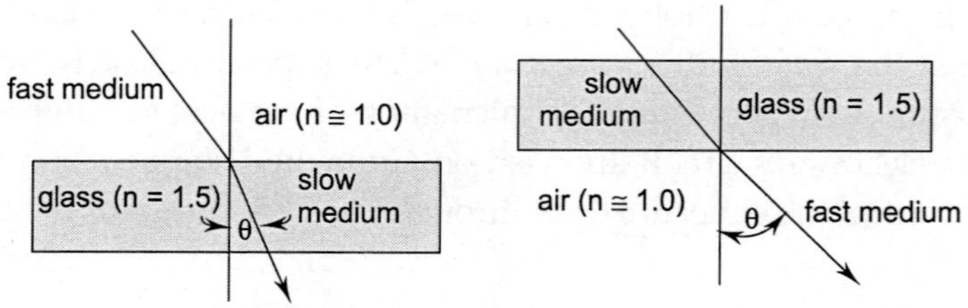

Fig. 8.12 *Refraction of Light Based on Two Mediums*

The *absolute index of refraction* compares the speed of light in a vacuum to the speed of light in the chosen medium. Therefore, it is only stated for one medium and the vacuum is assumed. The refraction of light is also dependent on the wavelength, l of the light ray that passes from one medium into another medium.

8.7 LENS

A *lens* is defined as an optical element that refracts light from an object passing through it to form an image of the object. We will consider an individual lens, but optical systems in photolithography consists of many different lenses and mirrors in the optical system. Lenses are important in the optical system of photolithography equipment because of the need to project the image of the reticle pattern on to the photoresist. This projection must be done with the proper resolution, dimensional control, and alignment. Lenses and refractive optics in photolithography steppers and step - and - scan system play a critical role in achieving the resolution needed to achieve the CD line widths in modern wafer fabrication.

A lens has either one or both surfaces as spherical-shaped, which defines whether the lens is converging or diverging. Lenses refract light in such a way that light rays converge or diverge on a principal focus.

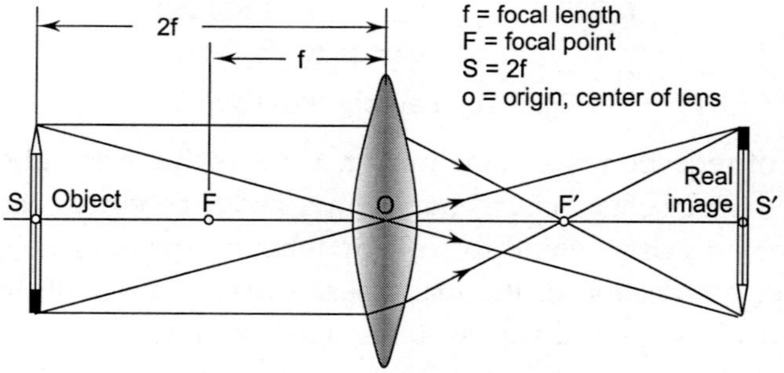

Fig. 8.13 *Converging Lens with Focal Point*

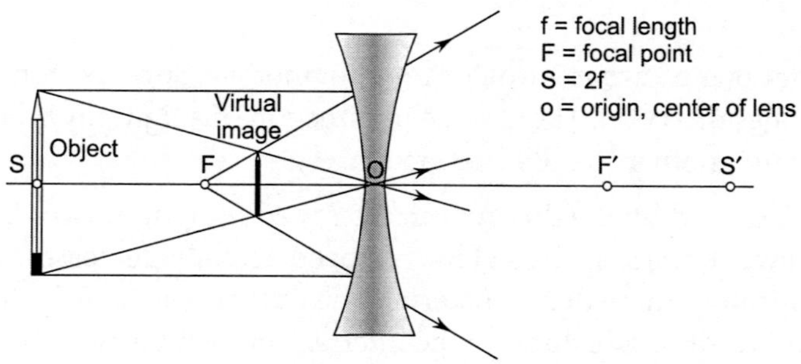

Fig. 8.14 *Laser Induced Lens Compaction*

The principal focus, or focal point, is where the light rays are refracted and will converge – the image carried by the light is sharp and in focus. The distance from the center of the lens to the focal point is called the focal length.

8.7.1 Lens Material

Lens have traditionally been made of glass. The lens material is an important variable as the exposure wavelength is continually reduced. For 248 nm wavelength DUV, a suitable lens material is fused silica. It has less light absorption at the DUV wavelengths. At the 193 nm DUV and 157 nm VUV wavelength, calcium fluoride (CaF_2) is being investigated as a possible candidate for lens material.

Traditional optical materials such as glass have greater absorbance and are more sensitive to laser damage at 248 nm KrF wavelength. Absorption causes a loss in exposure power and induces heat in the optics, which leads to refractive index changes and imaging problems. Thermal effects can also cause focus change to optical system heating from DUV laser light sources.

Another concern regarding lens materials is that a laser beam can create compaction damage. *Lens compaction* is a structural rearrangement of the lens material that causes densification of the lens material. Compaction occurs in lens materials, including fused silica. It can not well understood, but it occurs due to the total cumulative laser exposure and peak power density.

8.8 E-BEAM LITHOGRAPHY (EBL)

The direct write method of electron beam (e-beam) lithography is capable of transferring a high-resolution pattern directly to the reticle surface. In e-beam lithography, an electron source produces many electrons that are accelerated and focused in the shape of a beam toward the reticle. The electron beam is focused either magnetically or electrostatically and is scanned across the entire reticle or scanned only over the printed area to eventually transfer the pattern to the reticle.

To apply the e-beam resist on the reticle, the reticle is cleaned and is chrome film is spin coated with a suitable photoresist, then soft baked. The standard e-beam resist is positive tone. However, this resist is not suitable for submicron line widths. Alternative resists are being developed, such as chemically amplified resists. Advanced e-beam mask writers are capable of patterning 0.36 μm minimum feature sizes on a reticle.

Shorter line widths can be obtained in silicon wafers with e-beam system than with photolithography system. When a photolithography system is used with a 1 × size mask, diffraction effects can occur during wafer resist exposure. The diffraction pattern occurs, where the edges of the opaque mask regions are imaged on the wafer. The severity of the problem is related to the wavelength of the light source used for the exposure. A shorter wavelength exposure source reduces the diffraction effect. Some of the techniques used to overcome the diffraction problem involve the use of :

(*i*) Shorter-wavelength UV light sources.

(*ii*) Direct step mask aligners, in which 1 × size masks are not involved.

(*iii*) X-rays exposure sources used in conjunction with X-ray sensitive resists.

(*iv*) EBL systems to produce fine-line masks in conjunction with item (*ii*) or (*iii*).

(*v*) EBL system that directly expose resists on wafer surface.

It has been pointed out before a typical e-beam lithography system can scan only portion of a master mask (or wafer) at a given time. The mask is seated on an *x-y* table, which indexes to repeat a given circuit pattern until the entire mask (or wafer) is exposed. The positional accuracy of the *x-y* table is critical in this application. A laser interferometer is often used in a position control system to maximize the accuracy of the *x-y* table.

Since about 1960, scanning electron microscope (SEM) have been commercially available. In SEM, electrons emitted by a tungsten filament (or a lanthanum hexaboride, LaB_6, field emission source) are accelerated through a potential difference of a few kilovolts to 50 KV in vacuum. The electrons are then passed through magnetic focusing coils (electronstatic focusing is also used) and deflection coils to provide a faster scan pattern on a sample (See Fig. 8.15).

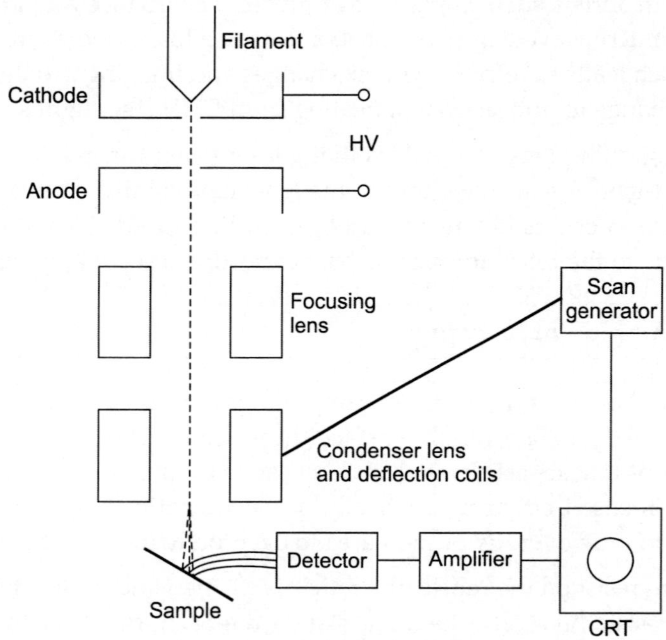

Fig. 8.15 *Block Diagram of SEM*

The electron beam serves as a very small probe, easily less than 100 Å in diameter. Various imaging techniques are possible. The most common is secondary electrons emission. Secondary electrons are ejected from the sample by the impinging primary electrons with energies typically less than 50 eV. These secondary electrons are detected and the quantity emitted at a particular point on the sample provides a level of contrast in a cathode-ray tube (CRT) or television (TV) image. The CRT or TV scan is synchronized with the scan of the electron beam. Variations in the quantity of secondary electrons emitted occur both as a result of sample topography and material differences.

Other imaging techniques used are cathodoluminesence, induced current, X-rays, and back scattering of primary electron. Back scattered electrons are detected by allowing them to create electron hole pairs, thus generating a current in a silicon detector. Various electron diffraction phenomena are also useful in characterizing materials.

It has been pointed out that the resolution capability of a lithographic process using ultraviolet light is limited by the wavelength of the light. This practical limit on line width is around 1 to 3 μm is microelectronics circuits, because of the necessity of maintaining clearly defined line edges. The wavelength of electrons at the voltages used in SEM's is of the order of 1 Å, so the limiting criterion is not wavelength, but the diameter of the electron beam. Electro beam diameters of 100 Å are routine in commercial SEM's. Beam diameters of 20 Å and less have been attained using high brightness field emission electron guns. This, however, requires a very high vacuum and consequently a very clean system. Pump down time is, consequently, substantially longer. In any case, it is clear that the possibilities for the use of e-beam fabrication of microelectronic circuits are extensive. The limit on line widths and packing density is not set by the electron beam writing process as much as it is by the resolution of the resist and the capability of aligning mask patterns on the Si wafer.

Even when lithographic techniques are used to align masks and expose resists on the wafer, production of masks by e-beam lithography (EBL) has distinct advantages. Superior line definition can be achieved in e-beam lithography on the master mask, thus providing a higher quality mask. Turn around time from the digitizing of a mask design to the computer. Controlled e-beam lithographic production of the mask can be substantially less than when photographic reduction is used. Additionally, a change is design simply requires a modification of a computer program. Masks can be checked on the SEM after development of the resist by secondary electron imaging, or the computer can be used to print out overlays so that alignment of successive masks can be checked.

Commercial e-beam lithographic systems are available and existing SEM's can be adapted to perform e-beam lithographic work. In most cases, the e-beam in existing machines is probably not sufficiently stable to achieve the highest resolution or bright enough to achieve sufficiently rapid exposure for production purposes, but considerable improvement over photolithography can be made, and such adapted machines provide a useful tool for experimental circuit work and e-beam lithography development. Such as in the area of e-beam resists. Many adapted machines make use of a flying spot scanner for controlling the position of the e-beam. In this arrangement, an image tube is masked with a positive or negative mask of the pattern to be defined in the electron resist. A CRT scans the mask-covered image tube and the output of the image tube is then utilized to turn the electron beam on and off at the appropriate location in its scan.

In other systems, the e-beam is controlled by a computer to write the pattern directly in the resist. Work has been done on multiple beam systems and imaging systems to achieve higher processing rates.

If the pattern is not directly defined on the wafer, as it is in such applications as bubble memories and charge coupled devices, a mask must be used to do the exposure on the wafer. To take full advantage of the small line widths achievable with e-beam lithography, the mask and patterns must be aligned with even greater accuracy. Standard optical techniques will not suffice for alignment or exposure. Exposure must be with short wave ultraviolet, X-rays, or electron beam to achieve adequate line definition at the wafer image. Alignment must be done automatically using one of the short-wavelength types of radiation mentioned above or by means of a sophisticated interferometric technique using visible or near visible ultraviolet light.

8.9 X-RAY LITHOGRAPHY

X-ray lithography is an established technology for imaging patterns on wafers with CD's below 100 nm (0.1 µm). An X-ray source projects X-rays onto a special mask that forms a pattern on a resist coated wafer. This technique is not widely used for wafer fabrication, primarily due to its high cost of ownership in comparison to optical lithography. One major chip maker has used X-ray lithography for patterning commercial chips throughout the 1990's, including advanced microprocessors on 200-mm diameter silicon wafers. The system components for X-ray lithography are :

(*i*) A mask that has pattern of X-ray absorbing materials on a material that transmits X-rays.
(*ii*) An X-ray source.
(*iii*) An X-ray resist.

The X-rays used in lithography are known as soft X-rays. They operate in a region of the electromagnetic spectrum from about 0.1 nm to about 10 nm (Fig. 8.16). These differ from the familiar X-rays used in medicine, which have a shorter wavelength and are referred to as hard X-rays.

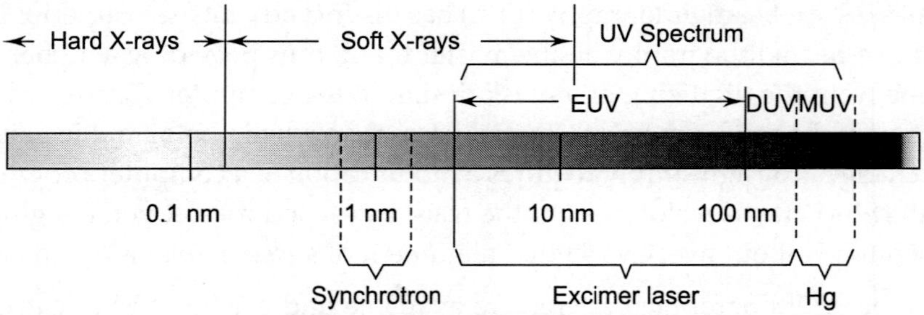

Fig. 8.16 *X-Ray Spectrum*

The most common X-ray source for lithography is known as a synchrotron. High-energy electrons are forced into closed curved paths by magnetic fields and made to accelerate which causes them to emit radiation. This action produces X-ray radiation that is very intense and reasonably collimated (parallel). A projection system based on special mirrors will project the X-rays on to the mask and resist coated wafer. Each synchrotron ring provides radiation for multiple X-ray steppers.

X-ray Mask

A typical X-ray mask structure is as shown in Fig. 8.17.

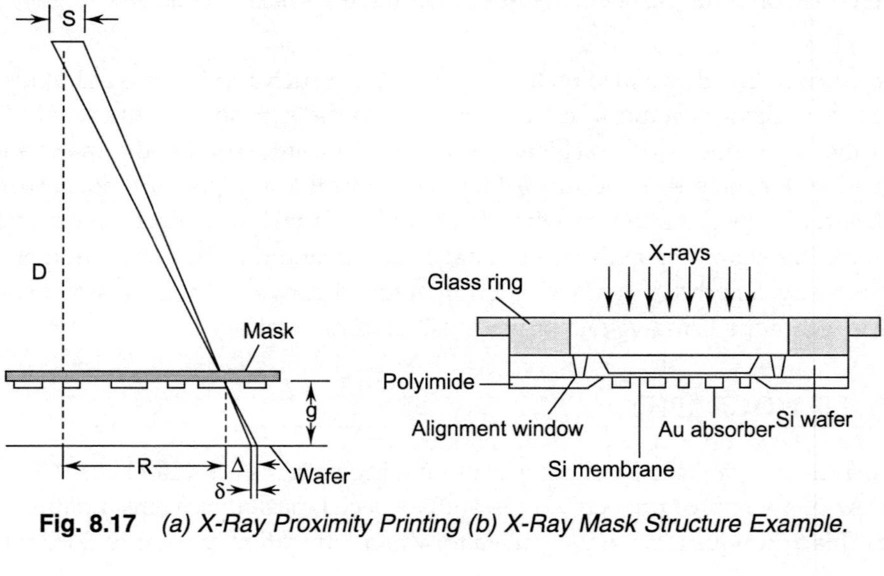

Fig. 8.17 *(a) X-Ray Proximity Printing (b) X-Ray Mask Structure Example.*

An X-ray mask consists of a transmissive membranes substrate and an absorber patterned on it. The ratio of metal are available that are fully transparent or fully absorbent, unlike the combination of glass and chromium in optical lithography. These thicknesses are determined by the transmission of the materials for the X-ray wavelength of interest. X-ray masks are made by electron lithography and use the technologies of mask making and direct writing. The pattern on the X-ray mask must be as perfect as a photo mask, but the dimension is the same as on the wafer because of proximity printing. For features down to 0.1 to 0.2 μm geometry it is harder to check the pattern integrity and to repair defects in X-ray than in optical lithography.

Of the heavy metals with large $\rho\ 2^4$ values, gold was once widely used, because it is relatively easily patterned by lift-off or electroplating. Tungsten and tantalum are used now, because they are easily etched by dry etching. The thicknesses of gold necessary for absorption of 90% of the incident X-ray flux are 0.7 μm, 0.5 μm, 0.2 μm and 0.08 μm for X-ray wavelength, 4.4 Å (PD_L) 8.3 Å (Al_K), 13.3 Å (Cu_L) or 44.8 Å (L_K) respectively.

In general, the metal is considerably thicker than the chromium layer (0.1 μm) on a photo mask. Methods for high resolution patterning of the gold include electroplating and ion milling. Electroplating produces excellent definition with vertical walls, but requires a vertical wall primary pattern in a resist that has a thickness equal to that of the metal to be plated. More often, a subtractive process has been employed in which a thinner resist layer is used to pattern a thin layer of refractive metal ; the refractive metal serves as a mask for ion-milling the underlying gold. With this method walls that depart from the vertical by 20° or less can be formed. The minimum line width attainable by ion milling 0.5 μm-thick gold is ~ 0.4 μm. For higher resolution, and where gold thickness can be reduced, longer wavelength such as the 12 Å Cu_L radiation may be used. Lines as well as 0.16 mm have been replicated with this type of radiation.

X-ray Sources

X-rays are produced by the interaction of incident electrons and a target material. The maximum X-ray energy is the energy E of the incident electrons. If E is greater than the excitation energy E_C of the characteristic lines of the atoms of the material, the X-ray spectrum will contain these lines.

X-ray generation by electron bombardment is very inefficient process ; most of the input power is converted into heat in the target. The X-ray flux is generally limited by the heat dissipation in the target. With electrons focused to a spot 1 mm in diameter on an aluminium target on a water cooled stem, 400 to 500 W is a typical upper limit for the input power. The X-ray power produced is only about 10 mW, and this power is distributed over a hemisphere. The X-ray power is proportional to $(E-E_C)$.

Another X-ray source, which is capable of an order of magnitude greater flux, is plasma discharge. There are several versions, but all functions by heating a plasma to a temperature high enough to produce X-ray radiation. The radiation consists of strong lines superimposed on a weak continuum. The source is pulsed at a low rate. In one embodiment, the source size is 2 mm. The repetition rate is 3 Hz. Special problems with such a source are reliability and contamination produced in the plasma chamber.

X-ray Projection

The use of X-ray projection lithography has been demonstrated. The key components of this system are X-ray mirror which consist of alternating layers of high index and low index material deposited on a smooth substrate. The choice of materials is determined by X-ray wavelength, and the individual layer thickness is determined by the wavelength and the incidence angle. The normal incidence reflectivity is better than 50% at $\lambda = 4.5$ nm. A laser-produced plasma source is adopted for the X-ray source. The system configuration is illustrated schematically in Fig. 8.18.

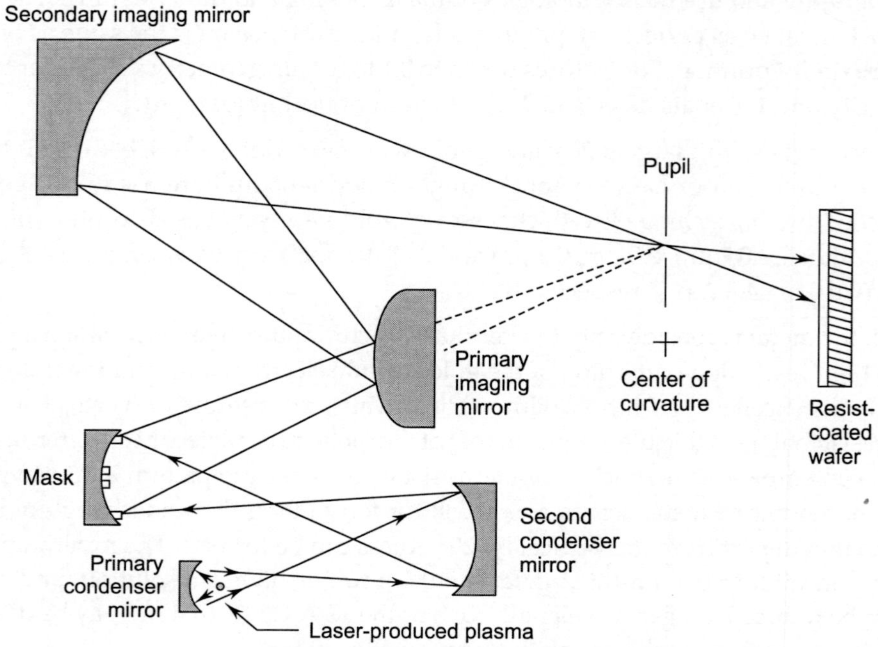

Fig. 8.18 *Schematic Illustration of an X-Ray Projection System*

The demagnification ratio is 1 : 5. The advantages, compared to proximity methods, are greater latitude with respect to placement of the mask pattern and less radiation-induced damage of the mask. Theoretical resolution is 8.9×10^3 line-pairs per millimeter, or a line width of 56 *nm*, based on $\frac{1}{\lambda t}$, where $\lambda = 4.5$ *nm* and t is 125. The image field size is as small as 5 *nm* in diameter. The exposure time is estimated roughly as 15 minutes for a 6-inch wafer.

8.10 ION PROJECTION LITHOGRAPHY (IPL)

Ion projection lithography is a technique in early development that uses ion beam to expose resist, either through a mask or by serial writing on the resist with a finely focused beam in Fig. 8.19.

If a mask is used, then a stitched approach is needed, using a broad beam to create small sub fields of exposed resist on the wafer surface. IPL uses multi electrode electrostatic optics to direct hydrogen or helium ions to the wafer. Ion transfer their energy more efficiently to the resist than electron beams, because ions have a larger mass. There are also fewer secondary electrons produced by ion bombardment and those produced have very low energy, which makes for less back scatter. Back scatter creates proximity effects that determines the limit to the minimum feature size for wafer patterning. Very high resolution is achievable from ion projection lithography. With 50 nm features sizes already demonstrated in a research environment.

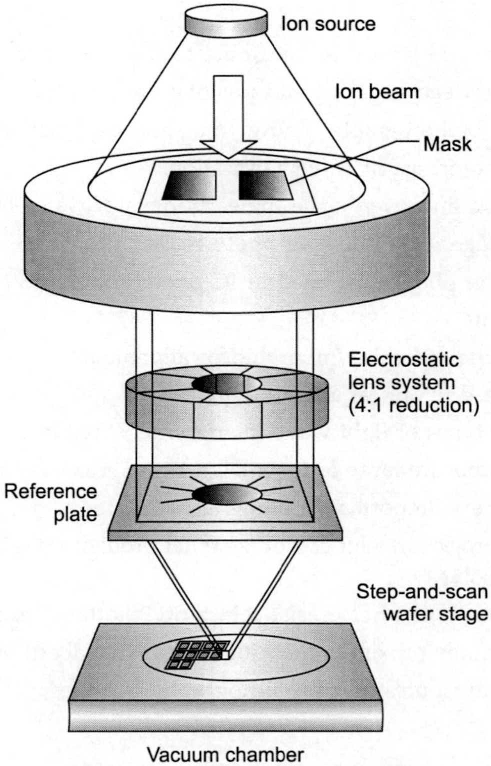

Fig. 8.19 *Ion Projection Lithography*

SUMMARY

Photolithography transfers a pattern from a mask to a UV light-sensitive photoresist on the wafer surface. Resolution is the smallest feature that can be patterned. Photolithography requires process latitude to consistently pattern wafers that meet the requirements. Negative lithography prints a pattern in the wafer that is opposite the mask pattern, where as positive lithography prints the same mask pattern in the wafer. Modern photolithography is based on optical lithography. Alignment and exposure are critical in lithography to meet critical submicron resolution requirement.

The continued success of optical lithography is based on improvements to all subsystems. Light is important to project the pattern on the resist. UV light at certain energy peaks creates the photochemical reaction in the resist. UV exposure sources are the mercury arc lamp for conventional lithography and the excimer laser the deep UV. Reflection and refraction are important parameters for manipulating the light through lens systems for image projection.

Lithography improvements have produced sub wavelength lithography, where patterning achieves a CD below the exposure light wavelength. Next-generation lithography is evaluating the lithography technology for eventually replacing optical lithography.

TUTORIAL SHEET

1. What is photolithography?
2. List the different or wavelengths used in photolithography and the name for each wavelength used between 436 and 157 nm.

3. Define Resolution.
4. What is overlay accuracy, and how this contribute to the mask overlay budget?
5. Explain the difference between negative and positive lithography.
6. List the eight steps of photolithography, and give a short explanation of each step.
7. Give two purposes of photoresist in wafer fabrication.
8. How has photoresist been improved since the early days of wafer fabrication?
9. What are two disadvantages of a negative photo resist?
10. What is the most common photoresist used in IC photolithography?
11. Describe opto lithography.
12. Explain why design ground rules are important for alignment?
13. Define light. Why is it need in optical lithography?
14. List and explain the two types of light wave interference. What is an optical filter?
15. What is a typical dose exposure latitude for a DUV resist? How is this measured?
16. What is optics, and Why is it important in lithography?
17. Explain how a scanning projection aligner works? What problems did the scanning projection aligner works? Aligner try to resolve.
18. What is one factor that reduces photo resist selectivity? Is this desirable or undesirable.
19. In *x*-ray lithography, why only proximity masking used? What about projection printing with *x*-rays?
20. What are some positive attributes of x-ray lithography?

9

ETCHING

9.1 INTRODUCTION

Etching is used in microchip fabrication to chemically remove layers from the surface of a wafer during manufacturing. Etching is a critically important process module, and every wafer undergoes many etching steps before it completes. For many etch steps, part of the wafer is protected from the etchant by a "masking" material which resists etching. In some cases, the masking material is photoresist which has been patterned using photolithography. Other situations require a more durable mask, such as silicon nitride.

In other words, the etching is based on removal of undesired material or layer by immersing the wafer into a solution which reacts with the exposed film to form by-products.

The two fundamental types of etchants are liquid phase (wet) and plasma-phase (dry). Each of these exist in several varieties. In general, there are two classes of etching processes.

1. Wet etching, where the material is dissolved when immersed in a chemical solution.
2. Dry etching, where the material is sputtered or dissolved using reactive ions or vapour phase etchant.

Before starting the discussion on various etching techniques. It is necessary to understands the some of useful parameters related to etching.

9.2 ETCHING PARAMETERS

1. **Selectivity.** If the etch is intended to make a cavity in a material, the depth of the cavity may be controlled approximately using the etching time and the known etch rate. More often, though, etching must entirely remove the top layer of a multilayer structure, without damaging the underlying or masking layers. The etching system's ability to do this depends on the ratio of etch rates in the two materials.
   The ability of the etchant to attack the desired material layer while at the same time not attacking other material present on the wafer is known as selectivity. Chemical Techniques are more selective than the physical means since chemical can be highly specific to their reactions.

The selective etching is frequently needed. Wet etching provides highest selectivity then plasma etching.

2. **Anisotropy.** The ability of the etchants to attack the layer in a directed manner is called anisotropy. In etching two different mechanisms exist, one is chemical etching which is isotropic and other is ionic etching which is directed or anisotropic in nature.

   Some etches undercut the masking layer and form cavities with sloping sidewalls. The distance of undercutting is called *bias*. Etchants with large bias are called *isotropic,* because they erode the substrate equally in all directions. Modern processes greatly prefer anisotropic etches, because they produce sharp, well-controlled features.

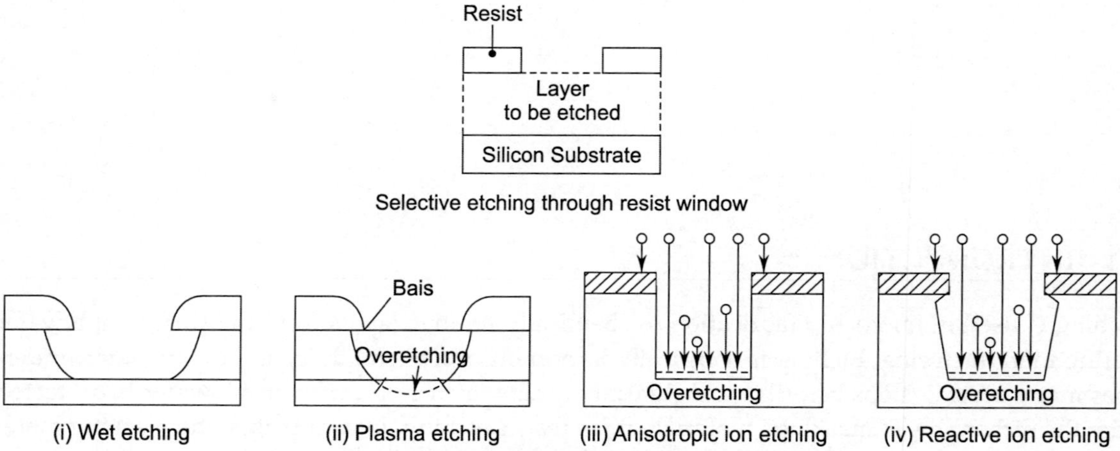

Fig. 9.1 *Anisotropicity in Different Types of Etching*

We can measure the anisotropy of etch process by using the relation.

$$A = 1 - \frac{V_H}{V_V}$$

where, V_V = Verticle etch rate

V_H = Horizontal etch rate

If $A = 0$, then the process will be isotropic

If $A = 1$, then the process will be an isotropic in nature, and it tells that lateral etch rate will be zero.

In the isotropic etching process the undercuts are formed in the device which effect the performance of the device. So, it is more important to control the verticle and horizontal etch rate. High selectivity implies no or little attack on the underlying surfaces. The selectivity factor must be high enough to ensure that photoresist is not removed before the etched layer.

3. **Uniformity.** Uniformity is related to the etch rate of the same material in different sections of the wafers. The etch rate should be same everywhere. The fluid flow patterns in the etching reactor may cause non-uniform etch rates. The system pressure influences etch rate and uniformity through a phenomenon called mean free path. This is the average distance a gas atom or molecules will travel before a collision with another particle. Low pressure is Preferred to achieve the unifomity, because at the higher pressure, there are many collisions that give the particle many directions, which causes loss of edge profile control.
4. **Throughput.** Throughput is a measure of the time, it takes to process a lot (batch) of wafers. In the batch process, the number of wafers processed together so the throughput is high in batch process than single wafer etching.

9.3 CLASSIFICATION OF ETCHING PROCESS

There are various methods to catch the unwanted materials. These methods are classified with the help of Tree Diagram in Fig. 9.2.

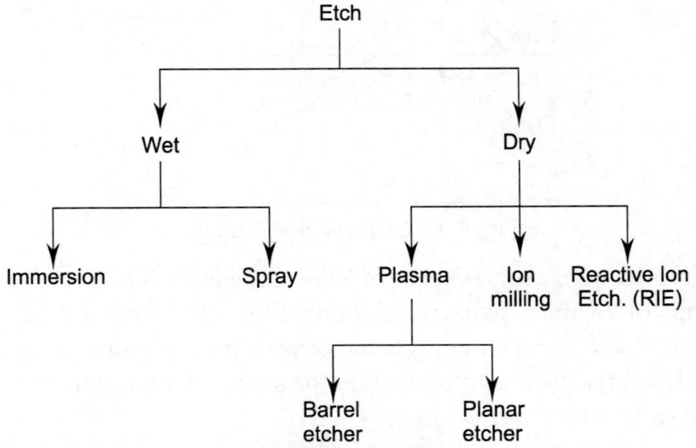

Fig. 9.2 *Classification of Etching Process*

In wafer fabrication, etching refers to a process by which material is removed from the wafer, *i.e.*, either from the silicon substrate itself or from any film or layer of material on the wafer. There are two major types of etching : dry etching and wet etching.

9.3.1 Wet Etching

Wet Etching is an etching process that utilizes liquid chemicals or etchants to remove materials from the wafer, usually in specific patterns defined by photoresist masks on the wafer. Materials not covered by these masks are 'etched away' by the chemicals while those covered by the masks are left almost intact. These masks were deposited on the wafer in an earlier wafer fabrication step known as lithography (as discussed in chapter 8).

A simple wet etching process may just consist of dissolution of the material to be removed in a liquid solvent, without changing the chemical nature of the dissolved material. However, a wet etching process involves one or more chemical reactions that consume the original reactants and produce new species.

A basic wet etching process may be broken down into three basic steps:

1. Diffusion of the etchant to the surface for removal;
2. Reaction between the etchant and the material being removed; and
3. Diffusion of the reaction by-products from the reacted surface as shown in Fig. 9.3.

Reduction-oxidation (redox) reactions are commonly encountered in wafer fabrication wet etching processes, *i.e.*, an oxide of the material to be etched is first formed, which is then dissolved, leading to the formation of new oxide, which is again dissolved, and so on until the material is consumed.

Wet etching is generally isotropic, *i.e.*, it proceeds in all directions at the same rate. An etching process that is not isotropic is referred to as 'anisotropic.' An etching process that proceeds in only one direction (*e.g.*, vertical only) is said to be 'completely anisotropic' (as discussed in section 9.2).

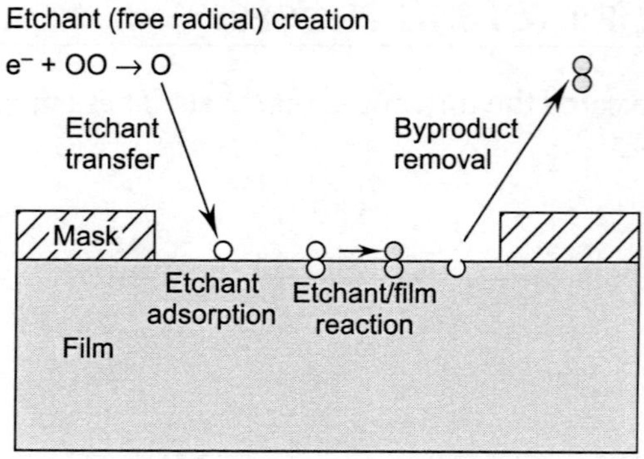

Fig. 9.3 *Steps of Wet Etching*

In semiconductor fabrication, a high degree of anisotropy is desired in etching, because it results in a more 'faithful' copy of the mask pattern, since only the material not directly under the mask are attacked by the etchant. Isotropic etchants, on the other hand, can etch away even the portion of material that is directly under the mask (usually in the shape of a quarter-circle), since its horizontal etching rate is the same as its vertical rate.

When an isotropic etchant eats away a portion of the material under the mask, the etched film is said to have 'undercut' the mask. The amount of 'undercutting' is a measure of an etching parameter known as the 'bias.' Bias is simply defined as the difference between the lateral dimensions of the etched image and the masked image. Thus, the mask used in etching must compensate for whatever bias an etchant is known to produce, in order to create the desired feature on the wafer.

Because of the isotropic nature of wet etching, it results in high bias values that are not practical for use in pattern images that have features measuring less than 3 microns. Thus, wafer patterns with features that are smaller than 3 microns must not be wet-etched, and should instead be subjected to other etching techniques that offer a higher degree of anisotropy.

Another important consideration in any etching process is the 'selectivity' of the etchant. An etchant not only attacks the material being removed, but the mask and the substrate (the surface under the material being etched) as well. The 'selectivity' of an etchant refers to its ability to remove only the material intended for etching, while leaving the mask and substrate materials intact.

Selectivity, S is measured as the ratio between the different etch rates of the etchant for different materials. Thus, a good etchant needs to have a high selectivity value with respect to both the mask (*Sfm*) and the substrate (*Sfs*), *i.e.*, its etching rate for the film being etched must be much higher than its etching rates for both the mask and the substrate.

Advantages

Despite the resolution limitations of wet etching, it has found widespread use because of its following advantages:

1. Low cost
2. High reliability
3. High throughput
4. Excellent selectivity in most cases with respect to both mask and substrate materials. Automated wet etching systems add even more advantages

5. Greater ease of use
6. Higher reproducibility
7. Better efficiency in the use of etchants.

Disadvantages

Of course, like any process, wet etching has its own disadvantages. These include the following :

1. Limited resolution
2. Higher safety risks due to the direct chemical exposure of the personnel
3. High cost of etchants in some cases
4. Problems related to the resist's loss of adhesion to the substrate
5. Problems related to the formation of bubbles which inhibit the etching process, where they are present and
6. Problems related to incomplete or non-uniform etching.

9.3.1.1 Immersion Wet Etching

In this method of etching the wafers are dipped into the etchant solution for some time after the masking. The etchants are chemicals which react with the unmasked material and remove it from the surface. This process is called immersion method. In this, the silicon (single-crystal or poly-crystalline) may be wet-etched using a mixture of nitric acid (HNO_3) and hydrofluoric acid (HF). The nitric acid consumes the silicon surface to form a layer of silicon dioxide, which in turn is dissolved away by the HF.

The overall reaction is as follows:

$$Si + HNO_3 + 6\,HF \rightarrow H_2SiF_6 + HNO_2 + H_2 + H_2O.$$

Silicon dioxide may, as mentioned above, be wet-etched using a variety of HF solutions.

The overall reaction for this is:

$$SiO_2 + 6\,HF \rightarrow H_2 + SiF_6 + 2H_2O.$$

Water-diluted HF with some buffering agents such as ammonium fluoride (NH_4F) is a commonly used SiO_2 etchant formulation.

Wet etching of aluminium and aluminium alloy layers may be achieved using slightly heated (35–45 deg °C) solutions of phosphoric acid, acetic acid, nitric acid, and water. Again, the nitric acid consumes some of the aluminium material to form an aluminium oxide layer. This oxide layer is then dissolved by the phosphoric acid and water, as more Al_2O_3 is formed simultaneously to keep the cycle going.

Other materials on the wafer may be wet-etched by using the appropriate etching solutions.

Disadvantages of Immersion Method

On major disadvantage of this method is that when aluminium and aluminium alloys layers are based on phosphoric acid, the by-product of the reaction forms a tiny bubbles of hydrogen (as shown in Fig. 9.4). These bubbles cling to the wafer surface and block the etch action. These bubble forms the bridges of aluminium that can cause electrical shortage between the two leads.

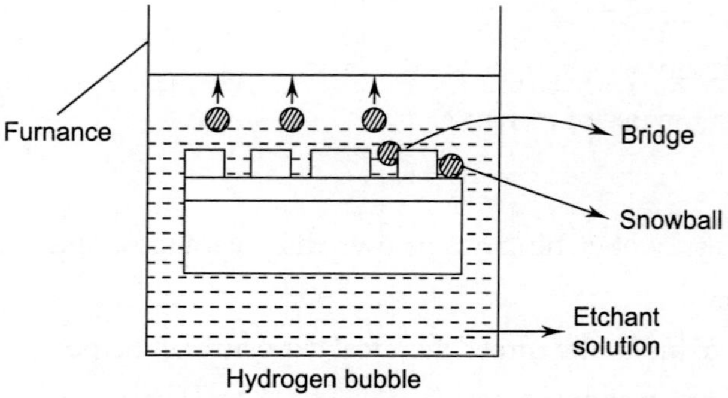

Fig. 9.4 *Hydrogen Bubble Blockage Etchant*

9.3.1.2 *Wet Spray Etching*

Wet spray etching offers advantages over immersion etching. This method minimize contamination from the etchants. This method is more controllable, because the etchants can be instantly removed from the process tank by opening the shuttle to the drain. Single wafer chunk spray system offers advantage of considerable uniform process.

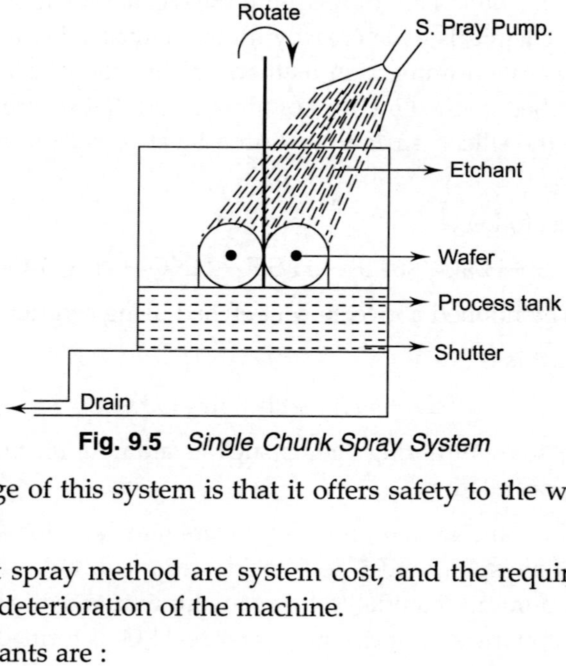

Fig. 9.5 *Single Chunk Spray System*

One another advantage of this system is that it offers safety to the worker, because the spray system is usually closed.

Disadvantages of wet spray method are system cost, and the requirement of Etch-Resistant materials to prevent the deterioration of the machine.

Some typical wet etchants are :

Table 9.1: Typical Wet Etchants

Material	*Gas*	*Etching rate*	*Mask*	*Selectivity*
Si (a-Si)	1. KOH 2. HNO_3+H_2O Hf	~ 6 – 600 nm/min (anisotropic) ~ 100 nm/min	Resist	> 50 : 1
SiO_2	1. HF 2. BHF	~ 10 – 1000 nm/min	Resist	> 50 : 1

Si_3N_4	1. HF 2. BHF 3. H_3PO_4	~ 100 nm/min ~ 100 nm/min ~ 10 nm/min	Resist SiO_2	> 50 : 1
GaAs	1. $H_2SO_4 + H_2O_2 + H_2O$ 2. $Br + CH_3OH$	~ 10 um/min	Resist	> 50 : 1
Au	1. $HCl + HNO_3$ 2. $KI + I_2 + H_2O$	~ 40 nm/min ~ 1 um/min	Resist	> 50 : 1
Al	1. $HCl + H_2O$ 2. NaOH	~ 500 nm/min	Resist	> 50 : 1

9.3.1.3 *Wet Etching Equipment*

The below Fig. 9.6 shows the wet etching equipment, it consist of wafer boat, probe with glass cover, thermometer, teflon cover, and the automatic controller. In this equipment, first the wafers are placed in the wafer boats, and teflon cover is put on the top to make the process isothermal, then etchants filled in the tank are rotated from the bottom and raise the temperature to the reaction temperature. This temperature is raised with the hot plates, and temperature also control the etchrate that is why a thermometer is placed inside the chamber to check it. This process can handle manually with control panel or automated with controller programming.

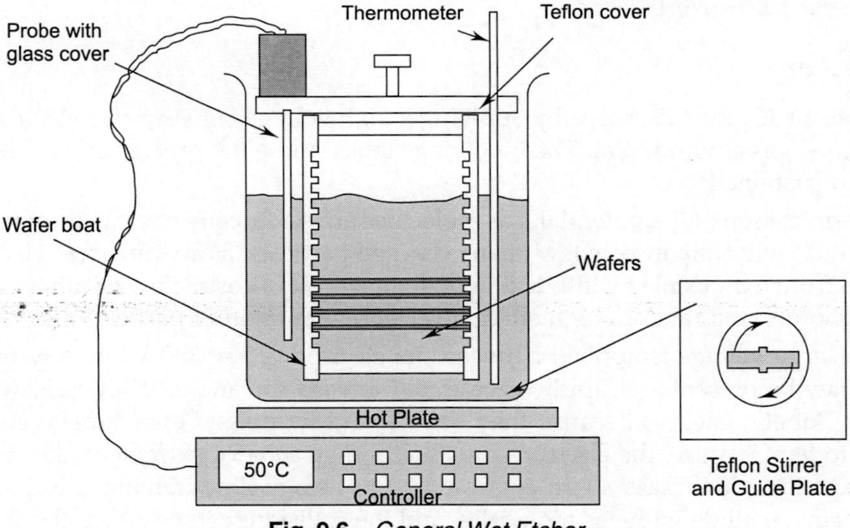

Fig. 9.6 *General Wet Etcher*

9.3.2 Dry Etching

Dry Etching is an etching but generates process that does not utilize any liquid chemicals or etchants to remove materials from the wafer, but generates only volatile by-products in the process. Dry etching may be accomplished by any of the following : 1. through chemical reactions that consume the material, using chemically reactive gases or plasma ; 2. physical removal of the material, usually by momentum transfer; or 3. a combination of both physical removal and chemical reactions.

Plasma etching is an example of a purely chemical dry etching technique. On the other hand, physical sputtering and ion beam milling are examples of purely physical dry etching techniques. Lastly, reactive ion etching is an example of dry etching that employs both physical and chemical processes.

Like wet etching, dry etching also follows the resist mask patterns on the wafer, *i.e.*, it only etches away materials that are not covered by mask material (and are therefore exposed to its etching species), while leaving areas covered by the masks almost (but not perfectly) intact. These masks were deposited on the wafer by an earlier wafer fabrication step known as 'lithography'.

9.3.3 Plasma Etching

Before starting the plasma etching, firstly, we will discuss, what is plasma and how it is exited for the etching.

Plasma is described as the fourth state of matter, the first three states (solid, liquid, gas) are known to us because they exist around us. Plasma through abundent else where in the universe. It occurs on the earth only under certain circumstance. As we know that if we transform solid into liquid, liquid into gas, it requires energy to transform. Like this plasma also require energy. The right amount of energy is applied to gas. This gas will transform into plasma, which is mixture of electrically charged and neutral particle, including atoms, ions and free radical.

Natural occurance of plasma is lighting and northern lighting. Man made plasma examples are neon signs and fluorocent tube.

9.3.3.1 Plasma Excitation

DC glow discharge

One can initiate a DC glow discharge by applying a high voltage between two electrodes in a low pressure (~ 1 torr) gas environment. The following steps outline the process by which a discharge is initiated and sustained:

1. Before application of the potential, gas molecules are electrically neutral and the gas at room temperature will contain very few of any charged particles. However, a free electron may be released from a molecule by interaction of, for example, a cosmic ray or other natural radiation, a photon, or a random high energy collision with another particle.
2. When a large voltage is applied between the electrodes, say 100 V/cm, any free electrons which may be present are rapidly accelerated toward the anode. They quickly attain high velocity (kinetic energy) because they have such low mass. Since, kinetic energy can be related to temperature, the electrons are "hot" - they achieve extremely high temperatures because of their low mass, in an environment of heavy, slow-moving cold gas molecules. They begin to collide with gas molecules, and the collisions can be either elastic or inelastic.
   (*a*) Elastic collisions deplete very little of the electron's energy and do not significantly influence the molecules because of the great mass difference between electrons and molecules: Mass of electron = 9.11×10^{-31} Kgm, Mass of Argon = 6.64×10^{-27} Kgm.
   (*b*) Inelastic collisions excite the molecules of gas or ionize completely removing an electron. (The excitation - relaxation processes are responsible for the glow).
3. As soon as, the electrons obtain enough energy to overcome the ionization energy of the gas molecules, inelastic collisions with neutral gas molecules or atoms create positive ions, and this releases more electrons into the system.
4. Newly produced electrons are accelerated toward the anode and the process cascades. With sufficient voltage, the gas rapidly becomes filled with positive and negative particles throughout its volume, *i.e.*, it becomes ionized.

5. Positive ions are accelerated toward the negative electrode (cathode). Collision with the cathode causes the emission of secondary electrons which are emitted from the cathode into the plasma.
6. Free electrons from secondary emission and from ionization are accelerated in the field to continue the above processes, and a steady state self-sustaining discharge is obtained.
7. The discharge reaches a steady state when electron generation and loss processes balance each other.

   Electrons are lost by : (*a*) drift and diffusion to the chamber walls, (*b*) recombination with positive ions, (*c*) attachment to neutral molecules to form negative ions.
8. The glow discharge, overall, must always remain neutral, although portions of it may be charged negatively or positively.
9. Most of the space between the electrodes will be filled with a soft glow as a result of the relaxation of excited species, however, certain dark spaces will be observed. The most significant dark space will be seen at the cathode. This dark space, known as the ion sheath, is a positive space charge region which is key to most dry etching processes.

The appropriate conditions of voltage, and gas pressure, a self-sustaining glow discharge can easily be obtained. A DC system must have current flow in the form of electron migration toward the anode and ion migration toward the cathode.

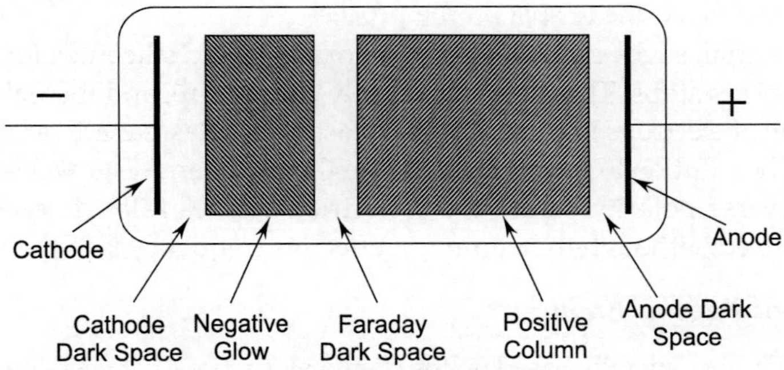

Fig. 9.7 *Plasma Excitation*

Cathode dark space and its importance

The cathode dark space is the most important region of a dc glow discharge and it is characterized by:

1. **Fewer electrons -** Highly mobile electrons are accelerated quickly away from the negative cathode.
2. **More positive ions -** As electrons escape, they leave a high concentration of positive ions. These are attracted to the negative cathode.
3. **Low conductivity -** Positive ions move slowly. The result is that the field is higher across the cathode dark space.
4. **Ions are accelerated -** When the ions drift into the field of the dark space, they are accelerated towards the cathode.
5. **Ions collide -** Ions collide with the cathode, and with neutral species on their way to the cathode. Therefore, the cathode is bombarded both with ions and with neutral species such as radicals and gas molecules. The ionic bombardment releases secondary electrons which sustain the glow discharge. At low pressures and high voltages, the bombardment can cause sputtering.
6. **Secondary electrons -** Secondary electrons produced at the cathode quickly accelerate to the ionization energy. As they leave the high field region, and encounter ionizing collisions, they

lose energy, and then only have enough energy to excite. This is the edge of the negative glow region.

7. **The dark space increases as pressure is reduced** - (longer mean free path, higher energies, fewer collisions). Eventually the dark space extends nearly across the chamber; the *m.f.p.* becomes too large to maintain a sufficient number of collisions and the glow is extinguished.
8. **Any conductive material** placed on the cathode will be subject to ion bombardment.
9. If a **non-conductive material** is placed on the cathode, the material charges up, and all the applied voltage exists across the material, and the plasma is extinguished.

Current flow, DC plasma

Net current flow must exist in DC plasma. Current travels through the electrodes, plasma and the external circuit.

(*a*) Electrons flow toward the anode, and enter the external circuit through the anode.

(*b*) Ions flow through the plasma toward the cathode. Since, they cannot travel through wires. When they strike the cathode, they accept an electron from the cathode, and are converted to neutral gas atoms or molecules.

(*c*) Collisions of fast moving electrons with gas molecules and with the anode provide a continuous source of new electrons to sustain the process.

If an insulating material, such as a dielectric or an oxide coated silicon wafer is placed on one of the electrodes, the process stops. The dielectric rapidly charges up, and the field on the gas drops, causing the generation of electrons to cease. For this reason, DC plasmas are rarely useful for etching most silicon wafers. The solution to this problem lies in the use of an AC power source of sufficiently high frequency to reverse polarity before the dielectric becomes fully charged. Typically, Radio Frequency (RF) is utilized, although the minimum effective frequency is in the range of > 100 KHz.

9.3.3.2 *Processes of Plasma Etching*

Plasma etching methods are generally based on the chemical reactions between the surface to be etched and an active species (reactive ion that comes from the plasma) that fall on the surface. To illustrate the etching process, silicon substrate etched in carbon - tetra fluoride (CF_4). Plasma is taken as an example. Figure 9.8 shows schematically the plasma etching process of the Si/CF_4 system.

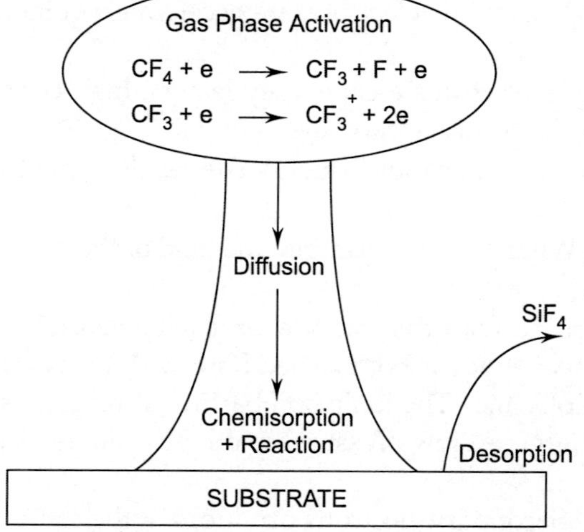

Fig. 9.8 *Schematic Illustration of Si Etching in CF_4 Plasma*

The following are the general steps of plasma etching:

(*i*) Impact a molecule gas (usually relatively inert) with an accelerated electron to generate reactive species, such as

$$CF_4 + e \Rightarrow CF_3 + F + e^*_{(dissociation)}$$

or $$CF_3 + e \Rightarrow CF_3 + e^* + e^*_{(ionization)}$$

(*ii*) Adsorb the sufficient radicals on the surface being etched, *e.g.*,

$$Si_{(surface)} + F_{(gas)} \Rightarrow Si\text{-}F_{(adsorbent)}$$

(*iii*) Choose proper chemistry so that the radicals will react with the wafer surface to form volatile by products, *e.g.*, by successive fluorination:

(*i*) $Si\text{-}F + F_{(gas)} \Rightarrow SiF_{2\,(adsorbent)}$

(*ii*) $SiF_2 + F_{(gas)} \Rightarrow SiF_{3\,(adsorbent)}$

(*iii*) $SiF_3 + F_{(gas)} \Rightarrow SiF_{4\,(adsorbent)}$

The final etch product formed must be volatile and stable so that it is not dissociated in the plasma and is easily desorbed in the next process.

(*iv*) Desorbs the volatile products from the surface,

$$SiF_{4\,(adsorbent)} \Rightarrow SiF_{4\,(gas)}$$

(*v*) And lastly, pump the volatile product out of the reactor.

9.3.3.3 *Effects of Ions in Plasma Etching*

1. **Increase the etch rate -** The etch rate is increased when the neutral flux and the ion beam both are applied to the substrate. This is shown in XeF_2 beam and Ar^+ beam directed the Si substrate that increase the etch rate by an order of magnitude. When neutral flux act together with ion beam.

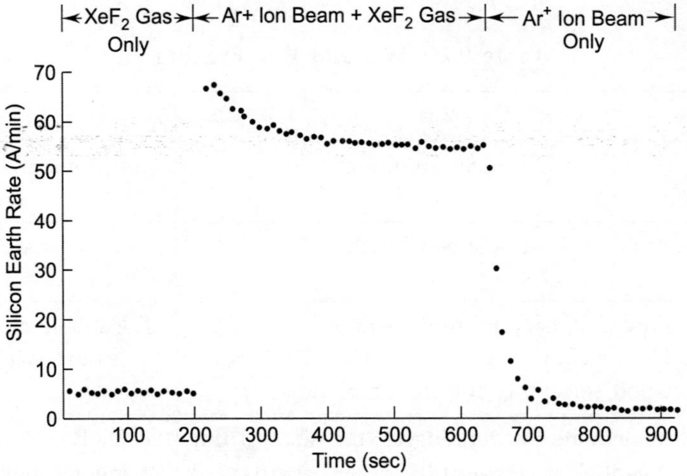

Fig. 9.9 *Increasing Etch Rate with XeF_2 and Ar^+ Beam*

2. **Attenuate the effects of rate limiting steps :** Energy in bombardment can also reduce the effect of rate limiting steps of all three basic steps of etching processes namely.

   (*a*) *Chemisorption of the incident reactive species on the surface being etched*. Ion-induced surface damage can increase the probability of incident reactive species to react with surface. This is done with fluorinated surface film and exposing reactive surface site.

$$\mathrm{Si + F + I^+ \Rightarrow SiF}$$
$$\mathrm{Si - F + F \Rightarrow SiF_2}.$$

I^+ is the impinging ion flux.

(b) *Formation of volatile product molecules*. This fromation in ion bombardment can hasten the formation of volatile spieces SiF_4 (in case of SiF, system) from partially reacted layer (Si $F_x \Rightarrow x \rightarrow 1, 2, 3, 4$) present on the surface at the time of ion impact.

(c) *Desorption of the product molecules from the surface.* Ion bombardment can efficiently sputter away the product molecules because of their low binding energy to the surface. Moreover, we have

$$\mathrm{SiF_{2\,(adsorbant)} + I^+ \Rightarrow Si\,F_{2\,(gas)}}$$

3. **Enhance anisotropy :** Anisotropy refers to directional etching where by the surface is etched vertically without undercutting the sides. Travelling ions can be directed perpendicular to the wafer surface. This will enhance the etchrate of the bottom surface. (as shown below)

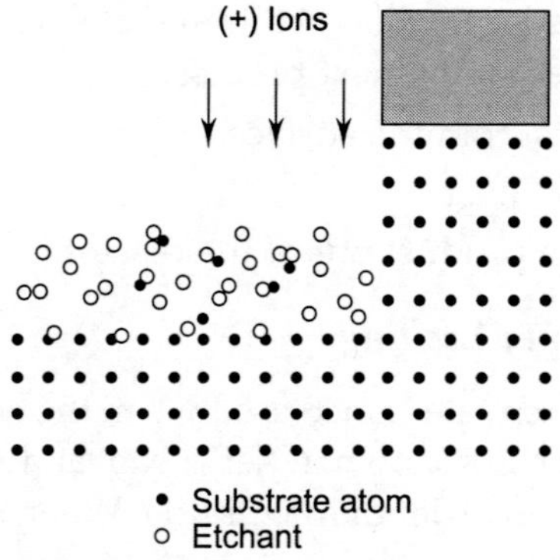

Fig. 9.10 *Etchrate Enhancement*

Table 9.2: Wet and Dry Etching

	Wet	*Dry*
Method	Chemical Solutions	Ion Bombardment or Chemical Reactive
Environment and Equipment	Atmosphere, Bath	Vacuum Chamber
Advantage	1. Low cost, easy to implement 2. High etching rate 3. Good selectivity for most materials	1. Capable of defining small feature size (< 100 nm)
Disadvantage	1. Inadequate for defining feature size < 1 um 2. Potential of chemical handling hazards 3. Wafer contamination issues	1. High cost, hard to implement 2. low throughput 3. Poor selectivity 4. Potential radiation damage
Directionality	Isotropic (Except for etching Crystalline Materials)	Anisotropic

We have Compared Chemical (wet) and Plasma (Dry) etching on the basis of various parameter both types of etching is done on the SiO_2 and noted the difference between the wet etching and Dry etching is given as in Table 9.3.

Table 9.3: Differences in Wet and Dry Etching

Chemical etching of SiO_2	***Plasma etching of SiO_2***
Etchant : HF + H_2O (for etching SiO_2.)	*Gases :* CF_4 + H_2 (or almost any other gas containing F).
Species in solution : F^-, HF^-, $H^+SiO_4^{2-}$, SiF_4, O_2 - whatever chemical reactions and dissociation produces.	*Species in plasma and on wafer :* CF_x^+ ($x \bullet 3$), and all kinds of unstable species not existent in wet chemistry. Carbon based polymers, produced in the plasma which may be deposited on parts of the wafer.
Basic processes : SiO_2 dissolves.	Etching of SiO_2, formation of polymers, deposition of polymers (and other stuff) and etching of the deposited stuff, occurs simultaneously.
Driving force for reactions : Only "chemistry", *i.e.,* reaction enthalpies or chemical potentials of the possible reactions; essentially equilibrium thermodynamics.	*Driving force for reactions :* "Chemistry", *i.e.,* reaction enthalpies or chemical potentials of the possible reactions, including the ones never observed for wet chemistry, near equilibrium, and non-equilibrium physical processes", *i.e.,* mechanical ablation of atoms by ions with high energies.
Energy for kinetics : Thermal energy only, *i.e.,* in the 1 eV range.	*Energy for kinetics :* Thermal energy, but also kinetic energy of ions obtained in an electrical field. High energies (several eV to hundreds of eV) are possible.
Anisotropy : None; except some possible {hkl} dependence of the etch rate in crystals.	*Anisotropy :* Two major mechanisms, 1. Ions may have a preferred direction of incidence on the wafer. 2. Sidewalls may become protected through preferred deposition of *e.g.,* polymers Completely isotropic etching is also possible.
Selectivity : Often extremely good.	*Selectivity :* Good for the chemical component, rather bad for the physical component of the etching mechanism. Total effect is open to optimization.

9.3.3.4 Mechanisms of Plasma Etching

The two main types of species involved in plasma etching are reactive chemical species and ions. The neutral species like free radicals and reactive species such as Cl_2 are mainly responsible for the chemical components in plasma etching processes. We will now discuss these mechanisms one by one :

(a) Chemical Etching

When the reactive neutral species act by themselves then mechanism is called chemical etching. Chemical etching of materials in plasma systems is commonly done by free radicals. These free radicals have incomplete bonding and they have unpaired electrons.

Examples are the fluorine free radical, F and neutral CF_3 free radical. Both of these can be produced by the reaction between free electrons in the plasma and CF_4

$$e^- + CF_4 \rightarrow CF_3 + F + e^-$$

These free radicals are highly reactive chemical species. These species reacts with the material to be etched, for example, Si. The by product should be volatile species that leave the surface, exposing more Si to be etched as shown below.

$$4F + Si \rightarrow SiF_{4\,(gas)}$$

(b) Sputtering Etching Mechanism

This mechanism is the ejection of material from a surface caused bombarment of energetic inert ions such Ar^+ and Xe^+ (as shown in figure). An important parameter is yield which is defined as the number of ejected surface atom per incoming ion at a given ion energy. One important aspect of sputtering is that yield changes with angle of ion flux as shown in Fig. 9.11(*b*). The maximum sputtering yields occors because fewer collision close to the surface are required to eject an atom as the flux angle is moved away form the surface normal. This preferential material removed often result in resists faceting.

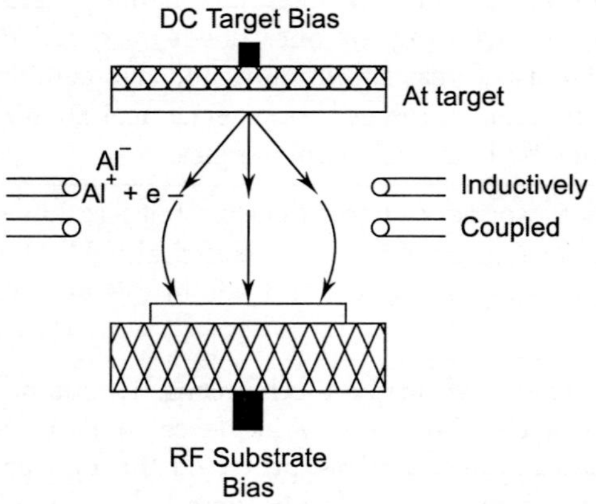

(a) Sputter Mechanism

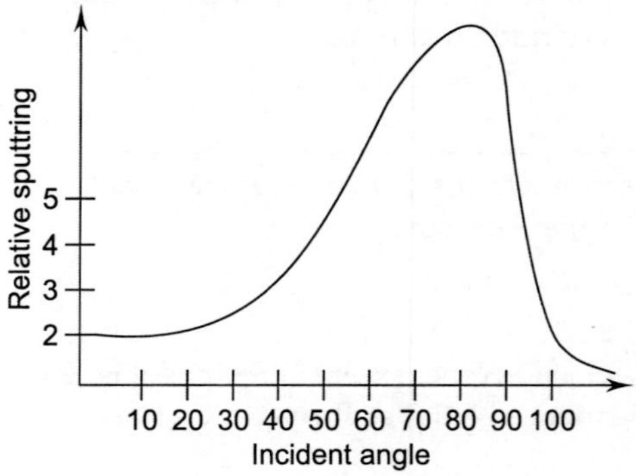

(b) Graph Shows Yield Changes with Angle of Ion Flux

Fig. 9.11

Disadvantage - It has low selectively, poor profile control, can create residue and particulates.

(c) Accelerated Ion-assisted Etching

In this process, like the sputtering process, ions are accelerated by the sheath potential. But, unlike the sputter etching, the purpose of these accelerated ions are not to sputter away the surface. They are to damage the surface only, leaving dangling bonds and dislocations in the surface. This is to modify the surface into a more reactive form so that the damaged surface will react with the neutral etchants more easily.

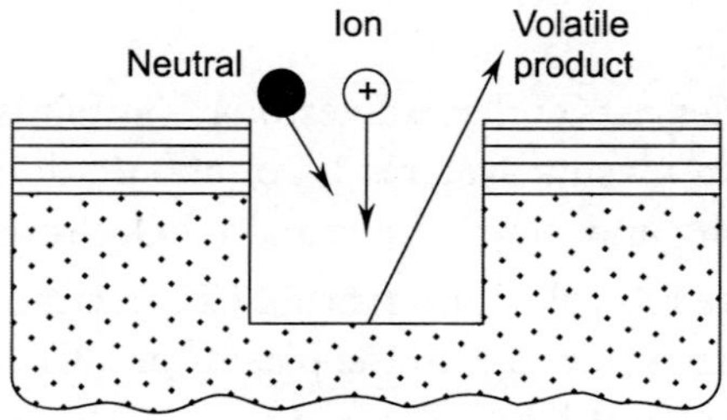

Fig. 9.12 *Accelerated Ion-Assisted Etching*

(d) ION Assisted Anisotropic Etching

Two mechanisms are proposed to explain the phenomenon of ion assisted anisotropy. Anisotropic etching is believed to result from a combination of physical and chemical removal processes. The ratio of vertical etch rate to horizontal etch rate may be increased either by reducing the horizontal rate or by increasing the vertical rate.

(*a*) *ION Induced Damage Mechanism*: In this model, bombarding ions have sufficient energy to break crystal bonds, making the film more accessible and the surface more reactive to the active chemical etchants. At the sidewalls, where there is essentially no ion bombardment, the etching process proceeds at the nominal chemical etch rates.

(*b*) *Surface Inhibitor Mechanism*: In some etch chemistries, the surface exposed to the plasma is likely to become coated with a chemisorbed film of etchant radicals and unsaturated species, which polymerize and adhere tenaciously to the material being etched. The resulting polymer coating inhibits the chemical reactions necessary to etch. Ion bombardment can cause the polymers to desorb, exposing horizontal surfaces to the etching gas. Vertical surfaces experience little or no bombardment, therefore etching in the horizontal direction can be completely blocked.

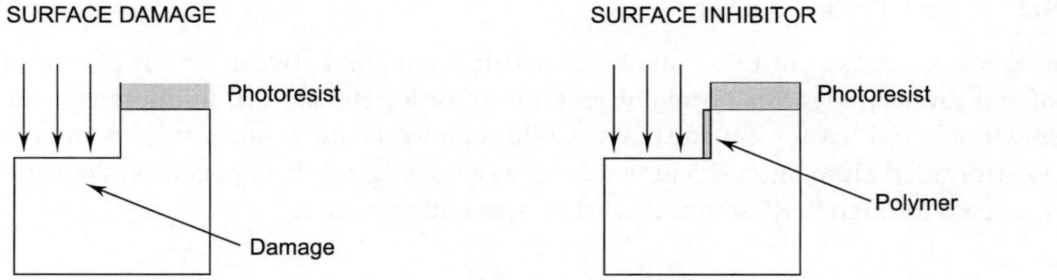

(a) Speeds Chemical Reaction Horizontal Surfaces *(b) Slow Chemical Reaction on Vertical Surfaces*

Fig. 9.13

9.4 TYPES OF ETCHERS

9.4.1 Planner Etcher

Planner Etcher consist of a chamber, with wafer holders, RF power supply arrangement, heaters, vacuum system of the etchants for etching process. The process of etching take place in the chamber, and chamber is grounded, because it makes the great difference between anode and cathode potential so that greater potential is obtained. It forms the large number of ions or reactive species which increases etch rate.

Firstly, gases (etchants) are fed inside from the chamber through the inlet point. The wafers are placed on the holder. Heaters provide the required amount of heat for the reaction process.

The reactive ions reacts with surface of wafers and etch the unwanted material and the etched materials are pumped out by vacuum system. Only one side of wafer is exposed to the plasma, the other side is protected from the unwanted etching and contamination. These type of reactors are simple in construction and operation, but there are some disadvantage of this system. First only few wafers can be loaded at a time. The number of wafers are directly prportional to the Area of planner etcher. If we increase the area of planner etcher, it is difficult to maintain the uniformity, and selectivity is also poor.

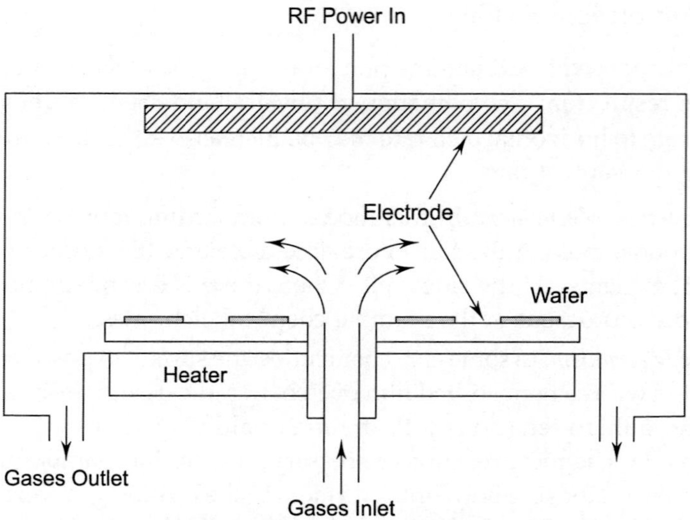

Fig. 9.14 *Schematic of a Planar Plasma Etcher*

9.4.2 Barrel Plasma Etcher

Barrel reactor consist of two hollow concentric cylinders between which plasma discharge is confined. In barrel reactor normally pressure of order 500 to 1000 millitorr is used. The outer cylinder of barrel reactor is made of quartz through which input gases an pass and inner cylinder is a perforratted aluminium shield which act as a floating anode of plasma system and allows the ions to pass through it. RF source is used for generation of ions.

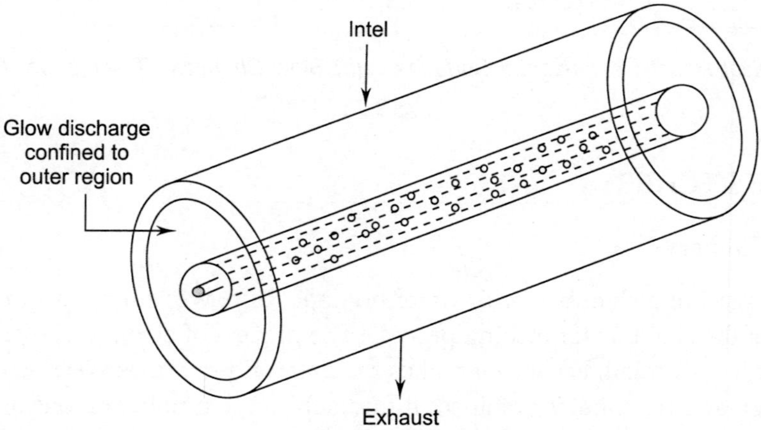

Fig. 9.15 *Barrel Reactor*

Wafer to be etched is placed vertically in wafer holder axially along the inner cylinder. A suitable inlet for input gases is provided in quartz tube. The perforratted Al mesh allow the reactive species to penetrate into inner cylinder inside which etching actually takes place.

Advantages

(*i*) It is cheaper than other methods.

(*ii*) Large number of wafer can be loaded at a time result in higher throughput.

9.4.3 Reactive Ion Etcher

In RIE, the substrate is placed inside a reactor in which several gases are introduced. A plasma is struck in the gas mixture using an RF power source, breaking the gas molecules into ions. The ions are accelerated towards, and reacts at the surface of the material being etched, forming another gaseous material. This is known as the chemical part of reactive ion etching. There is also a physical part which is similar in nature to the sputtering deposition process. If the ions have high enough energy, they can knock atoms out of the material to be etched without a chemical reaction. It is a very complex task to develop dry etch processes that balance chemical and physical etching, since there are many parameters to adjust. By changing the balance it is possible to influence the anisotropy of the etching. Since the chemical part is isotropic and the physical part highly anisotropic the combination can form sidewalls that have shapes from rouned to vertical. A schematic of a typical reactive ion etching system is shown in the Fig. 9.16.

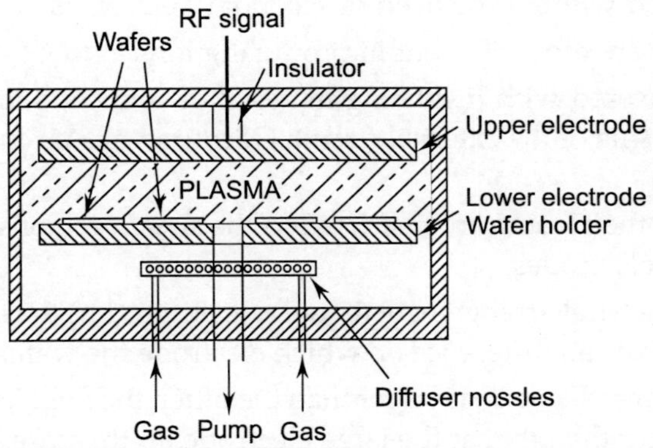

Fig. 9.16 *Typical Parallel-Plate Reactive Ion Etching System*

A typical (parallel plate) RIE system consists of a cylindrical vacuum chamber, with a wafer platter situated in the bottom portion of the chamber. The wafer plate, which is usually grounded, is electrically isolated from the rest of the chamber. Gas enters through small inlets in the top of the chamber, and exits to the vacuum pump system through the bottom. The types and amount of gas used vary depending upon the etch process; for instance, sulphur hexafluoride is commonly used for etching silicon. Gas pressure is typically maintained in a range between a few millitorr and a few hundred millitorr by adjusting gas flow rates and/or adjusting an exhaust orifice.

Plasma is initiated in the system by applying a strong RF (radio frequency) electromagnetic field to the wafer platter. The field is typically set to a frequency of 13.56 megahertz, applied at a new hundred watts. The oscillating electric field ionizes the gas molecules by stripping them of electrons, creating a plasma.

In each cycle of the field, the electrons are electrically accelerated up and down in the chamber, sometimes striking both the upper wall of the chamber and the wafer platter. At the same time, the much more massive ions move relatively little in response to the RF electric field. When electrons are absorbed into the chamber walls they are simply fed out to ground and do not alter the electronic state of the system. However, electrons absorbed into the wafer platter cause the platter to build up charge due to its DC isolation. This charge build up develops a large negative voltage on the platter, typically around a few hundred volts. The plasma itself develops a slightly positive charge due to the higher concentration of positive ions compared to free electrons.

Because of the large voltage difference, positive ions tend to drift toward the wafer platter, where they collide with the samples to be etched. The ions react chemically with the materials on the surface of the samples, but can also knock off (sputter) some material by transferring some of their kinetic energy. Due to the most vertical delivery of reactive ions, reactive ion etching can produce very anisotropic etch profiles, which contrast with the typically isotropic profiles of wet chemical etching.

Etch conditions in an RIE system depend strongly on the many process parameters, such as pressure, gas flows, and RF power. A modified version of RIE is deep reactive-ion etching, used to excavate deep features.

9.4.3.1 Advantages of RIE

1. Electrons absorb energy from the electric field, and achieve enough energy to ionize. They oscillate between the electrodes at the power supply frequency. The glow discharge can therefore be sustained without the need of secondary electrons.
2. Oscillating electrons are more efficient in producing ions at low (~ 10-3 torr) pressure.
3. Electrodes can be covered with insulating material or semiconductor. No net current need flow through the plasma or the external circuit. Of course, an AC current can be measured in the external circuit.
4. The electric fields in the plasma are restricted to the sheath regions and are such as to repel electrons from both electrodes.
5. In a physically symmetrical arrangement (equal size electrodes) the fields are identical at both electrodes, and it makes no difference on which electrode the wafer to be etched is placed.
6. In a chamber where one electrode is larger than the other, the field in the sheath at the smaller electrode will be larger than that at the large electrode. In this case, ionic bombardment will be greatest at the small electrode.
7. As the frequency increases, the minimum operating pressure needed to sustain a discharge decreases. The minimum operating pressure is less than one millitorr at 13.56 MHz. Most plasma etching operations are well above this lower pressure limit.

9.4.3.2 Inductively Coupled Olasma RIE Etcher

Other types of RIE systems exist, including inductively coupled plasma (ICP) RIE. In this type of system, the plasma is generated with an RF powered magnetic field. Very high plasma densities can be achieved, though etch profiles tend to be more isotropic.

A combination of parallel plate and inductively coupled plasma RIE is possible. In this system, the ICP is employed as a high density source of ions which increases the etch rate, whereas a separate RF bias is applied to the substrate (silicon wafer) to create directional electric fields near the substrate to achieve more anisotrolpic etch profiles.

The demands of etching small features in integrated circuits has driven the development of high density plasma sources. One of the most successful methods of creating an intense plasma is radio frequency induction. The plasma is sustained by an RF current (5-50 A) carried by an inductive element, usually an air-core coil. This current induces an electric field in the plasma region which is primarily tangential to the plasma boundary. The tangential component of the RF induction electric field may then penetrate the plasma region 1-2 cm and generate an intense plasma. Most significantly, the plasma is sustained without creating a high voltage sheath at the plasma-wall boundary. This improves the purity of the process and greatly increases the ion generation efficiency.

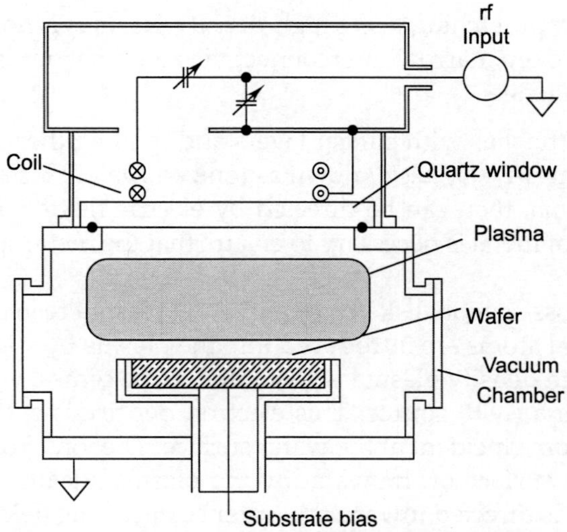

Fig. 9.17 *Inductively Coupled Plasma RIE Reactor*

In the ICP design shown above, we have developed a coil which significantly reduces the sputtering of contamination from the quartz window. The inductive element in an ICP typically has an RF voltage in excess of 1 kV during operation. This high voltage attracts ions from the plasma toward the coil, which then can sputter from the window. To prevent window erosion and subsequent process contamination, we have used a low-aspect ratio helical coil with the lower turn of the coil being electrically grounded. Strong electric fields originating on the upper turn of the coil are terminated on the lower turn, rather than extending into the plasma region. Sputtering of the window is significantly reduced and etching of polymer materials is documented to be residue-free.

9.4.3.3 Ionized Physical Vapour Deposition (I-PVD)

As the physical size of devices and interconnects on integrated circuits shrinks further and further, the aspect ratio of the metal interconnects increases. The aspect ratio is the via depth divided by the

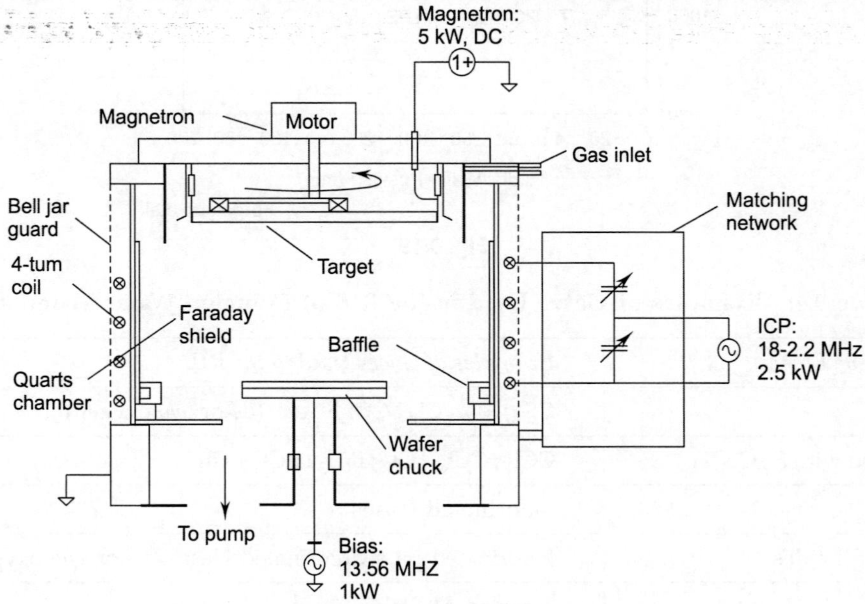

Fig. 9.18 *I-PVD Reactor*

via diameter. While anisotropic etching using high-density plasmas is now commonly practiced to create the via, refilling the deep, narrow interconnect vias with barrier layers and metals remains a substantial technical challenge.

One approach to lining the vias with barrier layers and refilling the vias with metals is Ionized Physical Vapour Deposition (I-PVD), also known as Ionized Metal Plasma (IMP). Metal atoms are ionized in an intense plasma, then can be directed by electric fields perpendicular to the wafer surface. This directed flux of metal is necessary to ensure that the metal species reach the bottom of deep vias.

The figure above is a cross-sectional sketch of an I-PVD plasma reactor. The inside diameter of the reactor is 450 mm. Metal atoms are introduced into the plasma by sputtering from the target at the top of the reactor. A high density plasma is generated in the central volume of the reactor by an Inductively Coupled Plasma (ICP) source. This electron density is sufficient to ionize approximately 80% of the metal atoms incident at the wafer surface. The ions from the plasma are accelerated and collimated at the surface of the wafer by the plasma sheath. The sheath is a region of intense electric field which is directed toward the wafer surface. The field strength is controlled by applying a radio frequency bias to the wafer chuck, as shown in Fig. 9.19.

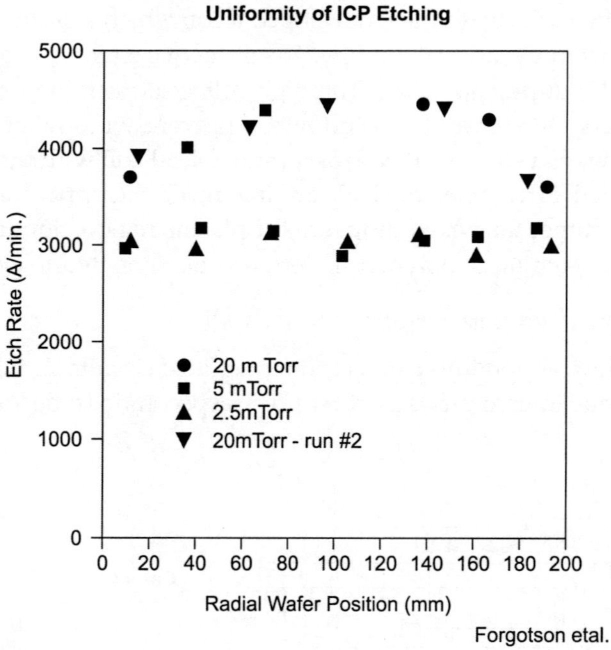

Fig. 9.19

Table 9.4: Examples of Gases Used in the RIE of Common Wafer Materials

Material to be Etched	*Examples of Gases Used in the RIE*
Polysilicon	CF_4 ; SF_6 ; Cl_2 ; CCl_3F ; etc. (*w/* or *w/o* oxygen)
Al ; Al doped with Si, Cu, Ti	CCl_4 ; $CCl_4 + Cl_2$; BCl_3 ; $BCl_3 + Cl_2$
Tungsten	Fluorinated Gases
Refractory Silicides	Fluorinated plus Chlorinated Gases (*w/* or *w/o* oxygen)
TiN ; TiC	Same as Al Etch

Uniformity of etching in the ICP is shown in the figure above. These etching tests were performed on 200 mm wafers that were spin-coated with polymer. The etch uniformity at low pressure (2.5 and 5 m Torr) is measured at better than 3% (standard deviation/average). At 20 m Torr, however, the etch rate is considerably higher in the centre of the wafer than at the edge. The RF bias was not used in this experiment. With the addition of a 30w bias to the wafer, the etch rate exceeds 1.5 μm/minute.

9.4.3.4 High Density Plasma Etch Systems

This etcher system consist of microwave supply (2.45 GHz), magnetic coils, wafer holder with RF supply (13.56 MHz) gas inlet point. In this system ions are generated in one section of the chamber, and then accelerated towards the target wafer surface; the density of the ions are high and controlled separately. A microwave discharge using Electron Cyclotron Resonance (ECR) to produce high density plasma, ECR uses a magnetic field to change the normal straight trajectories of electron into helical orbit or circular orbit. The etch rate is faster in this case.

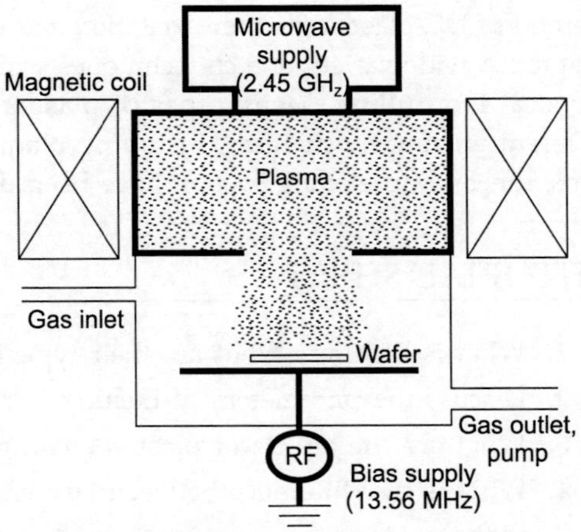

Fig. 9.20 *High Density Plasma (HDP) Etch Systems*

9.5 ION BEAM MILLING

This is an another type of dry-etch system (as shown in Fig. 9.21). It is a physical process ; the wafers are placed on a holder in a vacuum chamber, and argon gas is passed to the chamber. At the enterance, argon is subjected to a stream of high-energy electron from a set of cathode (–) anode (+) electrodes. The wafers are place on the – *ve* grounded holder, which attract the ionized + *ve* argon. When argon (ionized) transfer to the wafer surface, they exposed wafer layer and remove the amount of material from the wafer. The removed material is highly directional, which result in good definition of small opening. This process is also called ion beam milling. It has poor selectivity ; especially with photo resist layer.

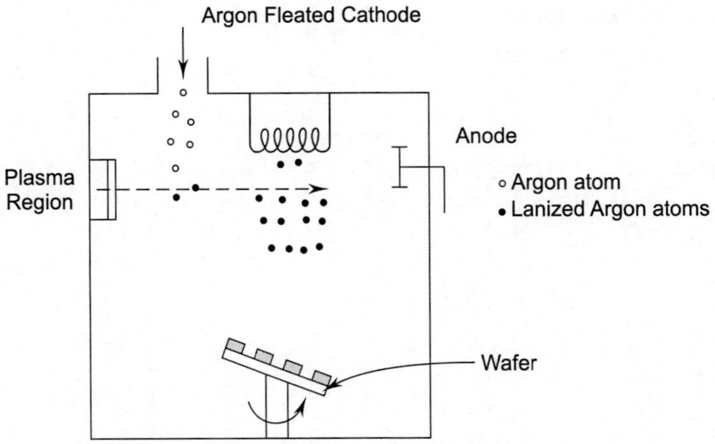

Fig. 9.21 *Ion Beam Milling*

SUMMARY

We have seen in this chapter, how etching is used to etch the pattern films by selectively removing material with help of mask layers. Anisotropy and throughput are the other etching parameter useful for etching process. Etching can either wet or dry, the former using liquid chemical enchants and the latter gas phase, plasma produced etches Plasma etching is an example of a purely chemical dry technique. The method of DC plasma glow or excitation and effects of ions in plasma etching is explained in detail in this chapter. A wide variety of etch techniques which ranging from the purely chemical (wet etching) to purely physical ion milling. Many types of plasma etching equipment have been developed that utilize the different etching mechanism, including planner, barrel, reactivation etching. High Density plasma systems, ion milling etching system have been described the end of chapter.

TUTORIAL SHEET

1. What is Etching? What are their types?
2. Discuss the parameters of Etching.
3. What are the effects of plasma etching?
4. What is the difference between Dry and Wet etching? Also write their advantages and Disadvantages.
5. Discuss the classification of various types of etches used for VLSI Fabrication process.
6. How cathode dark space is important region of DC glow Discharge?
7. Explain the process of plasma excitation and reactions involved in plasma etching process.
8. Explain the working of Planner and barrel etcher.
9. How reactive ion etcher works and Discuss their advantages?
10. Explain Inductively coupled plasma (RIE) Etcher with the help of Diagram.
11. How High Density plasma etchers works?
12. Explain the working of ion beans milling etch system.
13. Write a short note on:
    (*i*) Sputtering process
    (*ii*) Anisotropic etching
    (*iii*) Deposition of silicon domicile.
14. Discuss about the following terms:
    (*a*) Selectivity
    (*b*) Uniformity
    (*c*) Throughput.

10

Ion Implantation

10.1 INTRODUCTION

Ion implantation is the introduction of energetic, charged particles into a substrate such as silicon. The practical use of ion implantation in semiconductor technology has been mainly to change the electrical properties of the substrate. Typical ion energies are between 30 to 300 KeV, and typical ion does vary from 10^{11} to 10^{16} ions/cm^2. Note that the dose is expressed as the number of ions implanted into 1 cm^2 of the semiconductor surface area. The main advantages of ion-implantation are its more precise control and reproducibility of impurity doping and its lower processing temperature requirement compared to those of the diffusion process.

Ion implantation is an alternative to a deposition diffusion and is used to produce a shallow surface region of dopant atoms deposited into a silicon wafer. This technology has made significant in roads into diffusion technology in several area. In this process, a beam of impurity ions accelerated to kinetic energies in the range of several tens of KV is directed to the surface of the silicon. As the impurity atoms enter the crystal, they give up their energy to the lattice in collision and finally come to rest at some average penetration depth, called the projected range expressed in μm. Depending on the impurity and its implantation energy, the range in a given semiconductor may vary from a few hundred angstroms to about 1 μm.

The idea of doping semiconductor using ion implantation was patented by 'Shockley' at Bell Labs in 1954. Early work has been summarized by Gibbons, covering development up to 1968. An overview of work, since these can be found in the proceeding of the international conferences on ion implantation and the conferences on ion beam modification of materials. High dose and high energy implants have been used to form burried layers of different materials. New, low-energy machines have been designed to allow very shallow dopant layers to be formed, leading, in the limit, to ion beam deposition system. Focused ion beam can be used directly to write dopant patterns without the need for a masking layer. These new directions provide challenges and opportunities for equipment manufacture, process design, and process simulation.

10.2 THEORY OF ION IMPLANTATION

Ion implantation is a method for introducing controlled amounts of dopants into the silicon substrate to change its electronics properties. It is a physical process, there are no chemical reactions. A number of applications use ion implantation in modern wafer fabrication. The main application of ion implantation is the doping of semiconductor materials. Each doping application has specific requirements for the concentration of the dopant material and its depth. Ion implantation is preferred over diffusion in nearly all applications because of its ability to control dopant concentration and depth. It has become a critical process to meet the challenges of wafer fabrication with sub-quarter micron features sizes and larger diameter wafers.

10.2.1 Ion Implantation Equipment

The basic components of an ion implantation system are as shown in Fig. 10.1. The system consists of an ion source, a magnetic mass filter which separates unwanted ions from the beam, a charged path to accelerate the ions, electrostatic lenses and plates to focus the beam and provide X and Y beam deflection, and a target. The target is usually a metal disk that can hold up to 50 to 60 silicon wafers in concentric rings. The interior of the entire system is a vacuum chamber.

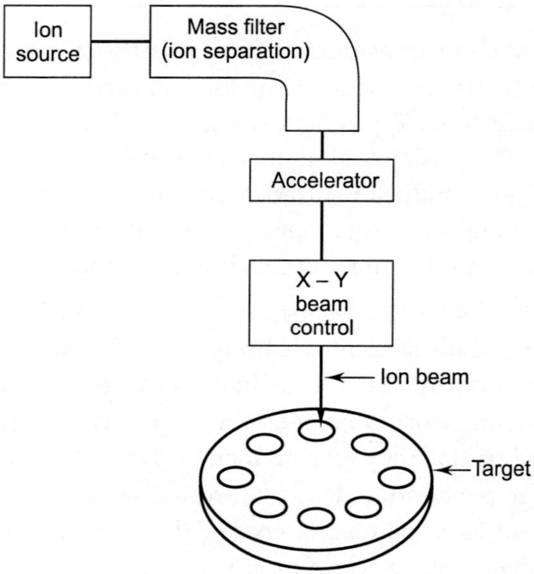

Fig. 10.1 *Basic Components of Ion Implantation*

The most commonly used ions are boron, phosphorus, and arsenic. Since, these elements are not gases at room temperature, molecular compound gases are used as ion sources. BF_3 or BCl_3 is used for boron ions. PH_3 or PF_3 is used for phosphorus ions. And AsH_3 or AsF_3 is used for arsenic ions. The ions are typically formed through bombardment by electrons from an arc discharge or a cold cathode source. The collision creates more than one kind of ion. The mass filter or ion separator use a magnetic field to deflect the ion beam. Since, ions with different masses are deflected to different degrees, the exit path from the magnetic field selects only the desired ions.

Many variations from the configuration of figure above are commercially available, depending on the desired application. In systems using high beam currents, the beam position may be fixed while scanning motion is provided by the target, which rotates with a simultaneous back-and-forth motion. In other systems, the beam scans in one direction while the target rotates.

Acceleration potentials range between 10 and 300 KeV for most implant system. The ion beam scan is typically focused to a diameter of about 1 cm at the target. The beam current, which can be accurately measured, ranges between about 10 μA and 2 μA. The number of implanted ions per unit area is called the ion dose ϕ. Typical ion doses range between 10^{11} and 10^{16} atoms/cm^2. The ion dose is related to the beam current I, the beam area A and the dose time t by the equation

$$\phi = \frac{tI}{qA}$$

where, q represents the charge per ion (1.6×10^{-19} C). If the beam current were set at 100 μA for 1.6 ms and beam area was 1 cm^2, the resultant dose would be

$$\phi = \frac{\left(1.6\times 10^{-3}\ \text{s}\right)\left(1\times 10^{-4}\ \text{A}\right)}{\left(1.6\times 10^{-19}\ \text{C}\right)\left(1\ \text{cm}^2\right)}$$
$$= 1\times 10^{12}\ \text{ions/cm}^2.$$

If the ion dose and the beam area are known, and the total target area to be scanned is known, an appropriate scan pattern and scan rate can be determined.

10.2.2 Range

The ion range is the total distance an ion travels in the silicon during ion implantation as shown in Fig. 10.2. The characterize range is necessary to understand energy. When the ions are accelerated due to electric potential difference to a high velocity, they gain energy. The ion energy is Kinetic Energy (KE) due to its motion, and it typically is expressed in units of Joules. However, for ion implantation, the energy is usually stated in terms of the number of the electronic charge times the difference in potential, or electron volts (eV). The equation that describes this energy is:

$$\text{KE} = nV$$

where, KE: energy in electron volts (eV)
n: charge state of the ion
V: voltage difference in volts.

For example, if ion with a single positive charge state moves in an electric field with a voltage difference of 100,000 volts (100 KV), then its energy is:

$$\begin{aligned}\text{KE} &= nV\\ &= (1)\,(100\ \text{KV})\\ &= 100\ \text{KV}.\end{aligned}$$

Higher implanter energy means the dopant atom will have an increased range and travel deeper into the wafer. Energy is important in implanters because of the need to control the range and, therefore, junction depth of the dopants. High-energy implanters have an energy greater than 200 KeV and may be as high as 2 to 3 MeV. High-energy implant is used in applications such as deep retrograde wells and retrograde triple wells. A retrograde well has higher doping concentration deeper in the well than at the surface. Ultra low-energy implanters have energies presently down to about 200 eV to enable doping at very shallow depths for source/drain applications.

There is a projected range, R_P, which is how far the implanted ions travel into the wafer, depending on the ion mass and energy, the target mass and the beam direction with respect to the wafer crystal structure. At the same time, not all ions come to stop simply at the projected range ; some stop at a shorter distance and others at a greater distance. Ions will also move in a lateral direction. Combining all these ion movements produces a distribution of distances travelled by the dopant atoms implanted in the wafer, referred to as straggle, or ΔR_P. The R_P (projected range) indicates how shallow or deep the junction depth can be formed, whereas the ΔR_P (straggle) represents the spread of the implanted species around R_P. As the implant energy of dopant atoms increases, range increases but the peak concentration of dopants decreases, because the range straggling increases. Projected range charts are available that predict the projected range for a given implant energy as shown in Fig. 10.3.

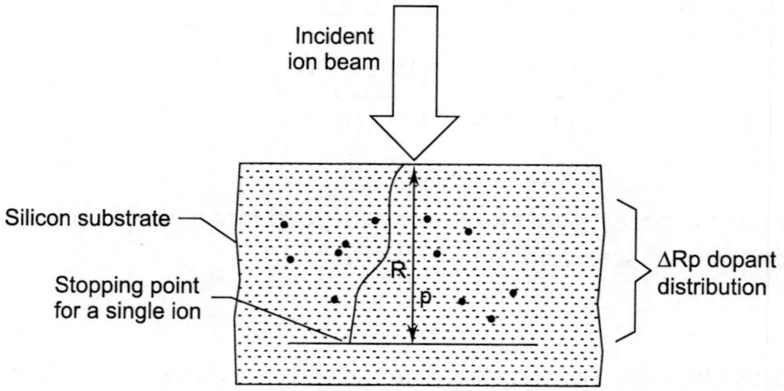

Fig. 10.2 *Range of Dopant Ion*

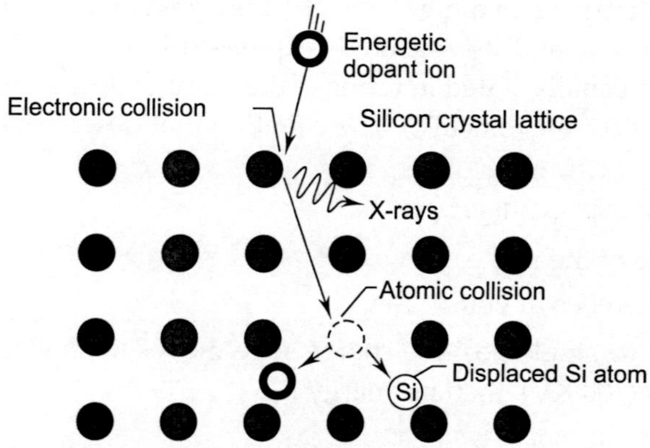

Fig. 10.3 *Energy Loss of an Implanted Dopant Atom*

10.2.3 Advantages of Ion Implant

The primary advantages of ion implantation are listed as below. These benefits effectively increase the process flexibility for fabricating wafers with sub-quarter micron technology.

1. *Precise control of dopant concentration*: Ability to precisely control the number of implanted dopants over a wide range of concentrations, from 10^{10} to 10^{17} ions/cm^2, and controlled to within ± 2% over this range. In contrast, diffusion can control dopant concentration to within 5 to 10% at high concentrations and becomes less effective at low concentrations.
2. *Good dopant uniformity*: Controls dopant uniformity by scanning the ion beam over a large wafer area, which is critical for smaller feature sizes on larger wafers.

3. *Good control of dopant penetration depth*: Controls the ions energy during the implant to achieve good controls of dopant penetration depth. This permits design flexibility. Such as buried layers, where the maximum dopant concentration is at buried layer and the minimum concentration is at the wafer surface.
4. *Produces a pure beam of ions*: Mass separation techniques produce a pure beam of ions free of contamination. Different impurities can be selected for implantation. It is done in a high vacuum that minimizes contaminants.
5. *Low temperature processing*: Implantation is performed at moderate temperature (< 125°C) which permits the use of a variety of masks, including photo resist.
6. *Ability to implant dopants through films*: Dopants can be implanted through films, such as an oxide or nitride. This condition allows adjustment of the threshold voltage of MOS transistors after the gate oxide is grown. It increases the flexibility to implant at various stages of the process.
7. *No solid solubility limit*: The dopant concentration during implantation is not limited by the solid solubility limit of the silicon wafer.

10.2.4 Disadvantages of Ion Implant

The major disadvantage of ion implantation is the damage done to the crystal structure when the energetic dopant ions bombard the silicon host atoms. As the high energy ions enter the crystal and colloide with substrate atoms, energy is transferred and some silicon atoms are displaced from the lattice. This action is referred to as radiation damage. Host or all this crystal damage is corrected with a high temperature anneal step.

Another drawback of implantation is the complexity of the implanter equipment. However, this drawback is offset by the implanter's capability for dose and depth control and overall process flexibility.

10.3 ION IMPLANTERS

An ion implant tool consists of the following five major subsystems:

- Ion source
- Extraction electrode and ion analyzer
- Acceleration column
- Scanning system
- Process chamber

The ion source and extraction assemblies are generally mounted in the same vacuum chamber. The ion source generates positive ions from a specific dopant gas or solid. The circular suppression electrode attached to the extraction assembly uses the electric field of a negative high voltage to draw the positive ions out of the ion source. Electromechanical controls on the extraction assembly aid in positioning the extraction electrode directly in front of the ion source plasma chamber. On the older ion implanters, these controls could be manually adjusted from the operator's control panel. Today, these controls are set by a process recipe and maintained by the tool's internal computer software.

10.3.1 Ion Source

The species to be implanted must be present in the beam as a charged particle, or ion. Because of their electrical charge, ions can be controlled and accelerated with electric and magnetic fields. Ions for implantation are generated in an *ion source*. A positive ion is formed from a dopant gas source or by vaporizing a solid. Typically B^+, P^+, As^+, and Sb^+ are produced by ionization of atoms or molecules. The most common feed material for a source of dopant atoms is a gas. Such as B_2H_6, BF_3, PH_3 and AsH_3. The gas is packaged in a relatively small cylinder (0.4 to 2.2 liters in size) and diluted with hydrogen to reduce the risks associated with inadvertent gas releases.

Another method of supplying the feed material is by heating and vaporizing a solid material, which is sometimes used to obtain arsenic (As) or phosphorus atoms from solid pellets. The solid material is vaporized at about 900°C and the volatile dopant atoms are transported to the ion source chamber. No dilute hydrogen is added, which permits higher maximum beam currents. A disadvantage of a solid source is the long set up time, which can be 40 to 180 minutes. Most of this required to heat and stabilize the vaporizer. There is also the need for more frequent maintenance due to the build-up of sputtered materials on the beam line components. However, many IC manufacturers prefer to use solid dopant because it is easier to deal with them from an environmental and safety standpoint.

10.3.2 Extraction and Ion Analyzer

The traditional implanter extraction system collects all the positive ions created inside the ion source and forms them into a beam. The ions are extracted through a slit in the source. They are repelled by the positive bias of chamber (anode) and attracted to the negative bias (cathode) of the extraction assembly as shown in Fig. 10.4.

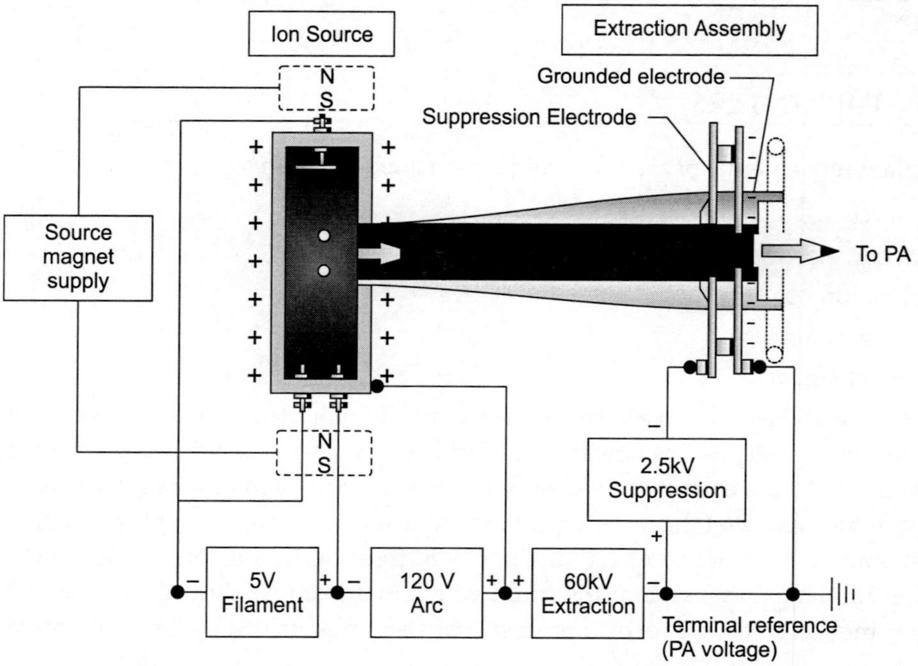

Fig. 10.4 *Interaction of Ion Source and Extraction Assemblies*

Since positive ions are like little magnets, each with a positive charge, they are attracted to a negative electric field. The higher the electric field, then the faster the ion will move. At the same time, the faster the ion moves than the higher is its kinetic energy and the further it will penetrate into the wafer when it strikes the surface. The negative bias of the extraction electrode also repels any stray electrons in the source plasma and forms the positive ions into a beam. A negatively biased *suppression electrode* is used to focus the ion beam into a parallel beam, the efficient beam to transport through the implanter. This condition is important for delivering high beam currents to wafers.

10.3.2.1 Mass Analyzer Magnet

The ions extracted from the source contain many different ion species and travel at a relatively high speed due to the acceleration provided by the extraction voltage. The species in the ion beam have different atomic mass units. In ion implanters, a magnetic ion analyzer separates the desired dopant ion from the main body of ion species. The analyzer magnet is shaped in a 90° angle. The magnetic field of the analyzer magnet causes the ion species to be deflected into an arc. A spectrum of ions is formed based on their unit (amu). For a given field strength, ions with heavy masses are not able to bend at the appropriate angle, while lighter ions bend too easily. There is one ion species, however, that bends just enough at the appropriate field strength to allow it to pass through the center of the analyzer magnet. This is the dopant species that eventually gets implanted into the wafer.

10.3.3 Acceleration Column

To achieve additional ion acceleration beyond the analyzer magnet, positive ions are accelerated in an electric field inside an *acceleration column* see Fig 10.5. This acceleration column is a linear design made of a series of electrodes separated by insulators, each with an increasing negative charge. As the positively charged ions enter the column, they start accelerating. In this manner, the total voltage difference in the electrode assembly and the acceleration column accelerates the dopant ions. The higher the total voltage, then the higher the velocity and, therefore, energy. Higher beam energy means the dopant ions are implanted deeper into the wafer, while lower beam energy is used for ultra shallow implants.

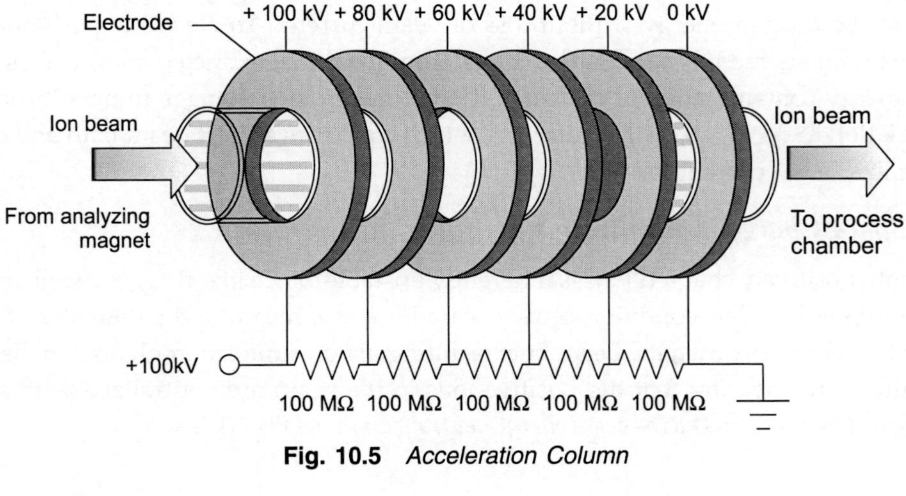

Fig. 10.5 *Acceleration Column*

10.3.3.1 Post Accelerator

There is additional beam focusing that takes place in the *post-accelerator* component of the acceleration column. It typically uses quadruple lens focusing based on four cylindrical poles in a two-lens system, employing electrostatic or magnetic repulsion to form the ion beam into a focused circular beam.

10.3.3.2 High-Current and High-Energy Beam

High-energy implanters are used to implant burried dopant layers. Such as for retrograde and triple wells. Most commercial high energy ion implanters employ *linear accelerator* technology to accelerate ions to high velocity and obtain beam energies in excess of 200 KeV and into the MeV range. High beam current (up to 25 mA) is desirable to reduce the implant time and increase wafer through put through the implanter. A linear accelerator is constructed as an alternating series of high-voltage (several 10 KV) electrodes and grounded quadruple focusing lenses. The timing of the voltage between each electrode and grounded quadruple is set to match the arrival of the dopant ions at each gap. The ions are accelerated through the gaps, with the final energies up to 20 times higher than the maximum acceleration in a traditional beam line acceleration column. This condition avoids the use of extremely high voltages. A traditional bending magnet is located at the end of the linear accelerator that serves as an energy analyzer to ensure that a pure, monoenergetic beam is formed for wafer implant. This added analysis magnet also aids in the removal of beam contaminants by removing masses other than the desired species.

10.3.3.3 High-Current and Low-Energy Beam

The need to form ultra shallow implants has promoted the development of high-current, low energy beam. A low-energy beam accelerates ions to a low velocity so that their projected range is shallow in the target wafer. Maintaining a well-focused beam with low-energy acceleration is difficult, because beam size is hard to control and usually increases with high-current and low-energy. Visualize this action by imagining a bowling ball rolled slowly down a bowling alley lane. It tends to veer to the side and into gutter much more easily than a bowling ball rolled straight and fast. A method used to achieve a high beam current in a low energy beam is *beam deceleration or decel*. With this approach, the beam is extracted from the ion source at a high energy and then decelerated down to the desired energy without loss of beam current. An electrode positioned farther down the beam line is used to decelerate the ions from the initial extracted energy. This approach may cause energy contamination of the beam if high-energy ions manage to pass through to the wafer. A variation to decel is the *differential lens*, which combines a short beam path and optimized optics minimize beam current-loss.

10.3.3.4 Space Charge Neutralization

A beam of only positively charged ions is inherently unstable. Since like charges repell one another with mutual repulsion. This condition causes beam blow-up, meaning the diameter of the beam increases as the ions pass down the beam line, resulting in non-uniform implantation. Beam blow-up is minimized by ensuring that the positive ions in the beam are neutralized with secondary electrons, a condition referred to as *space charge neutralization* see Fig 10.6.

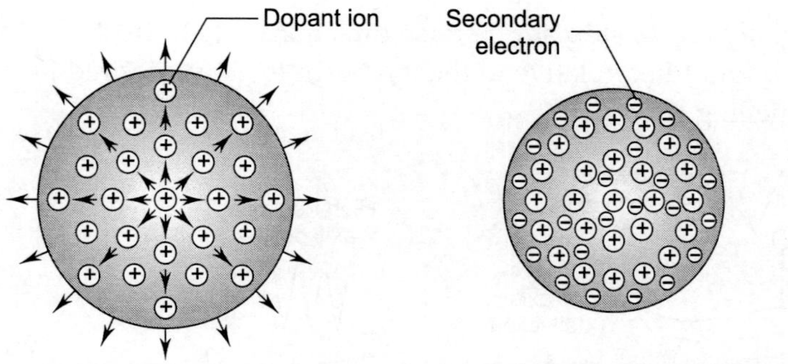

Fig. 10.6 *Space Charge Neutralization*

Secondary electrons are generated when high-energy dopant ions strike a surface as they pass along the beam line. The secondary electrons become trapped within the beam line due to the negative biased electrodes along the beam path, including the extraction electrode and the acceleration. The secondary electrons bounce back and forth between the negatively biased plants, unable to leave their specific region in the beam line. The trapped electrons serve to neutralize the beam and prevent or reduce beam blow-up.

10.3.4 Scanning System

The goal of the implanter is to form a highly pure ion beam from both a contamination and energy stand point. This pure beam is implanted into the target wafer. That is, the beam has only the desired dopant ions at a predetermined energy level. Because, the focused ion beam is typically small, on the order of 1 cm^2 for medium implanters and 3 cm^2 for high-current implanters, it must be scanned to cover the entire wafer. Wafer scanning is done either by moving the beam over a stationary wafer or moving the wafers through a stationary beam. Scanning plays a critical part in dose uniformity and repeatability.

Traditionally, low- to medium-current implanters keep the wafer stationary, while high current implanters will keep the ion beam stationary. The different types of scanning systems used in implanters are:

(*i*) Electrostatic scanning
(*ii*) Mechanical scanning
(*iii*) Hybrid scanning
(*iv*) Parallel scanning

10.3.4.1 Electrostatic Scanning

Electrostatic scanning deflects the ion beam across a stationary wafer by applying a specific controlling voltage to a set of X-Y electrodes (See Fig. 10.7). When the polarity of one side of the electrodes is made negative, the positive ion beam will deflect toward it. By properly positioning the two sets of electrodes and constantly varying the voltage, the beam is made to deflect across a wafer. This type of scanning could be compared to spray painting a surface, where the spray needs to be worked back and forth many times to uniformly coat the surface. An electrostatic scan moves the beam across the wafer (X-axis) 15,000 times in one second while it scans up and down (Y-axis) 1,200 times per second. Special care is made to ensure the scan is uniform at the outer edges of the wafer where the scan actually has to stop and reverse directions. In electrostatic scanning systems, the wafer can be twisted and tilted relative to the ion beam to active desired junction characteristics, and to reduce channelling effects.

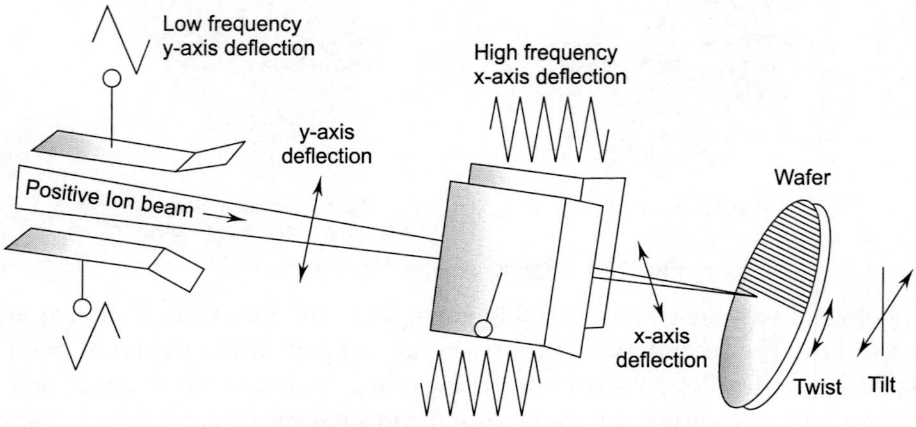

Fig. 10.7 *Electrostatic Ion Beam Scanning of Wafer*

Implanting a single wafer at a time is referred to as a serial process and is generally used for low-to-medium current implant applications. In a typical low current implant operation, a wafer might be scanned for 7 to 10 seconds to obtain uniform dose. Note that this same method of beam deflection can also be accomplished using an electromagnetic field approach instead of an electrostatic approach.

Because, the wafer is held fixed during electrostatic scanning with no elaborate mechanical assembly, there is less possibility of particulate contamination. Another advantage to this type of scanning is that electrons and neutrals are not deflected and are eliminated from the beam. The major disadvantage is that the ion beam does not strike perpendicular to the wafer, which leads to undesirable shadowing from the mask material that blocks implantation of the ion beam (See Fig. 10.8).

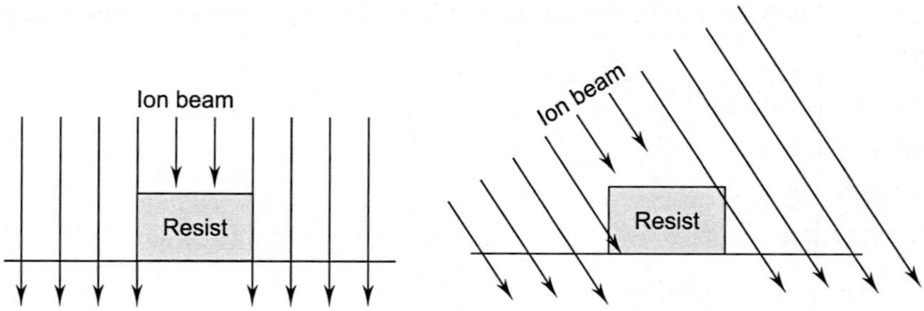

(a) Mechanical scanning with no tilt

(b) Electrostatic scanning with normal tilt

Fig. 10.8 *Implant Shadowing*

10.3.4.2 Mechanical Scanning

With mechanical scanning, the ion beam is fixed and the wafers are mechanically moved through the beam. This method is generally used for high-current implanters because electrostatic beam deflection is difficult at high currents and energies. The beam size may be 1 cm wide by 3 cm high. Mechanical scanning is done by rotating multiple wafers (*e.g.*, up to 25,200 mm wafers) fixed on the outer circumference of a large wheel assembly disk that is simultaneously moving up and down while rotating at 1000 to 1500 rpm (See Fig. 10.9).

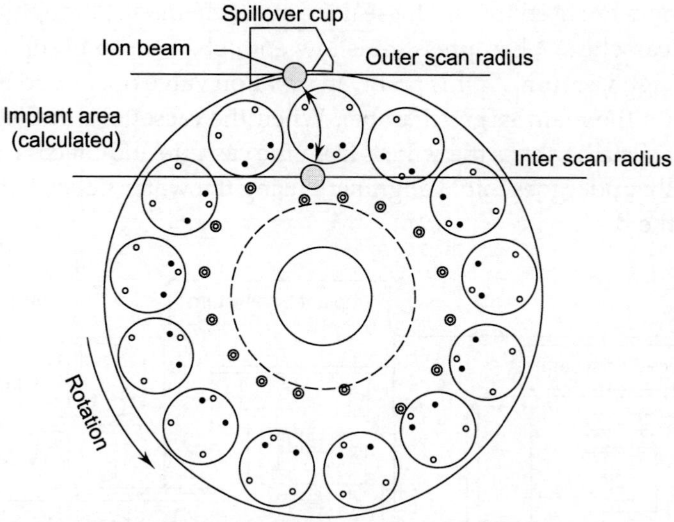

Fig. 10.9 *Mechanical Scanning of Implanted Wafers*

The disk diameter may be 5 feet or larger. While the disk is spinning, it is also moved up and down so that the beam scans from the outer edge to the inner edge of the wafer. The disk can also be tilted relative to the ion beam to prevent channelling through the interstitial spaces at the silicon lattice. Mechanical scanning is a batch process that effectively averages the beam power over a very large area. An advantage is the reduced wafer heating from the ion beam energy. However, there are potentially more particles generated from the mechanical assembly.

10.3.4.3 *Hybrid Scanning*

In the hybrid scanning system, the wafers are placed on a disk and rotated to scan in the y-axis. The beam is then electrostatically scanned in the X-direction. This method is generally used in a serial process for low-to-medium-current with a single wafer.

10.3.4.4 *Parallel Scanning*

In electrostatic scanning, the beam is not scanned perpendicular to the wafer surface, which causes shadowing. Parallel scanning is a method used to reduce shadowing and channelling by scanning the ion beam to within $< 0.5°$ of the wafer surface. In the parallel scanning method, the beam is first scanned electrostatically. Then, the beam passes through another set of magnets which corrects the beam angle and forces it to strike perpendicular to the wafer surface.

10.3.5 Process Chamber

The implantation of the ion beam into the wafer take place in the *process chamber*. The process chamber is a major implanter sub-assembly that includes the scanning system, *end station with* vacuum load locks for loading and unloading wafers, the wafer handler system, and the computer control system. There are also methods for dose monitoring and channelling control. The end station can be large if used for mechanical scanning. It is pumped down to the vacuum required for implantation with multiple roughing, turbo pumps, and cryo pumps to reach base pressure (*e.g.*, typically 10^{-6} for *r*).

Load/unload systems in the end station use robotic handling to move wafers between an input station and the scanning disk in the target chamber (See Fig. 10.10). Cassettes are loaded into the input rack and load locks are used to initially seal the input chamber. A roughing pump lowers the pressure around the cassettes. When pressure is low enough, a turbo pump continues the pump down until reaching high vacuum. At this point, an isolation valve opens and a robotic mechanism moves the cassette into the main target chamber. When the cassette is in the process chamber, a robotic wafer handler system moves the wafers from the cassette and places them on the scanning disk. The wafer usually undergoes initial alignment using the wafer flat or notch to ensure proper wafer orientation in the disk.

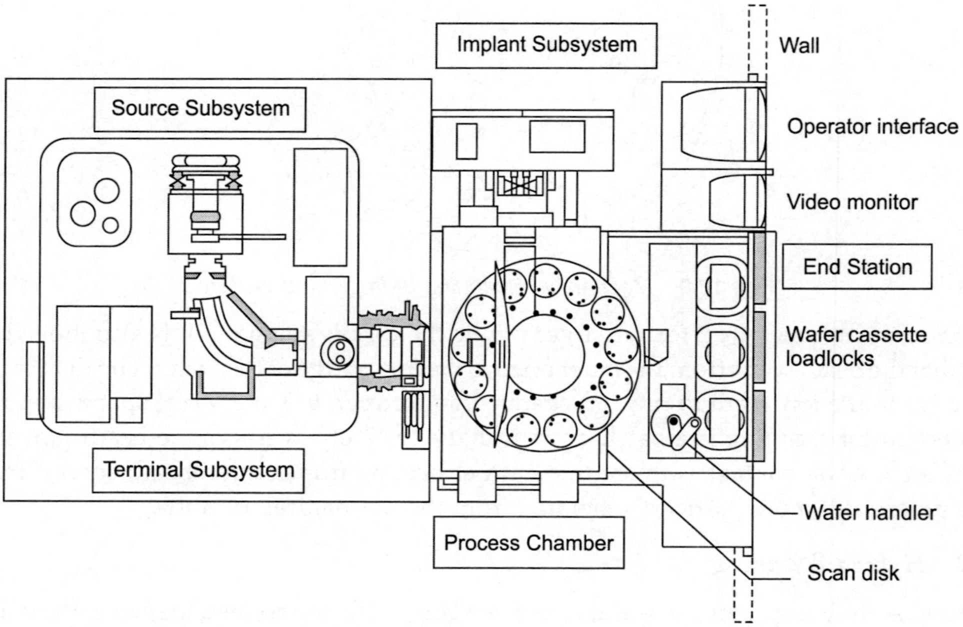

Fig. 10.10 *Wafer Handler for an Implant Process Chamber*

All subsystems are accessed by the technician through a central computer system. System status and information all obtained through the computer, including diagnostics and maintenance information.

10.4 ION IMPLANT TRENDS IN PROCESS INTEGRATION

Ion implantation requirements vary widely because of the different dopant requirement necessary for leading edge process. Examples of different implant process requirement for advanced MOS wafer fabrication are:

(*i*) Deep burried layer
(*ii*) Retrograde well
(*iii*) Punch through stoppers
(*iv*) Threshold voltage adjustment
(*v*) Source/drain implants
(*vi*) Silicon-On-Insulator (SOI).

10.4.1 Deep Burried Layers

Deep burried layers implanted in silicon with high energy ion implanters (> 200 KeV) are used in CMOS circuits for various reasons, such as the creation of reaction of retrograde wells (Fig. 10.11). A *triple well* has a burried implanted well layer beneath the standard n- and p- retrograde wells for improved device performance and packing density. An important reason for burried layers is to control latch-up, a reliability problem in CMOS circuits.

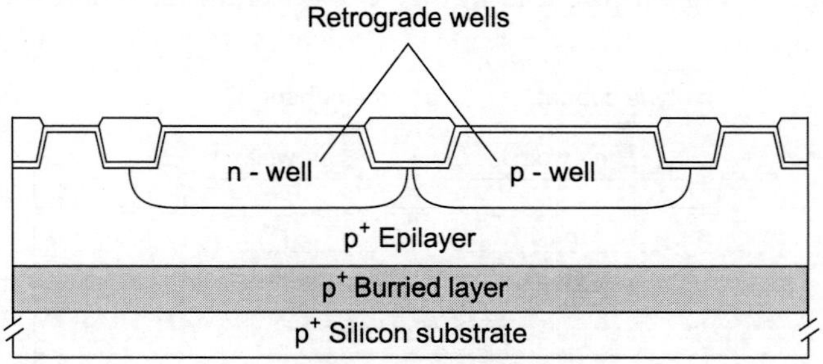

Fig. 10.11 *Burried Implanted Layer*

Latch up occurs when parasitic transistors cause the device to turn on unintentionally, leading to complete failure of the chip. Latch up is typically controlled in wafer fabrication by the use of an epitaxial silicon layer on the surface of the wafer. The low-resistance epilayer effectively shunts current to the ground plane. The epilayer also serve as an impurity getter. Because of the high cost of epilayer substrates, there is increased interest in the high-dose burried layer formed by ion implantation to replace the current shunting and impurity gettering of the epilayer.

10.4.2 Retrograde Wells

A well is a diffused area, where active devices (*e.g.*, MOSFET) are constructed. The *n*-type or *p*-type MOS transistor is placed in a well (or tub) of opposite conductivity type to form the semiconductor junction. An important design alternative for MOS devices is a *retrograde well*, which has the peak implanted dopant profile burried at a certain depth. This feature is shown in Fig. 10.12.

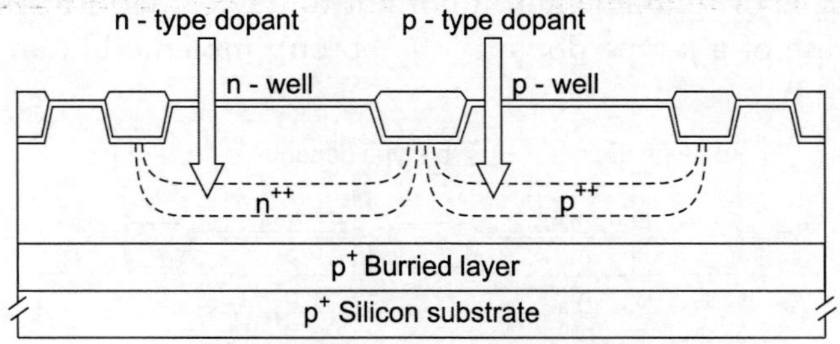

Fig. 10.12 *Retrograde Well*

Another term for a retrograde well is a vertically modulated well. A retrograted well is formed by a high-energy implant that produce high dopant concentration deep in the retrograde well to improve transistor immunity to problems such as a latch up and punch through.

10.4.3 Punch Through Stoppers

In submicron devices with short channel lengths, punch through stopper implants are necessary. *Punch through* is a condition, where the channel is shorted (punched through) with undesirable leakage current and leads to device failure. It occurs when the channel length is reduced for scaling (known as a short channel) and high electric fields occur at the drain end of the channel. In this case of an n-channel, the electrons moving from source to drain are accelerated by this high electric field and by energetic collision and create a free channel region and contact the source region field causing punch through. The punch through implant places dopants just below the active channel adjacent to the source and drain in order to carefully modify the well doping and prevent expansion of the drain depletion region into the lightly doped channel when the device is biased for operation (See Fig. 10.13).

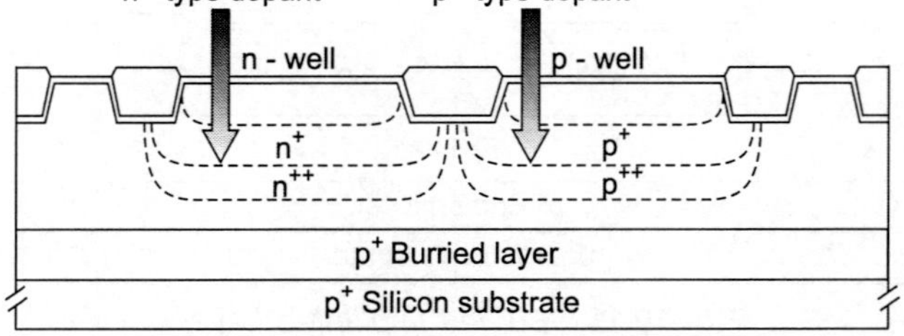

Fig. 10.13 *Punch Through Stop*

Precise control of the placement and dose of a punch through implant is necessary, with boron implant used for n-channel devices and phosphorus used for p-channel devices.

10.4.4 Threshold Voltage Adjustment

An MOS transistor consists of three distinct part : source, drain, gate. No current flows between the source and drain until the channel region underneath the gate becomes conductive when triggered by a voltage applied at the gate. The threshold voltage, V_{TH}, is the amount of voltage required to form the conductive channel between the source and drain that causes enhancement mode transistors to turn on. V_{TH} is very sensitive to the dopant concentration in the channel region. For optimum device performance, a dopant is implanted beneath the silicon layer to adjust the channel region to the required dopant concentration. This is called the MOS gate *threshold voltage implant adjustment* and is of fundamental important to device performance. To illustrate this adjustment take the case of a *p*-type dopant (*e.g.*, boron) implanted in an n-channel region to adjust V_{TH} (See Fig. 10.14).

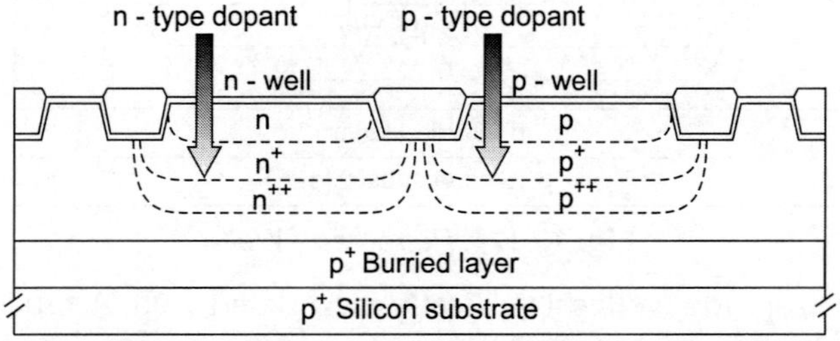

Fig. 10.14 *Implant for Threshold Voltage Adjustment*

The increase in the *p*-type dopant concentration will causes a change in the necessary V_{TH} (*e.g.*, V_{TH} increases from 0.9 V to 1.0 V). Threshold voltage adjustment with ion implantation was the first width spread use of implanters in wafer fabrication due to the implantation's ability to produce uniform, repeatable dopant concentrations.

10.4.5 Source/Drain Implants

Source/Drain (S/D) implants form highly doped regions (10^{20} to 10^{21} ions/cm^3) that interface with the lightly doped active channel and well regions (10^{16} to 10^{17} ions/cm^3) of an MOS transistor. S/D regions are doped to have the opposite conductivity type as the well that surrounds them. Arsenic implants are often used to form *n*-type source/drain for n-channel (*n*MOS) transistors, and boron or BF_2 implants are usually used to form *p*-type source/drain for *p*-channel (*p*MOS) transistors. By using implantation with its excellent control of doping placement, the lateral diffusion of the source/drain dopant ions into the channel region is minimized.

10.4.6 Silicon-On-Insulator (SOI)

A form of vertical device isolation that is becoming more important in advanced wafer fabrication is *silicon-on-insulator* (SOI). SOI technology is based on an insulator layer (oxide) burried within the silicon that effectively isolates the devices on the silicon surface. It was developed in the 1970's but has not been widely used due to process complexity and cost. SOI offers several advantages for deep submicron CMOS applications, including the complete elimination of latch-up, reduced electric fields to minimize hot carriers, and reduced parasitic capacitance. The concept of forming single-crystal silicon on an oxide layer (or other insulator material) is difficult because the dielectric material's crystalline properties are so different from pure silicon. If not properly controlled, this difference in crystal structure can lead to crystal defects in the silicon that affect device performance. The most widely used SOI technology is SIMOX.

10.4.7 SIMOX

At the time of this writing, SIMOX is the leading commercial process for SOI. In the *SIMOX (Separation by Implanted Oxygen)* process, a well-defined horizontal oxide layer is burried in the silicon wafer. This is done by implanting a high concentration of oxygen atoms into the wafer, traditionally using a high-energy implanter (*e.g.*, a 200 KeV oxygen implanter). The implant step is followed by a high-temperature thermal anneal (*e.g.*, 1300°C) to react the burried oxygen with silicon and form a continuous SiO_2 layer under the thin silicon surface.

This burried oxide layer is typically about 50 to 500 nm thick and serves as an excellent device isolation layer. The anneal process also regenerates the crystalline quality of the silicon layer remaining over the oxide. It should also be noted that there are new SIMOX techniques in development using low-energy, low-dose oxygen implanters that produce burried layers with improved dielectric properties.

10.5 ANNEALING

Ion implantation damages the silicon lattice by knocking atoms out of the lattice structure. With high dose, the implanted layer actually becomes amorphous. Furthermore, the implanted ions rarely enter the lattice structure sites of the silicon, instead stopping in interstitial sites outside the lattice. These interstitial dopants are electrically inactive until activated by a high-temperature *annealing* step. Annealing heats the implanted silicon substrate to repair crystalline damage and electrically activate dopants by moving the atoms into crystal lattice sites (See Fig. 10.15).

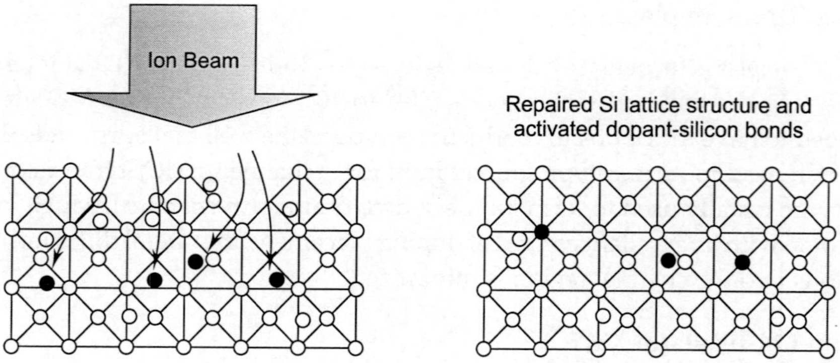

(a) Damaged Si lattice during implant (b) Si lattice after annealing

Fig. 10.15 *Annealing of Silicon Crystal*

Crystal damage repair is done at 500°C and dopant activation occurs at about 950°C. Electrical activation of dopants occurs as a function of time and temperature, with longer times and higher temperature increasing the dopant activation. There are two basic methods for heating the implanted wafer for anneal :

(*i*) Furnace anneal ; and

(*ii*) Rapid thermal anneal (RTA).

10.5.1 Furnace Anneal

Furnace anneal is the traditional method of annealing by heating the wafer in the hot wall furnace in the range of 800 to 1000°C for approximately 30 minutes.

At this temperature, the silicon atoms move back into the lattice sites and the dopant atoms enter substitutional sites in the lattice. However, annealing cycles at this time and temperature can cause extensive dopant diffusion and is undesirable for advanced IC wafer fabrication.

10.5.2 Rapid Thermal Anneal (RTA)

Rapid thermal anneal (RTA) anneals the wafer by using an extremely fast ramp and short dwell time at the target temperature (typically 1000°C). Annealing of implanted wafers is usually done in a single-wafer rapid thermal processor (RTP) with argon (Ar) or nitrogen (N_2) flowing into the chamber. A fast ramp and short dwell time is optimized to anneal the wafer to restore lattice damage and electrically activate the dopant while minimizing dopant diffusion in the silicon. RTA also minimizes a phenomenon known as transient enhanced diffusion. RTA is the optimal anneal method to achieve acceptable junction depth control in shallow implants.

Investigations have been done into spike anneals for RTA. The ramp-up rate is very fast at up to 150°C/sec. This speed permits a soak time reduction to about 1 second. Uniform ramp rates during a spike anneal are critical because much of the thermal exposure occurs during the ramp.

The RTA equipment design is similar to the rapid thermal processor previously discussed. There are different types of heating, with a common method being tungsten halogen lamps positioned on both sides of the wafer. Multiple optical pyrometers are often used to measure temperature across the wafer to ensure uniformity.

10.6 CHANNELLING

Single-crystal silicon atoms exhibit long range order, lying in rows and planes throughout the crystal. *Channelling* occurs when the implanted ions are not slowed by collision with silicon atoms and instead pass through the interstitial area of the crystal. The undesirable aspect of channelling is that the expected projected range has a very large spread, especially for shallow implants at low ion beam energies. Dopant ions passing through lattice channels do not encounter electronic or nuclear stopping (*i.e.*, have little energy loss) and therefore penetrate deeper than those ions that are stopped by the lattice atoms. There are four ways that channelling is controlled during implant :

(*i*) Wafer tilt
(*ii*) Screen oxide layer
(*iii*) Preamorphization of the silicon.
(*iv*) Using dopants with greater amu's.

10.6.1 Wafer Tilt

The most widely used method to reduce channelling during ion implantation is wafer tilt. This method orients the wafer at an angle relative to the incident ion beam. A common angle for <100> wafer is 7° from the perpendicular and ensures that a collision occurs within a short distance of the dopant ion entering the silicon. The tilt angle is set during the scanning process. In this manner, the crystal lattice presents a dense orientation to the incident beam to obtain better control of the projected range on the implanted ion. Note that channelling behaviour is different at low energies (< 1 KeV) for ultra shallow junction, where tilt has little effect on reducing channelling. Furthermore, tilt angles increase shadowing effects and may lead to asymmetric device behaviour. It is also necessary sometime to twist (or rotate in the vertical plane) the wafer surface. For the <110> crystal orientation, twisting the wafer 15° to 35° can help reduce channelling.

There are situations, where high-current implantation of shallow source/drain extension (*e.g.*, lightly doped drain) may require high-angle implants under the photoresist mask to obtain a shallower depth. Modern serial process (single-wafer) implanters underneath wafer features and onto trench side walls. Trench side wall doping is required for applications such as DRAM capacitor cell in order to form a conductive capacitor plate along the side wall.

10.6.2 Screen Oxide Layer

In some cases, a thin oxide layer (10 to 40 nm) is grown or deposited as a *screen oxide layer* on the wafer surface prior to implantation. It is also sometimes referred to as a *sacrificial oxide,* since it is deposited for the implantation process and then removed. Implanting through this amorphous oxide layer randomizes the directions of the ions as they enter the silicon lattice to reduce the channelling effects.

Randomization occurs due to the dopant ion impining on the SiO_2 molecules. Investigations indicates that a screen oxide is not always effective at reducing channelling and lead to dose uniformity problems. Variables that affect how successful a screen oxide is to control channelling are the implant energy, dopant species, the screen oxide thickness, and the orientation of the incident ion beam.

10.6.3 Preamorphization

A technique for reducing channelling is preamorphization of the single-crystal silicon lattice with an electrically inactive species, usually Si^+. This step is done prior to implantation and serves to destroy the single crystal structure in a thin layer of the silicon surface. The subsequent implant with the dopant ion into the amorphous crystal structure leads to minimal channelling. The implantation is followed by a thermal anneal to correct the lattice damage. Investigations show that preamorphization is especially important for shallow profiles at energy level below 1 KeV. The biggest disadvantage of preamorphization is that it involves an additional implant step that increases production cycle times.

10.6.4 Particles

Ion implantation is sensitive to particulate contamination. A particle located on the wafer surface can block the incident ion beam and cause improper implantation (See Fig. 10.16).

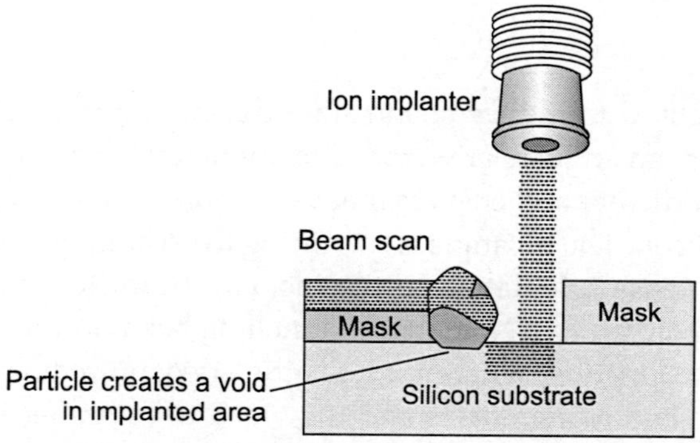

Fig. 10.16 *Implantation Damage from Particulate Contamination*

High current implanter create more particles due to erosion of beam line components. Implanter suppliers design the target chamber to reduce the high temperature from the high. Current beam and minimize particle formation.

Particles are usually monitored or unpatterned wafers to calculate the particles per wafer pass and take corrective action when necessary.

10.7 APPLICATIONS

Ion implantation is commonly used in the fabrication of both bipolar and MOS (metal oxide semiconductors) integrated circuits.

⇒ Some of the standard MOS applications of ion implantation are as follows:

1. It provides a means of accurately adjusting the threshold voltage of MOS devices. Historically, this has been the most important application because it made possible the low-cost production of *n*-channel MOS circuits.
2. It is commonly used as an alternative to predeposition in the fabrication of MOS circuits.
3. The resistivity of the starting silicon can be fixed by ion implantation.
4. In MOS devices, the overlap of the gate electrode over the drain region creates a parasitic feedback capacitance and it deteriorates the frequency response. Through ion-implantation source and drain can be built up to the gate edge after the diffusion has been conducted.

5. *p*-well can be formed by ion implantation in CMOS.
6. In MOS technology process control of low dose levels is required. The unique masking possibilities with ion implantation techniques allow considerable flexibility in the process.

⇒ Some *bipolar* applications of ion implantation are as follows:

1. It is used to form high-value resistor. Conventionally, resistors are formed with transistors bases and have sheet resistance of about 200 Ω/O. Large resistor values thus require large areas. With ion implantation, sheet resistance on the order of 100 KΩ/O can be obtained.
2. It can be used as an alternative to predeposition.
3. Accurate impurity profiles can be obtained for special-purpose bipolar devices. Through accurate profile control, ion implantation is often used to "fine tune" device parameters.
4. In avalanche transit time diode consisting of $p^+ - n - n^+$ structures p^+ diffusion is done into an n-epitaxial layer. Ion implantation is used for the p^+ region so that the width of the n layer can be controlled by epitaxial growth process.
5. By Ion-implantation burried layer can be formed easily.
6. Ion-implantation is useful in selective placement of doped regions on the same slice. High performance vertical pnp transistors in the same microcircuit as npn devices can be fabricated.
7. Emitter push effect as in diffusion process can be avoided by ion implantation.
8. No lateral diffusion take place here when very small emitter region formed by this method.

SUMMARY

Ion implantation is a doping process which introduces a dopant into silicon to modify its electronic properties. Ion implantation replace the diffusion because ion implantation has many advantages over it. Diffusion is the movement of one material through another to a concentration gradient. Ion implantation is a physical process that uses an implanter tool. Two important ion implant parameters are dose and range. The beam current is used to define the dose. The range is the total distance a dopant travels in the wafer and is related to energy and the mass of the dopant ions. The implanted ions are eventually stopped in the wafer due to collision or interaction with silicon atoms. There are six major subsystems of an implanter tool : ion source, extraction assembly and ion analyzer, acceleration column, scanning system and target chamber. The ion source generates ions for implantation. The extraction assembly extractions from the source and the analyzer magnet separates the desired dopant from the ion species. The accelerator accelerates the dopants, which are scanned on the wafer with the scanning system in the target chamber.

Wafer are annealed in either a hot-wall furnace or RTP to repair crystal damage and electrically activate dopants. Channelling is the passage of dopants through interstitial crystal areas, which leads to non-uniform junction depth. It is controlled by wafer tilt, preamorphization or screen oxide layer.

TUTORIAL SHEET

1. Give a short description of ion implantation.
2. Explain the common dopant processes in wafer fabrication.
3. What are the steps typically followed by ion implantation process?
4. Describe a doped region and dopant profile.
5. Give a general explanation of an ion implanter.

6. List the advantages of an ion implanter over diffusion.
7. What is the major disadvantage of ion implantation and how is it corrected?
8. Define dose. State and explain the dose equation.
9. What is range? Describe the relationship between energy and range.
10. List four classes of ion implanters, and give a short description of each one.
11. Describe the two primary energy loss mechanisms during implantation.
12. What is the purpose of an ion source? What is the most common source feed material?
13. Discuss the purpose of the extraction assembly and explain its operation.
14. What is the purpose of suppression electrode?
15. How does the acceleration column increase ion beam energy?
16. What does the beam energy define ? What is dose related to in ion implant?
17. Why is a high-current and low energy beam desirable? How is a high beam current achieved in a low energy beam?
18. List and give a explanation of the four types of scanning system.
19. Which type of scanning is used for low current application? High-current applications?
20. What is the purpose of annealing? Is it preferred to perform a furnace anneal or an RTA.
21. Describe channelling. List and give explanation of the three mechanisms for controlling channelling.
22. Why is SIMOX a potentially beneficial device fabrication technology? What is SOI?

11

METALLIZATION

11.1 INTRODUCTION

Metallization is the fabrication step in which proper interconnection of circuit elements is made. In this process, the metal layers electrically interconnect the various device structures fabricated on the silicon substrate. Aluminium is a popular metal used to interconnect various devices, both to make ohmic contact to the devices and connect these bonding pads on the chip edge. It is more suitable for this purpose because of its low resistivity and its adhersion compatibility with SiO_2. It can be easily vacuum deposited, and has high conductivity. In addition to pure aluminium, alloys of aluminium are used to form IC interconnections for different performance related reasons. For example, small amount of copper is added to reduce the potential for electromigration effects (in which the current applied to the devices induces mass transport of metal). Small amounts of silicon are also added to aluminium metallization to reduce the formation of metal "spikes" that occur over contact holes (discussed in the later part of chapter).

In the bipolar transistor, the central region (semiconductor) base is directly in contact with metal and the emitter and collector are also directly in contact with the metal. The metal plays very important role in all the three regions in determining the current flow characteristic. The applications of metallization can be divided into the three parts : (*i*) gate ; (*ii*) contact ; (*iii*) interconnects. In interconnections metallization, fine line metal patterns are used to connect the thousands of MOSFET or BJT devices in IC. For gate, polysilicon film is used for metallization and interconnection of MOS devices. All metallization directly in contact with semiconductor is called contact metallization. There are some desired properties of metals listed below for metallization. The reliability and stability of the material are two most important desired properties throughout the process and the operation of the devices.

The desired properties of metallization for Integrated Circuits are:

(*i*) Easy to form
(*ii*) Easy to etch
(*iii*) Should be oxidizable and stable in oxidizing ambients
(*iv*) Low resistivity

(*v*) Mechanical stability : good adhereness and low stress
(*vi*) No reaction with find metal (aluminium)
(*vii*) Should not contaminate devices, wafer, or working apparatus.
(*viii*) Surface smoothness
(*ix*) Stability throughout processing, including high temperature sinter, dry or wet oxidation gettering etc.
(*x*) Good device characteristics and lifetimes.
(*xi*) Environmentally safe material during processing and actual use, and recyclable.
(*xii*) Low cost.

11.2 SINGLE LEVEL METALLIZATION

In the single level metallization process, the small holes called contact holes are etched through the surface layer (*a*) and open the area on the device where contact is to be made for the interconnection (*b*) and a thin layer of conducting material is deposited on the wafer with the help of vacuum evaporation, CVD or sputtering techniques (*c*). The unwanted material portions of this layer are etched by photomasking method or by lift off. This step gives the thin lines of the metal that are known as interconnections of metal lines (*d*). After the step (*d*), the heat treatment, step called alloying is performed after metal patterning to ensure good electrical contacts between the metal and the wafer surface.

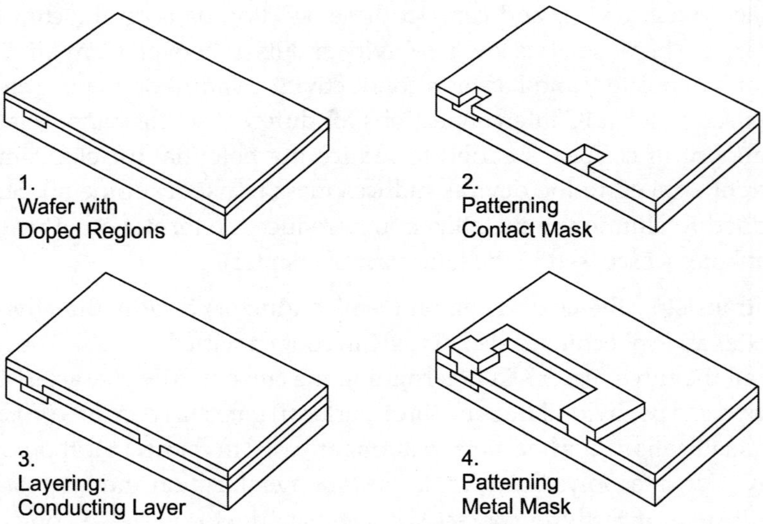

Fig. 11.1 *Single Level Metallization*

11.3 MULTILEVEL METALLIZATION

As integral circuits feature sizes continue to scale well below 0.25 microns, active device counts are reaching hundreds of millions. The amount of interconnect among the devices tends to increase linearly with : the number of transistors and club area is often limited by physical interconnect area. So due to this limitation, the interconnections dimensions are scaled down with device size and it results to move the technology of the interconnections to multilayer connections. In this, the connection of the various devices are made on the different layers and these layers are inter connected by vias, but this process creates problems with signal integrating and interconnect delay.

The Fig. (11.2) shows typical two metal layer structure of a device. This starts with a burried layer which is formed by silicidation of the silicon surface to produce a lower electrical resistance between surface and the next layer then a dielectric material layer is deposited over it, to provide electrical isolation between the metal layer, that is why it is called intermetallic dielectric layer (IDL). This dielectric material may be silicon dioxide, silicon nitride, or a polymide film and a mask is used to etch the area for the new contact hole called vias or plugs, down to the first metal lays and then a conducting material is filled into these vias or holes and a conducting path is created from the first level of metal layer to the next layer. The IPL/plug/metal deposition/patterning sequence is repeated for subsequent layers. A multilevel metal system is more costly and requires greater attention for planarization of the surface of the wafer and of low yield.

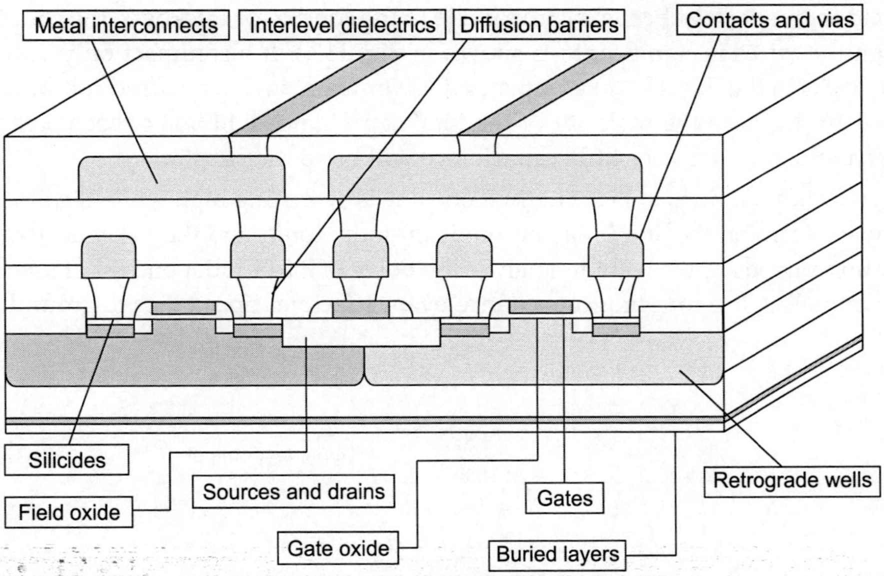

Fig. 11.2 *Multilevel Metallization*

11.4 EFFECTS OF MULTILEVEL INTERCONNECTIONS

As the levels of the metallization increases, the density of the wires also increases with increase in the number of devices on each chip. In these IC's, the wires are electrostatically coupled to nearest neighbours on the same layer, as well as above and below it as shown in figure.

If we scale all the dimensions of the IC by S, the capacitance per unit length among the wires remains unchanged. The resistance is increased by $\frac{1}{S^2}$ factor for per unit length of the wire. So, RC (factor) per unit length is increased by same factor if all of the lengths of interconnect are scaled then the total RC for the interconnect remains unchanged. But the device size reduced, and operating speed of the gates is improving, the delays of the interconnects between the gates remain fixed. This can translate into an increase in interconnect delay, relative to the driver delay, even when lengths are scaled.

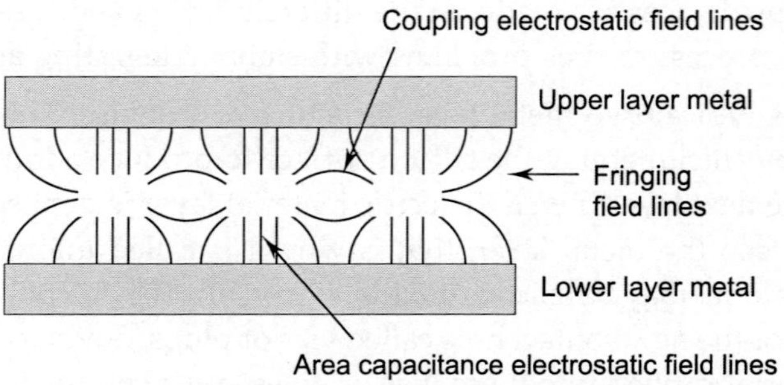

Fig. 11.3 *Multi-level Metal Interconnect Cross-section. Wires are Alternated in Direction from Layer-to Layer*

11.4.1 Coupling Capacitance and Inductance Effects

When the conductor spacings become comparable to the conductor thickness, the coupling capacitance between the wires is significant (as shown in Fig. 11.4). It has impact on performance and signal integrity. As in the Fig. 11.4(*a*) a upper and lower metal layers are used for the current flow or signal flow. In this case the majority of the total capacitance will be between signal wires for multilevel technology and a very little capacitance will be to the substrate.

Consider two coupled lines in Fig. 11.4(*b*) if one line is switching high while the other is switching low, the wave form on the line 1 may become non-monotonic, and the delay is increased. If, for delay calculation purposes, we want to analyze the delay of line 1 independent of line 2. Then, we can consider modeling the coupling capacitance by an effective capacitance to ground.

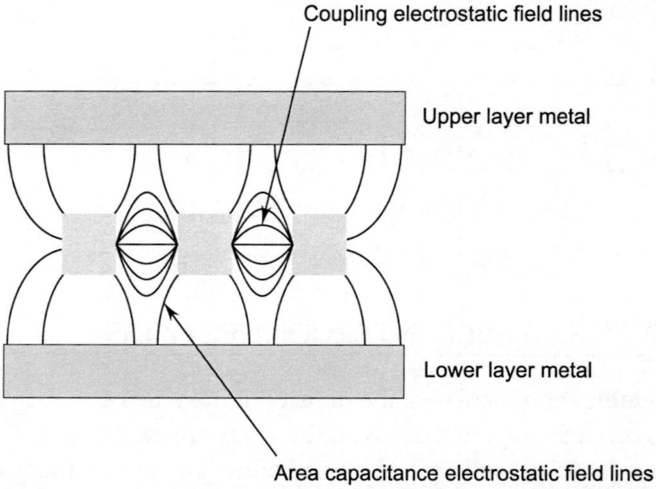

Fig. 11.4 (a) *Coupling Capacitance between the Same Layer Wires*

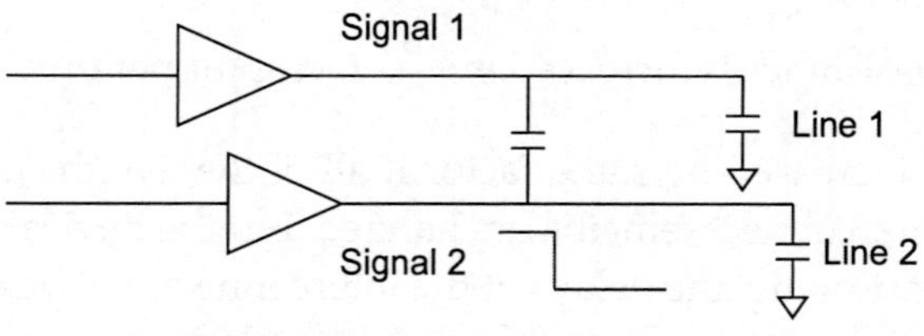

Fig. 11.4 (b) *Two Coupled Lines*

The internal inductance of on-chip metallic wires needs to be considered because the skin depth at the frequencies that we consider are comparable to the wire thickness or width. Most electrical current flow inside the wires due to lack of highly conductive ground plane on the chip or we can say the upper layer of the metallization. In this the mutual coupling between wires cover very long ranges and decrease very slowly with the increase of spacing. The resistivity of the lightly doped substrate layer is about two million times larger than aluminium and substrate is too far away from the high buses or clock wires.

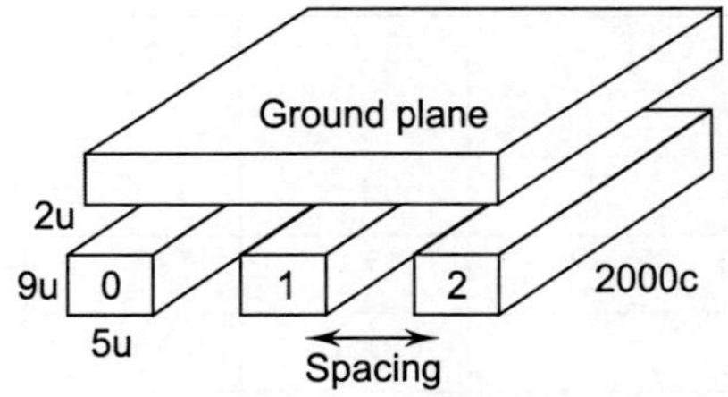

Fig. 11.5 *Coupling Inductance*

11.4.2 Local Interconnects and Global Interconnects

When technology trends are investigated, usually important distinction is made between two types of interconnects wires. Local wires, which connect gates or cells within blocks and with scaling these wires gets shorter. Global wires connects together and usually span a significant portion of the chip. They do not shrink when devices get smaller and even tend to increase with increasing chip sizes.

Local wires do not have significant RC interconnect delays. But, the total capacitance can be dominated by interconnect capacitance since wire capacitance does not decrease with scaling, whereas device capacitance does. Furthermore, when C dominates coupling can be a big factor and in effect, it can scale the capacitances several times nominal value. This affect both local and global wires.

11.5 METALLIZATION MATERIALS AND THEIR PROPERTIES

It is very difficult to select a metal for metallization of devices in the IC fabrication technology. The selected metal should have all the desired properties, which satisfy the condition for metallization. The primary metallization was aluminium for the development of VLSI-level circuits.

Choice of Metals

The selection of metallization materials depends on the requirement or applications for the Interconnections contacts, Diffusion barrier, metallic contact etc., (applications and the number of choices available are shown in Table 11.1)

Table 11.1: Choice of Metals for Matallization

S. No.	*Applications*	*Choices*
1.	Gates, interconnections and contacts.	Polysilicon, silicides, nitrides, refractory metals, Al or Cu and Alloys
2.	Diffusion barrier	Nitrides, carbides, borider T_1-W alloy,
3.	Selectively formed metallization on silicon.	Al, tungsten, silicides
4.	Top level	Al.

Some metals and their properties for metallization are given as :

Table 11.2: Metals Propertirs

Property/Metal	*Al*	*W*	*Cu*	*Ag*	*Au*
Resistivity	2.66	5.65	1.67	1.59	2.35
Youngs modulus ($\times 10^{-11}$ dyn cm^{-2})	7.06	41.1	12.98	8.27	7.85
Thermal conductivity (Wcm^{-1})	2.358	1.74	3.98	4.25	3.15
Coefficient of thermal expansion CTE $\times 10^6$ (°C^{-1})	23.5	4.5	17	19.1	14.2
M.P.(°C)	660	3387	1085	962	1064
Specific heat capacity (Jkg-1K-1)	917	138	38	234	132
Corrosion in air	Good	Good	Poor	Poor	Excellent
Adhesion to SiO_2	Good	Poor	Poor	Poor	Poor
Delay	3.7	7.8	2.3	2.2	3.2
Thermal stress per degree for films on Si ($\times 10^7$ dyn cm^{-2} °C^{-1})	2.1	0.8	2.5	1.9	1.2

11.5.1 Aluminium

Aluminium is very suitable for the purpose of metallization which has low resistivity and adhersion compatability with SiO_2. Aluminium alloys such as Al-Cu preferred over the pure aluminium for metallization because these inhibit problems like electromigration and junction spiking (to be discussed later in the chapter). One another disadvantage of aluminium as the metallization material is its low melting temperature (660 degree °C) and the low Aluminium silicon (Al-Si) eutectic temperature 577 deg C. At this temperature, aluminium and silicon dissolve into each other. This type of eutectic formation occurs when two materials heated in contact with each other melt at temperature much lower than their individiual melting temperature. It starts at 450°C and this restricts the maximum processing temperature once the Al layer has been deposited. (See Fig. 11.6).

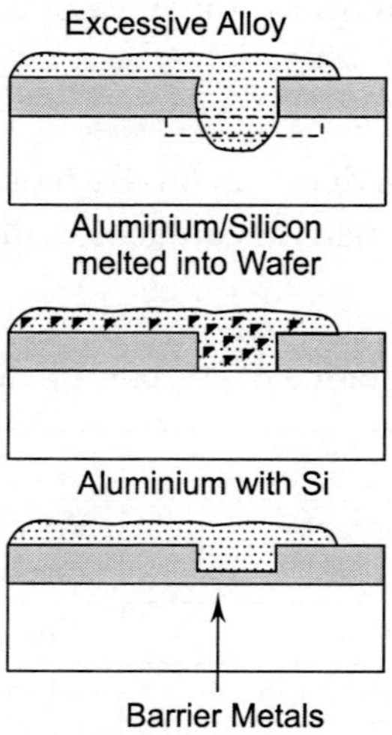

Fig. 11.6 *Eutectic Alloying of Aluminium*

Aluminium metal layers are usually deposited through Physical Vapour Deposition (PVD) by sputtering. Sputtering may be described as a series of four steps :

1. High energy ions are generated and are bombard a target (the source of material for deposition as shown in Fig. 11.7(*a*).
2. The ions sputter (eject) atoms from the target.
3. The sputtered atoms reach the substrate.
4. The sputtered atoms condense and form a thin film over the substrate.

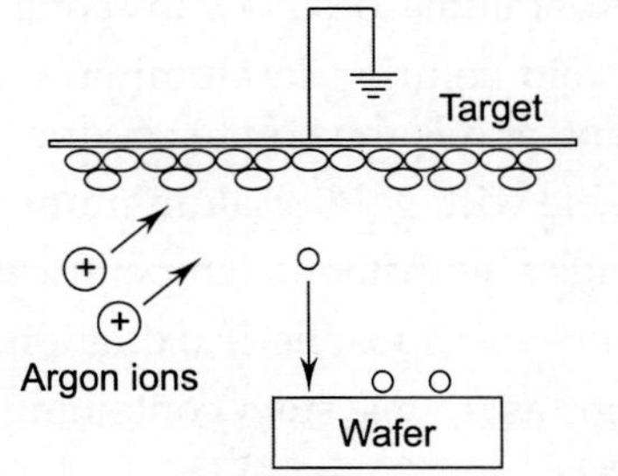

Fig. 11.7 (*a*) *Sputtering Process*

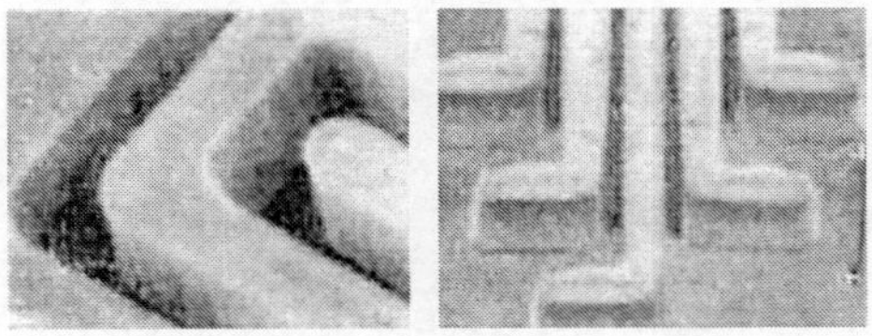

Fig. 11.7 (*b*) *Metal Lines After Deposition*

Upon exposure to oxygen, Al readily forms a native thin oxide on its surface (Al_2O_3). Even at ambient temperature. The presence of such oxide layer can increase the contact resistance of Al layer. It can also inhibit the sputtering of an Al target or etching of an Al thin film, resulting in processing difficulties.

Aluminium can also suffer from corrosion. All it take in the presence of a corrosive contaminant and moisture. For instance, if phosphorous doped silicon dioxide is deposited over aluminium lines, phosphoric acid can result. If moisture in gress through the glass occurs, the acid will lead to corrosion of the Al lines. The two ways of avoiding this problem are rinse the wafers immediately in water or to proceed, in "situ with O_2 plasma etch, which removes the resist and passivate the aluminium surfaces.

Advantages of Aluminium

The advantages of aluminium as a metallization system are listed below.

1. Aluminium can be used in a single-metal system which provides considerable simplicity compared to multimetal systems.
2. Aluminium is a low-cost source material.
3. Aluminium films have a high electrical conductivity, close to that of the bulk material.
4. Aluminium evaporation by a resistance heated tungsten coil is simple.
5. Aluminium adheres well to oxides.
6. Aluminium layers delineate well by photolithographic techniques.
7. Aluminium is easily etched without dissolving SiO_2 or elemental silicon.

8. Aluminium reacts with residual SiO_2 in contact cuts.
9. Aluminium forms low-resistance contacts to n+-type silicon, and to p-type silicon.
10. The appreciable solid solubility of silicon in aluminium presents the possibility of dissolution of silicon by a solid-state diffusion-type reaction. Al-Si alloys can be used to adjust the amount of penetration at alloying and reduce it compared to that of pure aluminium.
11. There are no compounds in the Al-Si system : thus brittle compounds are not possible, and void formation (Kirkendall effect) has not been a problem at Al-Si interfaces.
12. Silicon in solid solution in aluminium does not substantially lower its electrical conductivity.
13. Silicon recrystallized from solid solution in aluminium contains a substantial amount of aluminium as a p-type dopant, and is thus quite conductive.
14. Aluminium is readily bondable with gold or aluminium wire.
15. Aluminium is ductile and thus a withstands temperature cycling.
16. Aluminium has high resistance to atmospheric oxidation.
17. A single-metal system, in contrast to a system containing several dissimilar metals, cannot result in galvanic action effects.
18. The lowest possible melting point in the Al-Si system is the Al-Si eutectic (577°C).
19. Aluminium is a satisfactory metallization material for radiation-hardened integrated circuits.

Disadvantages of Aluminium

The disadvantages of aluminium metallization are listed below.

In many cases, methods exist for overcoming the limitations, and these are briefly described.

1. Aluminium is not readily chemically vapour plated in Al^+
2. Aluminium is not readily electroplated.
3. Electromigration can form voids in aluminium at current densities somewhat lower than those required in other metals.
4. Electromigration may form spikes which cause shorts through deposited dielectric in multi-level arrays.
5. Aluminium galvanically corrodes with dissimilar metals.
6. Aluminium recrystallizes at fairly low temperatures, causing hillocks of metal which could break through a dielectric or be difficult to cover.
7. The reaction of aluminium with SiO_2 can be significant at temperatures of the order of 500°C.
8. Aluminium forms compounds with gold which tend to be brittle and can result in decreased electrical conductivity.
9. Gold wire bonded to aluminium results in reliability problems due to Kirkendall effects form which is of high conductivity.
10. Aluminium is a soft metal and is easily scratched.
11. Silicon dissolved during contact alloying may precipitate at grain boundaries in the aluminium and cause reliability problems. In addition, it may present a bonding problem.
12. Aluminium is typically under tensile stress as deposited ; other metals are, however, also under stress as deposited.
13. The thermal coefficient of expansion of aluminium is substantially greater than that of silicon; this, in turn, results in considerable stress in the aluminium during temperature cycling of integrated circuits.

14. In multilevel-metallization structures, the problem of making good aluminium-to-aluminium contacts has not been completely solved.
15. Aluminium is electronegative and corrodes in electrolytes.
16. Aluminium is not directly solderable by conventional soldering techniques.
17. Aluminium has a higher work function than some metals, and thus results in higher thresold voltages in MOS devices.
18. Aluminium dissolves to a small extent in solutions used to etch vias in deposited dielectrics in multilevel metallized structures.

11.5.2 Copper

Another metal used for the metallization is copper. It is a better conductor than aluminium with resistance of 1.7 μΩ-cm, compared to a 3.1 μΩ-cm value for Al. All the current circuits are developed with copper metallization and low-k dielectric. It increases performance and reduces the number of metal layers required. Mostly, it can be deposited at low temperatures and is resistant to electromigration and copper based circuits will operate four times faster than the Al. To keep the copper from migrating into dielectric and poisoning it, a diffusion barrier was used, which lined the trench walls between the copper and the substrate. Motorola used titanium nitride as a barrier other possible barrier material include tatalum alloys and tatalum nitride.

IBM and Motorola produced their copper wired chips by electroplating the copper over the diffusion barrier. Although, ion-resisted technique can be used either in conjuction with electroplating or itself, it has several advantages over electroplating. It can produce thinner, more uniform layers of metals in a variety of architecture. It can be used in narrower trenches with higher depth to width aspect ratios. It can fill trenches from the bottom up ; automatically eliminating uneven deposition that can lead to voids in the metal lines or it can produce conformal thin films that mirror the shape of the patterned wafer.

Film consisting of multiple layers are readily deposited using different cathode materials. Copper, tantatam, tantalum nitride and a variety of other materials can be applied in this way. Copper metallization, for example, may begin by depositing a conformal film of tantalum 20 to 50 nm thick. Ions of copper are then deposited on the tantalum layer. The process can be halted when the new material has formed a thin conformal coating, or the deposition can be continued until the trench is filled completely. Another possibility is to use the thin copper layer as a "seed layer" and fill the trench electrochemically.

Advantages of Copper:

1. It has higher conductivity and reduced in resistivity.
2. Superior resistance to electromigration.
3. Higher ultimate tensile strength.
4. Tighter packing density (narrow line)
5. Higher melting point, and thermal conductivity
6. Fewer process steps – 20 to 30% fewer steps with damascene technique.

Disadvantages:

(*i*) Lack of feasible dry etching
(*ii*) Poor adhersion to dielectric material such as SiO_2 and low k polymers.
(*iii*) Lack of self passivating oxide similar to Al_2O_3 on Al.

Table 11.3: Comparison of Properties/Processes Between Al and Cu

Property / Process	*Al*	*Cu*
Resistivity (μΩ-cm)	2.65 (3.2 for Al-0.5% Cu)	1.678
Electromigration resistance	Low	High
Corrosion resistance (in air)	High	Low
CVD processing	Yes	No
CMP (chemical mechanical planarization) processing	Yes	Yes

Major Challenges in Using Copper Interconnects

1. Copper diffuses quickly into oxides and silicon.
2. Copper cannot be easily patterned using regular plasma etching techniques.
3. Copper oxidizes quickly in air at low temperatures (< 200°C) and does not form a protective layer to stop further oxidation.
4. Copper affects SiO_2 properties
5. Copper affects most insulators properties
6. Copper adheres poorly to bottom and side ILD.

11.5.3 Low Dielectric Constant

As integrated circuit (I_c) dimensions continue to decrease, RC delays, cross talk noise power dissipation of the interconnect structures become limiting factors of VLSI integrated circuits. Materials with low dielectric constant are being developed to replace silicon dioxide as ILD (Inter-level dielectric). The number of low-k dielectric materials have been developed to meet VLSI circuits needs. The major categories are oxide based materials like poly-alyene-ethers (PAE) or Hydrido-organic. Silo xane-spin on process can provide great. Uniformity and planarity. It is also less expensive than CVD processes. Some of the low-k dielectric materials are given in Table 11.4.

Table 11.4: Low-k Materials.

Metal system	*Low - k material*
Aluminium	parylene
	HSQ
	methyl silesequioxane
	F-doped oxide
	F-doped amorphous carbon
	Parylene-F (AF_4)
	Xerogel
Gold	polyimide
	BCB
	Xerogel

Barrier Layer

In the contacts, the barrier layer of $TiSi_2$ is used in metallization as shown in Fig. 11.8. It also reacts with Al, but at higher temperature of 500°C and refractory metal compound TiN, becomes more popular barrier layer. $TiSi_2$ provide good adhersion and make good electrical contact to the Si. The TiN serves as a barrier to interaction between the Aluminium (Al) and both the $TiSi_2$ and Si. TiN can provide better step coverage with CVD method of deposition. The TiN can be deposited using $TiCl_4$ with NH_3, H_2/N_2 or NH_3/H_2.

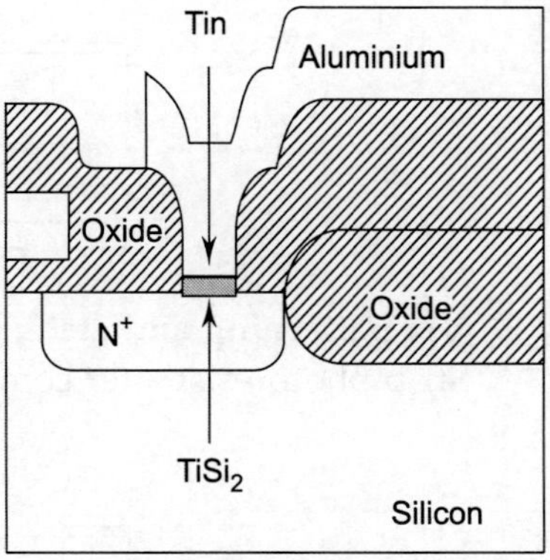

Fig. 11.8 *$TiSi_2$/TiN Barrier Layer between Si and Al*

$$6TiCl_4 + 8NH_3 \longrightarrow 6TiN + 24HCl + N_2$$
$$2TiCl_4 + N_2 + 4H_2 \longrightarrow 2TiN + 8HCl$$
$$2TiCl_4 + 2NH_3 + H_2 \longrightarrow 2TiN + 8HCl$$

The reactions with NH_3 take place at 400°C to 700°C temperature.

The Essential Properties of Barrier Metal

1. Good diffusion barrier properties so that the diffusivity of the two interface materials (*e.g.*, tungsten and silicon) is low at the sintering temperature (sintering refers to joining of the materials by thermal means).
2. High electrical conductivity with low ohmic contact resistance.
3. Good adhersion between the semiconductor and metal.
4. Resistance to electromigration.
5. Stability when thin and at high temperature.
7. Resistance to corrosion and oxidation.

11.5.4 Dual-Damascene Method

Dual-Damascene methods are currently used to etch the trenches, fill them electrolytically, then mechanically polish away the excess metal using a chemically active slurry. Etching and filling narrow structures with high aspect ratios will be especially difficult for dual-Damascene architectures.

Copper Metallization using Dual Damascene

1. **SiO_2 deposition :** ILD oxide deposition with PECVD to desired thickness for via. There is no critical gap fill therefore PECVD is acceptable.

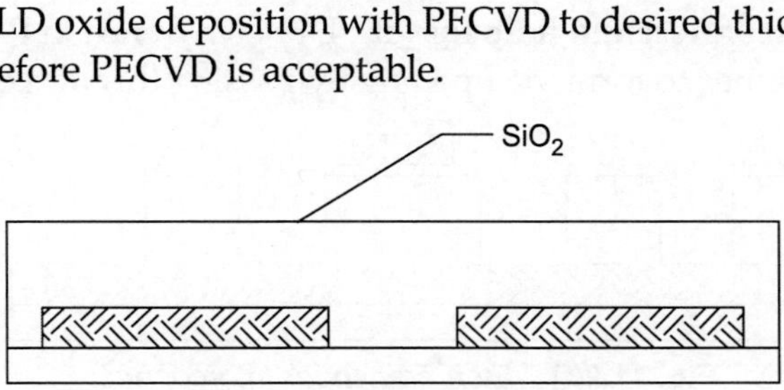

Fig. 11.9(a) *SiO_2 Deposition*

2. **SiN etch stop deposition :** Thin (250 Å) SiN etch stop is deposited on ILD oxide. The SiN needs to be dense and pinholefree; therefore HDPCVD is used.

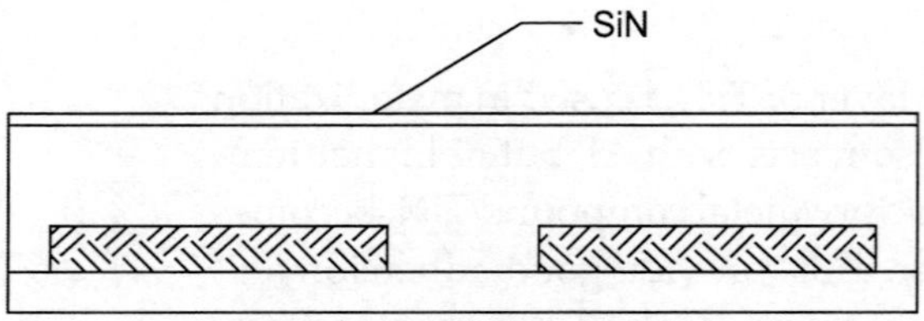

Fig. 11.9(b) *SiN Etch Stop Deposition*

3. **Via patterning and etch :** Photolithography to pattern and dry etch via openings into SiN. Strip photoresist after completion of etch.

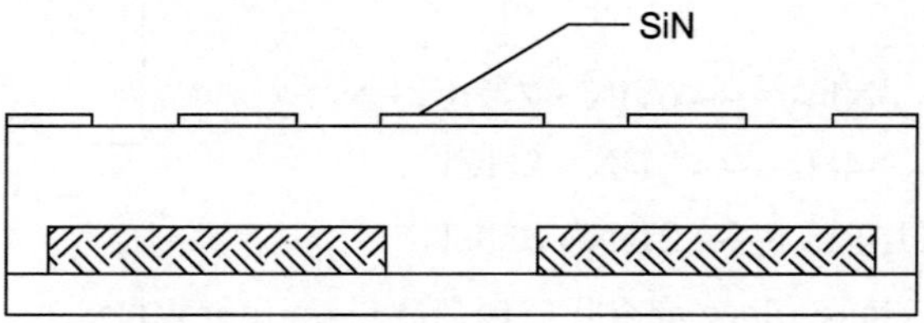

Fig. 11.9(c) *Via Patterning and Etch*

4. **Deposit remaining SiO_2 :** PECVD oxide deposition for remaining ILD oxide.

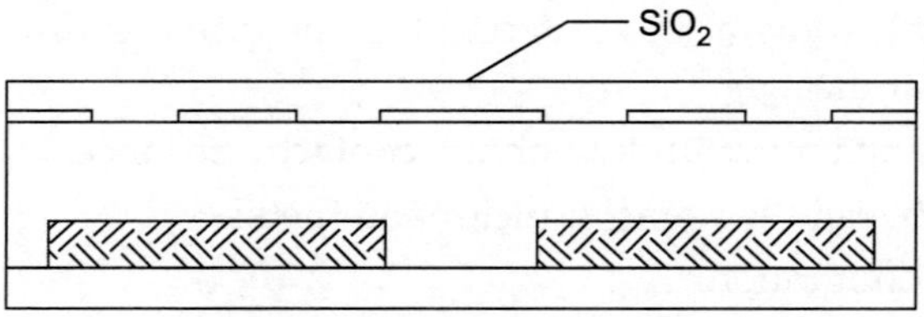

Fig. 11.9(d) *Deposit Remaining SiO_2*

5. **Interconnect patterning :** Photolithography to pattern SiO_2 trench with resist. Previously patterned via openings are located in trench.

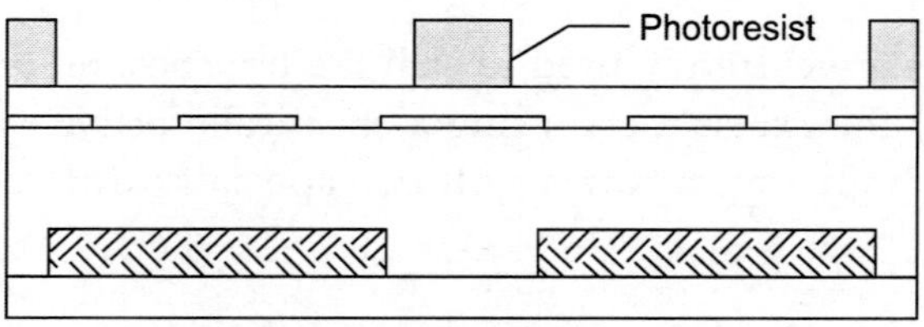

Fig. 11.9(e) *Interconnect Patterning*

6. **Etch trench for interconnect and hole for via :** Dry etch trench in ILD oxide, stopping on the SiN layer. Etch continues to form via opening by passing through opening in SiN.

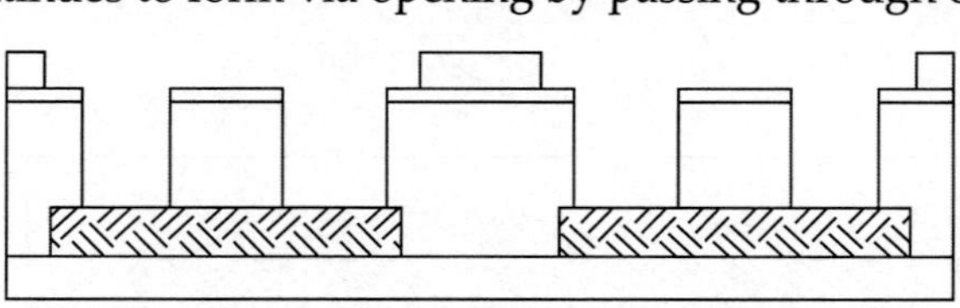

Fig. 11.9(f) *Etch Trench for Interconnect*

7. **Deposit barrier metal :** Deposit Ta or TaN diffusion layer with ionized PVD on bottom and sidewalls of trench and via.

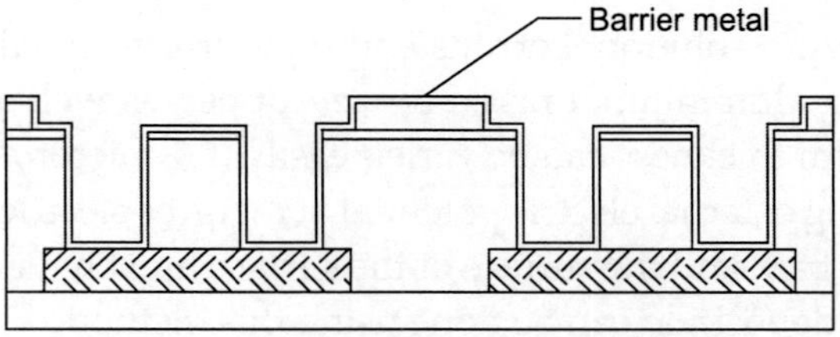

Fig. 11.9(g) *Deposited Barrier Metal*

8. **Deposit Cu seed layer :** Deposit continuous Cu seed layer with CVD. The layer must be uniform and free of pinholes.

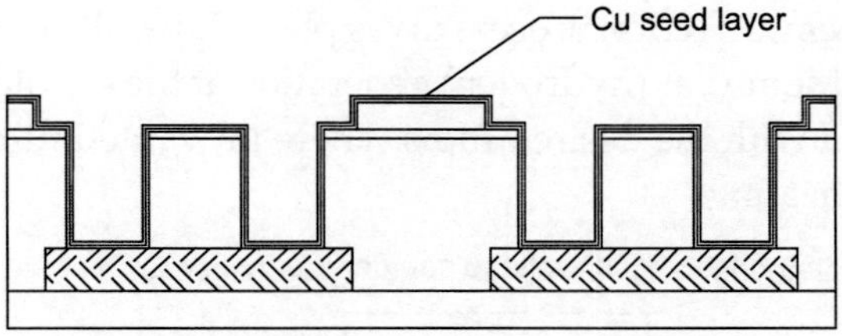

Fig. 11.9(h) *Deposit Cu Seed Layer*

9. **Deposit Cu fill :** Deposit Cu fill electrochemical deposition (ECD). Fill both via opening and trench.

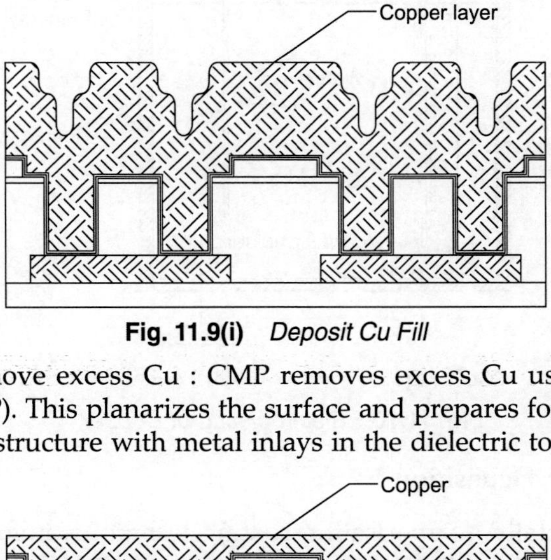

Fig. 11.9(i) *Deposit Cu Fill*

10. **Process step :** Remove excess Cu : CMP removes excess Cu using chemical mechanical planarization (CMP). This planarizes the surface and prepares for next level. The resulting surface is a planar structure with metal inlays in the dielectric to form the circuitry.

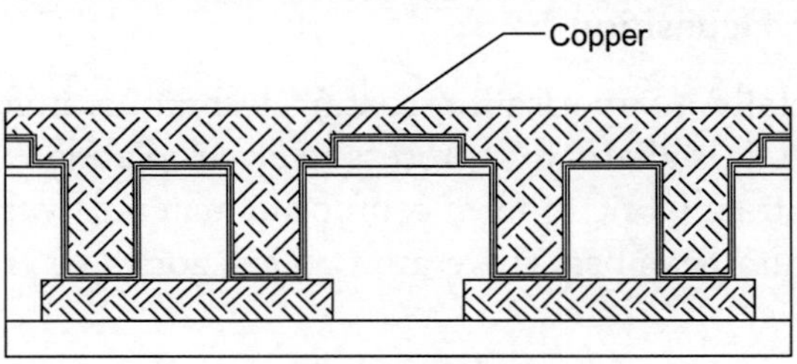

Fig. 11.9(j) *CMP Process*

11.6 METHODS OF METALLIZATION

11.6.1 Electrochemical Plating or Polishing (ECP)

The electroplating is achieved by passing an electrical current through solution containing dissolved metal ions and metal object to be plated. The metal object serves as the cathode in an electrochemical cell, attracting metal ions from solution. Ferrous and non-ferrous metal objects are plated with a variety of metals, including aluminium, brass, bronze, copper, as well as precious, such as gold ; platinum and silver. The film thickness can be varied easily (< 1 micron, > 100 microns). Control of deposition achieved by using external electric potential, it requires electrical contacts to the substrate while immersed in a liquid bath and the surface of the surface must have an electrically conducting coating (seed layer) before deposition can be done using this method.

Working : The electroplating of copper is shown in Fig. 11.10. In this process, the substrate is placed in a liquid solution of copper sulphate ($CuSO_4$) and is connected to the cathode (– ve pole). When an electrical potential is applied between a conducting area on the substrate and counter electrode in the liquid, a chemical redox process takes place. It results in the formation of a layer of copper on the substrate and some gas (hydrogen) generation at the counter electrode (anode). Some non-uniform films are build with the desired films. These unwanted films are removed with etching or bevel removal or annealing.

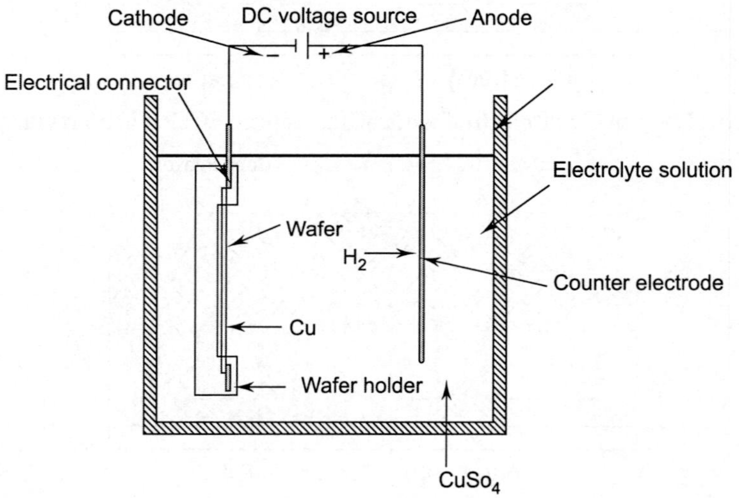

$$2Cu^{++} + 2H_2O \longrightarrow 2Cu(s) + O_2(G)$$

Fig. 11.10 *Electroplating of Copper*

11.6.2 Physical Vapour Deposition

This method is fundamentally a vaporisation coating technique, involving the transfer of the material on an atomic level. It is alternative process to electroplating. This is used to deposit the thin film fibers and to infiltrate fabric to form composite materials vapour deposition processes usually take place in vacuum chamber. There are two categories of physical vapour deposition processes :

(*i*) Evaporation
(*ii*) Sputtering

(*i*) **Evaporation :** The evaporation is the most common method of thin film deposition, the three steps of the evaporation process for metallization are :

(*a*) The solid aluminium must be changed into a gaseous vapour.
(*b*) The gaseous aluminium must be transported to the substrate.
(*c*) The gaseous aluminium must condense on the substrate.

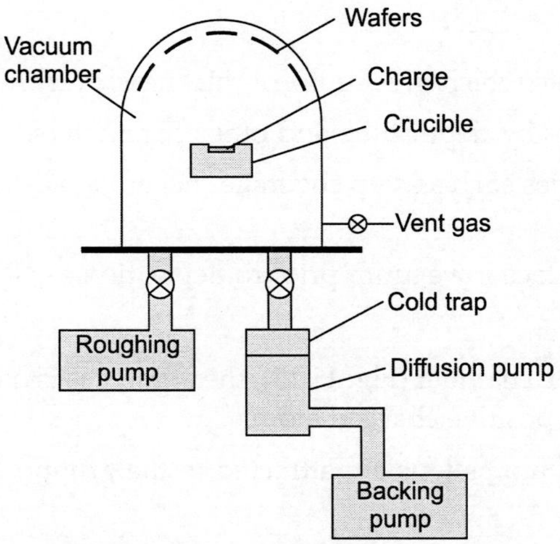

Fig. 11.11 *Evaporation Process*

The process chamber is shown in Fig. 11.11. The chamber can be a quartz bell jar, in which filament is heated by thermal resistance, and vacuum of 5×10^{-5} torr is established for the process. As the temperature rises the metal in the crucible chamber starts to melt, then the current through the filament is further increased, untill the metal (crucible) vapourize. The vapourized metal is set on the substrate with the help of masks. The metal vapour then condenses on substrate of semiconductor wafer, forming a desired metal layer.

We can use also e-beam in place of filament for heating or vapourization. The e-beam evaporation uses a focussed beam of electrons to heat the metal for deposition. The metal is kept in a water cooled crucible and exposed to e-beam, causing it to vapourize and condense on wafers.

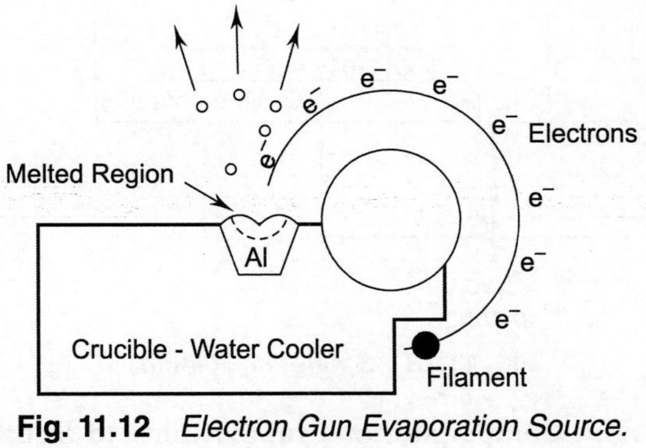

Fig. 11.12 *Electron Gun Evaporation Source.*

The system shown in the figure is called e-beam gun. This evaporation source consists of a water cooled. Crucible with a center cavity to hold Al. At side of crucible is a filament (high temperature). A high current is passed through the filament, which boils of electrons. The – ve electrons bent 180° by a magnet arrangement and these electrons strike the center of cavity. The high energy electrons create a pool of liquid Al in center of cavity. Al evaporate from the melted region and deposits on wafer in the holders at the top of chamber in Fig. 11.12. The crucible water maintains outer walls of the charge in solid state, thus presenting contamination.

(*ii*) **Sputtering :** In general sputtering is a term used to describe the mechanism in which atoms are ejected from the surface of material when that surface is struck by sufficiently energetic particles for the metallization Al-alloys, Ti, TiW Nickel, Cobalt and Gold etc., is used.

Need of Sputtering :

(*i*) It uses large-area-targets which give uniform thickness over the wafer.

(*ii*) It controls the thickness by depth-time and other parameters.

(*iii*) It controls film properties such as step coverage (negative bias), grain structure (wafer temperature) etc.

(*iv*) Sputter cleaned the surface in vacuum prior to deposition.

Steps of Sputtering

1. Ions are generated in the chamber (Fig. 11.13), the argon gas is usually used for this purpose. The ions are ionized to positive charged atom.
2. The positivety charged atoms/ions are attracted to the ground target. Ions start sputtering the target atom.
3. The sputtered atoms/molecules scatter in the chamber with some coming to rest on wafer.
4. Atoms condense and form a thin film or metallic lines for interconnections :

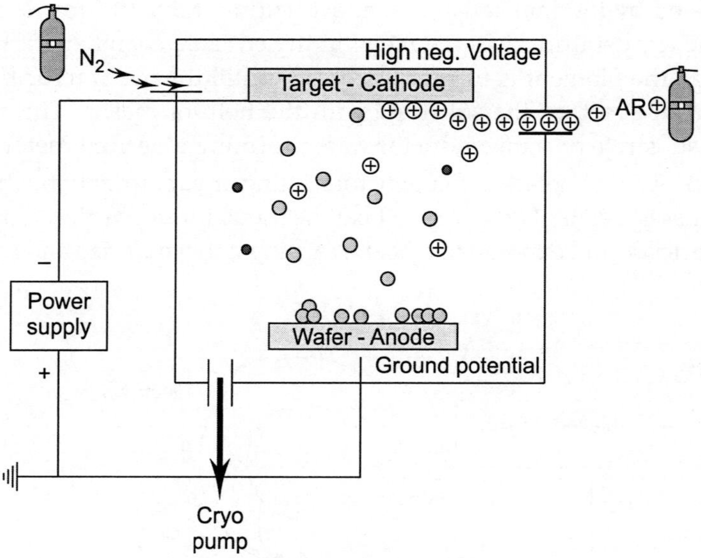

Fig. 11.13 *Sputtering Reactor*

Table 11.5: Comparison of Evaporation and Sputtering

Evaporation	*Sputtering*
1. low energy atoms	higher energy atoms
2. high vacuum path	low vacuum, plasma path
• few collisions	• many collisions
• line of sight deposition	• less line of sight deposition
• little gas in film	• gas in film
3. larger grain size	smaller grain size
4. fewer grain orientation	many grain orientation
5. poor adhersion	better adhersion

11.6.3 Chemical Vapour Deposition

Chemical vapour deposition (CVD) is the process of chemically reacting a non-volatile compound of a material to be deposited, with other gases, to produce a non volatile solid that deposit automatically on a suitably placed substrate. CVD offers coating that are conformal, has good step coverage and can coat a large number of wafers at a time. It has the ability to produce a large variety of films like polysilicon, dielectric SiO_2 and SiN and coating of metals, semiconductor. The reactor used for the metallization is same as discussed in the thin film deposition chapters.

In general, for an arbitary metal M, the reaction is as follow :

$$2MCl + H_2 \longrightarrow 2M + 2HCl$$

11.7 CHEMICAL MECHANICAL PLANARIZATION (CMP)

Chemical Mechanical Planarization (CMP) is a process that can remove topography from silicon oxide, metal and polysiliconsurfaces. It is the preferred planarization step utilized in deep sub-micron IC manufacturing. More recent scaling of transistor critical dimension has required the use of CMP for applications such as shallow trench isolation (STI) and trenched metal interconnection (Cu damascene). CMP has also been utilized for fabrication and assembly of Micro Electro-Mechanical System (MEMS).

In principle, CMP is a process of smoothing and planing surfaces with the combination of chemical and mechanical forces. It can, in a way, be thought of as a hybrid of chemical etching and free abrasive polishing. Mechanical grinding alone may theoretically achieve planarization but the surface damage is high as compared to CMP. Chemistry alone, on the other hand, cannot attain planarization because most chemical reactions are isotropic. However, the removal and planarization mechanism is much more complicated than just considering chemical and mechanical effects separately. CMP makes use of the fact that high points on the wafer would be subjected to higher pressures from the pad as compared to lower points, hence, enhancing the removal rates there and achieving planarization.

CMP is most widely utilized in back-end IC manufacturing. In this process, technology thin layers of metal and dielectric materials are used in the formation of the electrical interconnections between the active components of a circuit (*e.g.*, transistors, as formed in the front-end processing). As shown in Fig. 11.14, the interconnect is manufactured by depositing thin films of materials, and selectively removing or changing the properties of these materials in certain areas. A new level of thin film is deposited on top of old films and the process is repeated many times until the interconnect is complete. The goal of the CMP process is to planarize step heights caused by the deposition of thin films over existing no planar features, so that further levels may be added onto a flat surface.

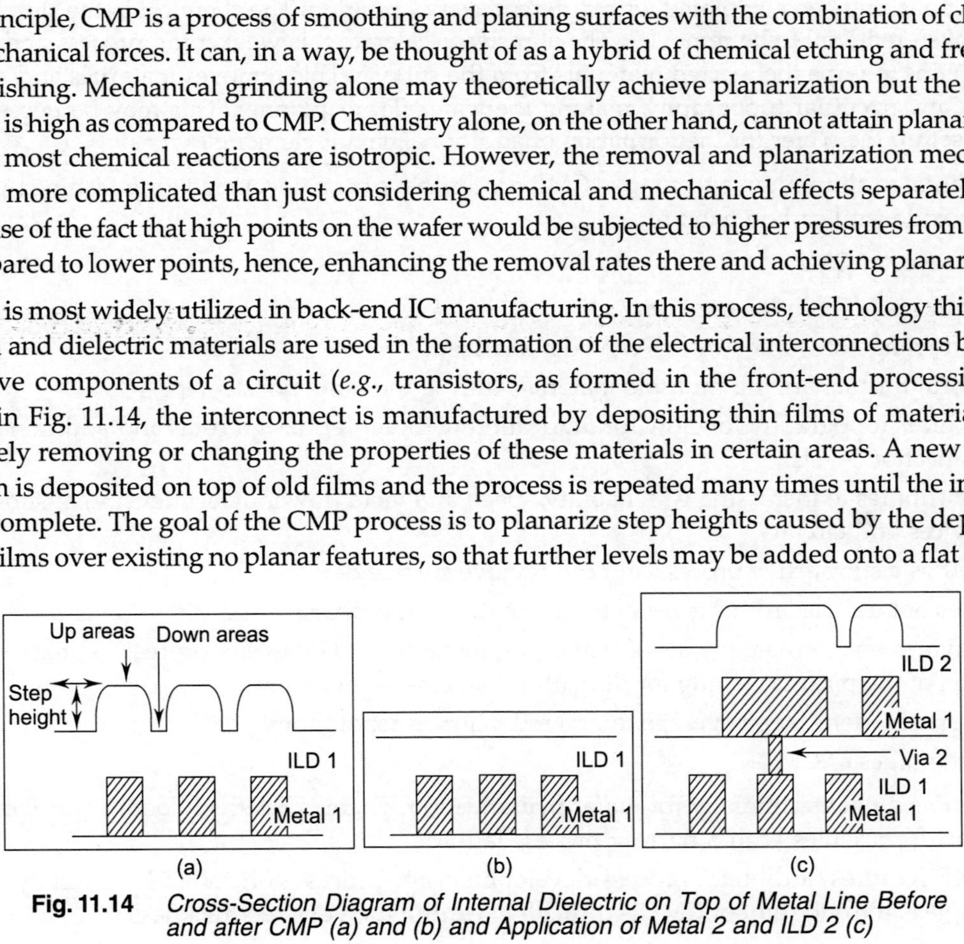

Fig. 11.14 *Cross-Section Diagram of Internal Dielectric on Top of Metal Line Before and after CMP (a) and (b) and Application of Metal 2 and ILD 2 (c)*

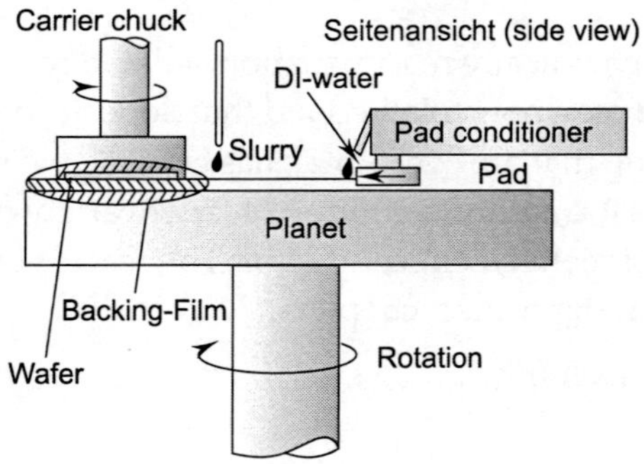

Fig. 11.15(a) *Chemical Mechanical Polishing (CMP)*

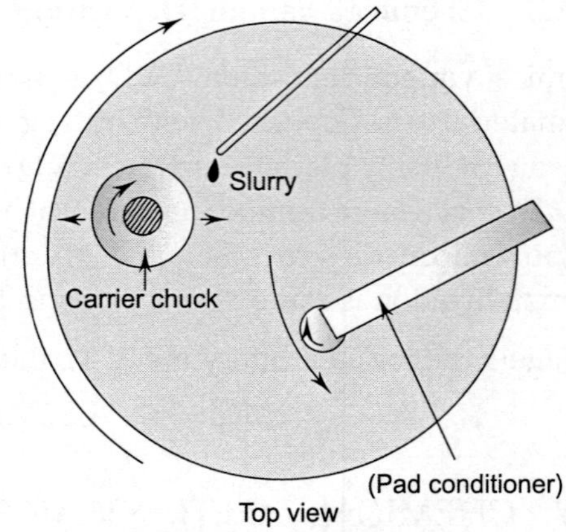

Fig. 11.15(b) *Top View of CMP*

Working of CMP Equipment

The equipment as shown in figure above consists of planet, carrier, slurry, DI water and rotation arrangement. The process uses an abrasive and corrosive chemical slurry in conjuction with the polishing pad and retaining ring, typically of a greater diameter than wafer. The pad and the wafer are pressed together by a dynamic polishing head and held in place by a plastic retaining ring. The dynamic polishing head is rotated with different axes of rotation. The chemical in the slurry reacts with weaken material to be removed. The abrasive accelerate this weakening process and polishing pad helps to wipe the reacted materials from the surface. This removes materials and tends to even out any irregular topography, making the wafer flat or planner. This may be necessary in order to set up the wafer for the formation of additional circuit elements.

In addition to aluminium and copper, CMP process have been developed for polishing tungsten, silicon dioxide, and carbon nano tubs.

Advantages of CMP

1. Achieves global planarization.
2. Wide range of wafer surfaces can be planarized.
3. Useful for planarizing multiple materials during the same polish step.
4. Reduces topography to allow for fabrication with tighter design rules and additional interconnection levels.
5. Contributes to increasing IC reliability, speed and yield (lower defect density) of sub-0.5 μm devices and circuits.
6. CMP is a subtractive process and can remove surface defects.
7. Does not use hazardous gases common in dry etch process.
8. Provides an alternate means of patterning metal (*e.g.*, Damascene process), eliminating the need of the plasma etching for difficult-to-etch metal and alloys.
9. Improves metal step coverage due to reduction in topography.

Disadvantages of CMP

1. CMP is a new technology for wafer planarization. There is relatively poor control over the process variables with a narrow process latitude.
2. CMP requires additional process development for process control and metrology. An example is the endpoint of CMP is difficult to control for a desired thickness.

3. New types of defects from CMP can affect die yield. These defects become more critical for sub-0.25 μm feature sizes.
4. CMP is expensive to operate because of costly equipment and consumables. CMP process materials require high maintenance and frequent replacement of chemicals and parts.

11.8 CMOS/PROCESS STEPS OF THE METALLIZATION

Step 1 : Metal interconnect 1: Deposit layer of metal (Aluminium or copper) over the entire wafer as shown in Fig. 11.16 (*a*).

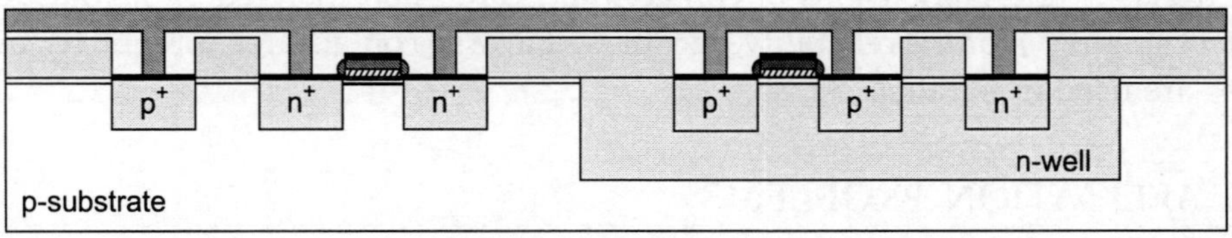

Fig. 11.16 (a) *Metal Interconnect*

Step 2 : Metal interconnect 1 : Photolithography sequence using Metal 1 mask and Metal selectively etched as shown in Fig. 11.16 (*b*).

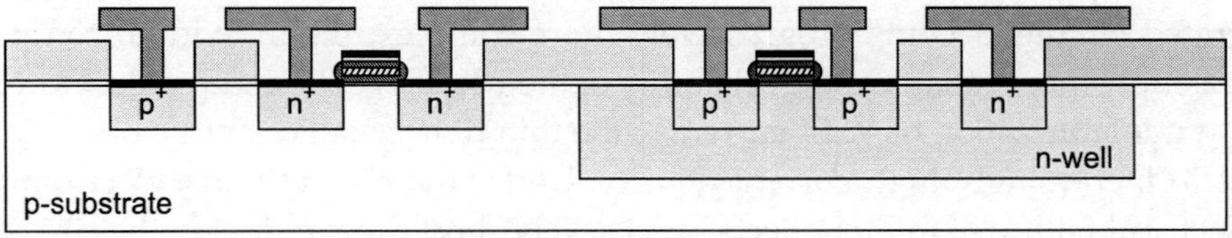

Fig. 11.16 (b) *Metal Interconnect*

Step 3 : Via windows : Firstly cover wafer with a layer of SiN_3 and make the Vias mask using the "via mask" for Lithograpy and then etched by using Plasma etching as shown in Fig. 11.16 (*c*).

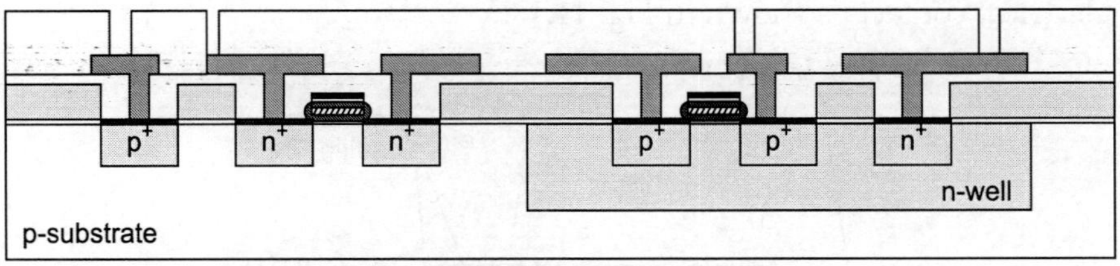

Fig. 11.16 (c) *Via Windows.*

Step 4 : Metal interconnect 2: Deposit layer of metal over the entire wafer as shown in Fig. 11.16(*d*).

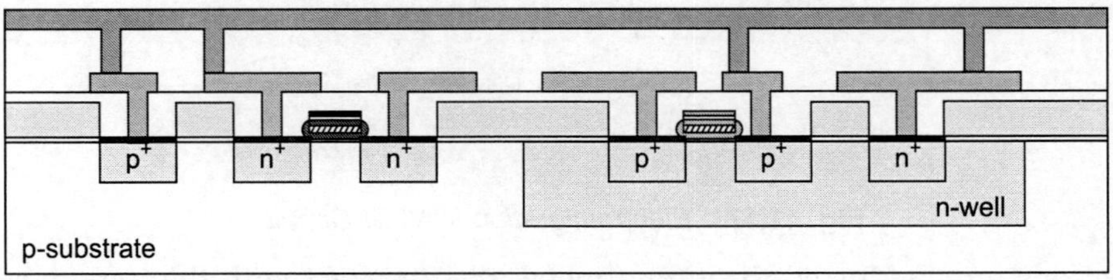

Fig. 11.16 (d) *Metal Interconnect 2.*

Step 5 : Metal interconnect 2 : Photolithography sequence using "Metal 2 mask" and Metal selectively etched as shown in Fig. 11.16(*e*).

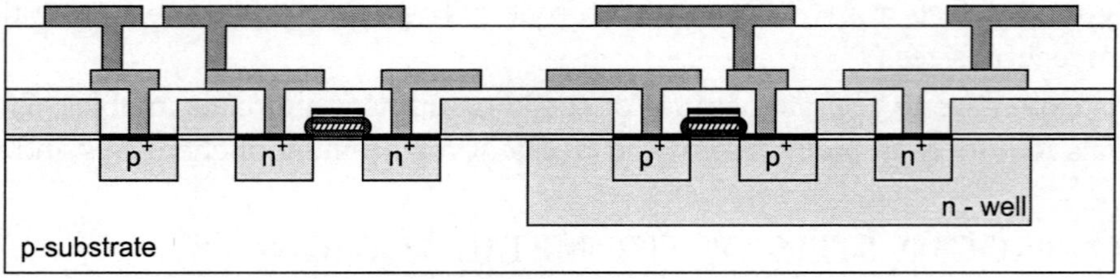

Fig. 11.16 (e) *Metal Interconnect 2.*

In the CMOS metallization, two masks required for each additional metal layer : "via n mask" and "metal n mask" and Dimensions of contacts/vias cannot be changed by layout designer (to avoid "contact spiking") for the reliability purposes. If a large contact area is required, many small contacts/vias are used in parallel.

11.9 METALLIZATION PROBLEMS

11.9.1 Electromigration

Electromigration is generally considered to be the result of momentum transfer from the electrons, which move in the applied field, to the ions which make up the lattice of the interconnect material. As the integrated circuits become progressively more complex, the individual component must become increasingly more reliable if the reliability of the whole is to be acceptable. However, due to the continuing miniaturization of VLSI currents ; the thin film metallic conductors or interconnects are subjected to increasingly high current density. Under these conditions, electromigration can lead to the electrical failure of interconnects in relatively short time, reducing the circuit lifetime to an unacceptable level.

Electromigration causes several kinds of failure in the narrow interconnect. The most familiars are void failure along the length of the line and diffusive displacement at the terminals of the line that destroy electrical contact as shown in Fig. 11.17.

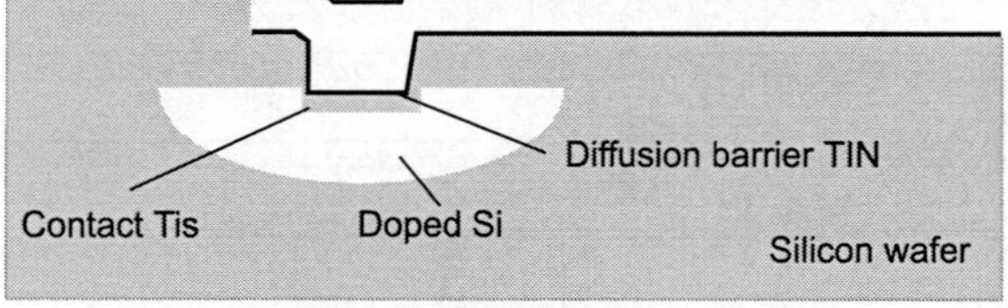

Fig. 11.17 *Electromigration in VLSI Circuit*

When electrons are conducted through a metal, they interact with the imperfections in the lattice and scatter (Fig. 11.18). Scattering occurs whenever an atom is out of the place for any reasons : Thermal energy produces scattering by causing atoms to vibrate. This is the source of resistance of metals. The higher the temperature, the more out of place the atom is, the greater the scattering and the greater the resistivity. We need electron scattering electromigration does not occur in semiconductors. But, in some semiconductor materials if they are also heavily doped that they exhibit metallic conduction. The main two driving forces for the electromigration is (*a*) Direct force – Direct action of the external field on the charge of the migrating ion (*b*) Wind force scattering of the conduction electrons by the metal atom under consideration :

$$F_{total} = F_{direct} + F_{wind} = Z^* eE$$

for simplicity, the term "electron wind force" often refer to the net affect of these two electrical forces. The Fig. 11.18 shows the forces.

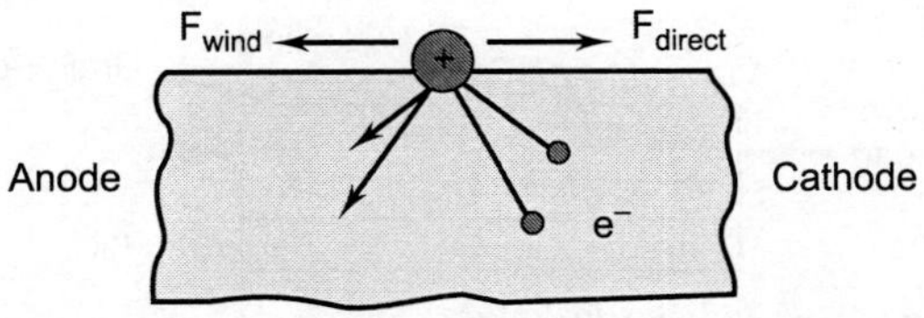

Fig. 11.18 *The Driving Force for Electromigration*

F_{direct} : Electrostatic interaction with the field.
F_{wind} : Momentum transfer from the carriers (e-).

Electromigration can be prevented by :

1. Proper design of the device such that the current densities in all parts of the circuits are practically limited.
2. Good selection and deposition of the passivation or thin films placed over the metal lines in order to limit extrusions caused by electromigration.
3. Increasing the grain sizes of the metal lines such that these become comparable to their widths.

11.9.2 Junction Spiking

The solubilities of silicon and aluminium are 0.25 wt% at 400°C, 0.5 wt% at 450°C and 0.8 wt% at 500°C. Therefore, wherever aluminium contacts silicon, the silicon will dissolve into aluminium during annealing at 450°C. In this way, aluminium starts to penetrate the silicon at few points, where spikes like structure is formed as shown in Fig. 11.19.

The one method to solve this problem by adding aluminium in the silicon by co-evaporation and the alloy of Al-Si is used instead of silicon and another solution is to introduce the barrier metal layer between the Al and the silicon substrate TiN (as discussed before).

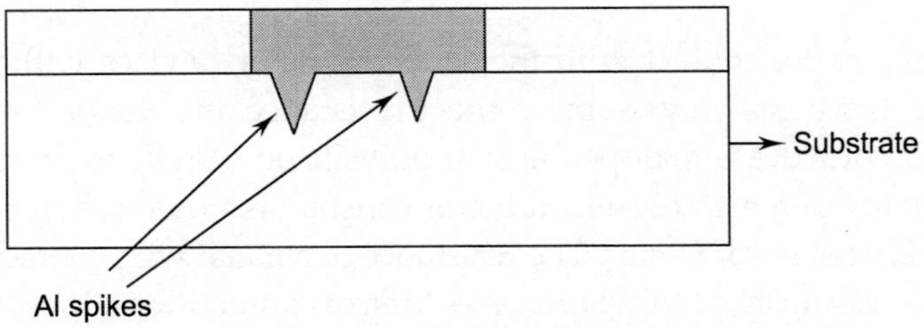

Fig. 11.19 *Junction Spiking*

11.9.3 Stress Migration

Stress migration take place due to difference between coefficient of thermal expansion for Al and silicon. (*i.e.*, Al – 23 × 10^{-6} °C^{-1} and Si – 2.6 × 10^{-6} °C^{-1}). If Al is deposited on Si substrate at low temperature and structure is heated, high compressive stresses in the aluminium can be generated due to these stresses the small hills or hillocks are formed as shown in Fig. 11.20 give The Al moves primarily along the grain boundaries of the aluminium. Since diffusion is usually faster along grain boundaries than through the interior of the grains. In the similar manner, the voids are formed under tensile stress.

The problem of stress migration can be solved by addition of elements that have limited solubility, for example, Cu is added in the Al in which the Cu atoms segregate and precipitate preferentially along the grain boundaries suppressing hillock formation.

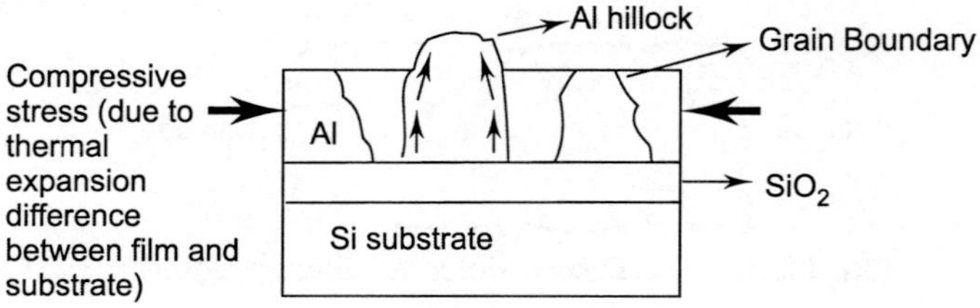

Fig. 11.20 *Stress Migration*

11.9.4 Metal Corrosion

Aluminium grows a passivating oxide in the presence of air and is naturally protected against corrosion. In VLSI circuits, we use aluminium which contains small amount of Cu, and this alloy of Al-Cu has no passivating oxide more valuerable to corrosion.

The corrosion comes from four sources, these sources are :

(*a*) Cl transported through the plastic packaging and passivation material.

(*b*) Cl from etching compound and etching by products.

(*c*) Phosphorus acid formed from excess of phosphorus in the phosphosilicate glass.

(*d*) Electro-chemical corrosion from dissimilar material.

$$Cl^- + H_2O \longrightarrow HCl + OH^-$$

$$3HCl + Al \longrightarrow AlCl_3 + 1.5H_2$$

$$AlCl_3 + 3H_2O \longrightarrow Al\,(OH)_3 + 3HCl.$$

The above reactions shows how the corrosion is formed in Al metallization.

11.9.5 Contact Issues

Ideally, the metal wires make contact with the semiconductor devices without adding parasitic resistance. In practice, there is always some contact resistance and the best one can achieve is to minimize it. A common practice is to dope the Si or polysilicon heavily to produce a low resistance ohmic contact. The quality of a metal-semiconductor constant is measured by specific contact resistance ρ_{Si}. Which has dimensions of Ω-cm^2. The resistance of a metal wire contacing the source, drain or gate of a mos devices is simply ρ_{sc}/A where, A $\rightarrow$ Area of contact. To reduce the contact resistance is to increase the metal semiconductor contact area without increasing the contact window size.

11.9.6 Keyholes Void in ILD

Keyholes void are formed during the fabrication of Integrated circuits. When the Plasma Enhanced CVD is used to form multilevel metallization. The Dielectric material is deposited on the Al contacts for Interlevel Dielectric (ILD). PECVD is used to fill the gaps between the two metal lines a void can form in the centre as shown in Fig. 11.21(*a*) when the planarization is done, the keyholes are formed due to the voids as shown in Fig. 11.21(*b*) and then another metallization layer is deposited on it as shown in Fig. 11.21(*c*). But, it affects the surface smoothness.

To avoid this problem, a multistep process is used in which the deposition / etching / deposition process is repeated. The Argon gas (Ar) etch step is used to cut the corners of the ILD and form a less severe geometry for next PECVD process. The repeated steps fill the gaps in a height/space ratio of about 1.0 without any void as shown in Fig. 11.22.

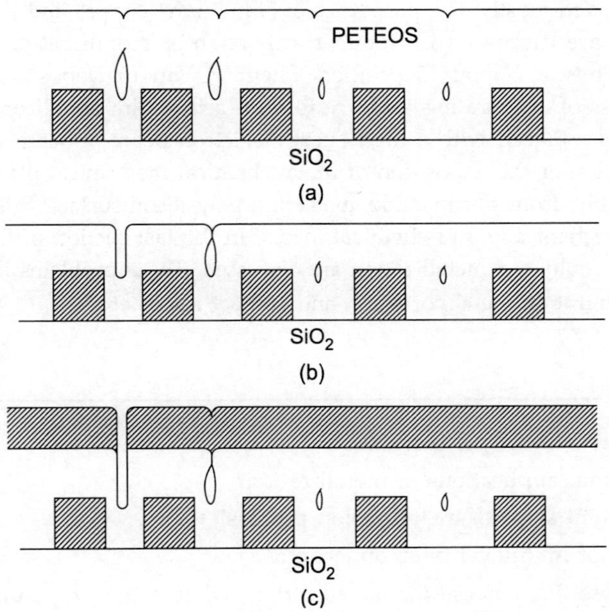

Fig. 11.21

Effects of keyholes in ILD on metal step coverage : (*a*) ILD before planarization ; (*b*) after ILD planarization, some keyholes open into trenches ; (*c*) after metal deposition.

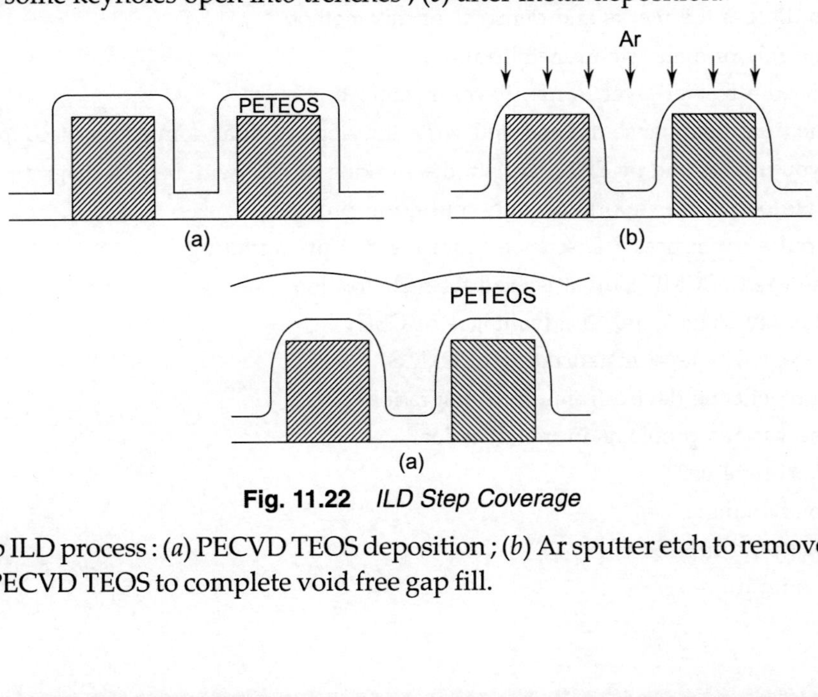

Fig. 11.22 *ILD Step Coverage*

A multistep ILD process : (*a*) PECVD TEOS deposition ; (*b*) Ar sputter etch to remove overhang ; (*c*) new layer of PECVD TEOS to complete void free gap fill.

SUMMARY

The interconnection of circuits are made with metallization process in which the metallization material is used to interconnect the various device structures fabricated on the silicon substrate metallization includes variety of component : contact to active area global interconnects, local interconnect, Via between interconnect levels, first level dielectrics and intermetal dielectrics to physically and electrically separate the different interconnect single level, multi-level interconnection and effects of multi-level interconnection are discussed. The choice of metal selection for metallization depends on applications like for gates, interconnection and contacts polysilicon, silicides nitrides, Al or Cu and Alloys, for diffused barriers nitrides, carbides borider T - w allow is selected mostly. The properties of aluminium, copper and low dielectric constant and benefits and limitations are discussed in detail in this chapter metallization layer the deposited by Electrochemical Plating, Physical Vapour Deposition, Chemical Vapour Deposition methods. The evaporation is the physical process of evaporating a source to form a thin film an silicon substrate electron beam evaporation, a filament is replaced with a high energy electron beam. e-beam uses a focussed beam of electrons to heat the metal after the deposition of metal chemical mechanical planarization (CMP) process is used to remove topography from silicon oxide, metal and polysilicon surface. It is used for smoothing and planning surfaces with mechanically and chemical forces. In the last section of the chapter, the problems related to metallization or multilevel metallization are discussed. These problems includes Electromigration, Junction spiking, stress migration metal corrosion and contact issues etc.

TUTORIAL SHEET

1. What is metallization and why it is needed?
2. What are the different applications of metallization?
3. What are the different steps of metallization process?
4. Discuss the effects of multilevel interconnections.
5. How will you choose the material for metallization? Discuss the properties of these materials.
6. Why aluminium is preferred for metallization and what are its limitations?
7. Compare the various properties of copper with aluminium.
8. Why multilevel interconnections are needed. What are the steps to fabricate multilayer metallization? Also discuss the merits and demerits of this method.
9. Discuss the role of dielectric in metallization.
10. Why we need a barrier layer in multilevel metallization?
11. Discuss the dual dominance method and write the steps of dual dominance of copper.
12. What do you understand by ECP? Explain the working of ECP with help of diagram.
13. Differentiate between evaporation and sputtering process.
14. What is sputtering process? Discuss e-beam method of sputtering.
15. Discuss the need of CMP. How it is used for metallization?
16. Discuss the advantages and disadvantages of CMP.
17. Write the steps of 2 level metallization of CMOS.
18. How we prevent the device from electromigration?
19. Discuss the various problems in metallization.
20. Write a short note on:
    (*i*) Junction Spiking;
    (*ii*) Keyholes/void in ILD;
    (*iii*) Stress migration.

12

PROCESS INTEGRATION (DEVICE FABRICATION)

12.1 INTRODUCTION

An integrated circuit consists of several patterned layers of materials that are used to form transistors and provide electrical interconnections for the circuit. In the modern process, the minimum feature size is less than about 90 nm, which allows for a tremendous packing density. Individual chip with more than 100 million FET's are becoming common place.

The techniques needs to fabricate silicon chips of this sophistication have been characterized as being the most studied element on earth. Now, that we have an understanding of the physical structure of CMOS integrated circuits we may process to studying how the circuits are fabricated in the manufacturing process. Our treatment will focus on those aspects of silicon chip fabrication that are important to VLSI design.

Today, IC's are complex devices with millions of individual components that are manufactured on a single silicon crystal and interconnected by tiny wires that are microns wide. The process of forming these devices out of a single piece of silicon crystal such as *p-n* junction fabrication. BJT, *n*MOS CMOS, *n*-well, *p*-well and twin process are described in this chapter. The major IC technologies that are used today are *n*-channel metal-oxide-semiconductor (NMOS). Complementary metal-oxide-semi conductor (CMOS) and bipolar. The purpose of this chapter is to present the fundamentals of integrating silicon processing steps to create silicon devices.

Silicon CMOS technology has been the dominant fabrication process for relatively high performance and cost effective VLSI circuits. The revolutionary nature of this development is indicated by the way in which the number transistors integrated in the circuits on a single chip has grown. Such progress is highlighted recent products such as high speed chips in which it is possible to process number of functions per second. In order to improve on this throughput rate it will be necessary to improve the technology, both in terms of scaling and processing.

12.2 SILICON PROCESSING

Silicon integrated circuits are created on larger circular sheets of silicon called wafers. They are typically 100 – 300 mm in diameter, and about 0.4 – 0.7 mm thick. A larger silicon circuit is about

1 cm on a side so that many individual circuit can be made on a single wafer. The location of a circuit is called a die site, with the number of sites per wafer depending upon the size of each site and relative to the overall surface area of the wafer. (Fig. 12.1) portrays a wafer with individual sites.

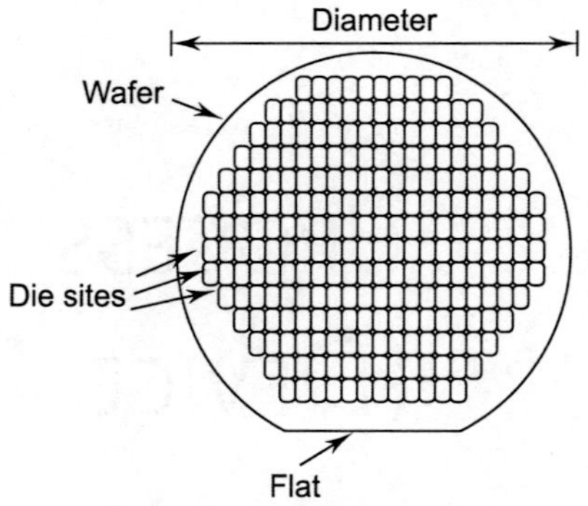

Fig. 12.1 *Silicon Wafer Showing Die Sites*

The flat is used as a reference plane to form an imaginary grid that is used to place the individual sites. Some wafers will have additional flats that are coded to provide information about the crystal orientation at a glance.

Starting with a base polishing surface, the wafer is subjected to thousands of individual steps in the manufacturing processes. The most important steps in the sequence are for creating and patterning the layers of materials needs in the CMOS structures. Most of the remaining steps are cleaning and rinsing of wafer. The manufacturing capacity of a chip factory is usually measured by the number of wafer starts per week *i.e.,* how many fresh wafers are introduced into the fabrication sequence. Wafers are processed in groups, and it takes several weeks for a lot to make it through the entire processing line.

Unfortunately, not every site on the wafer turns out to be a functional circuit. This is due to many factors that may arise in the manufacturing line that are inherent due to the complexity of the processing sequence. To describe this problem, we introduce the concept of the fabrication yield Y such as

$$Y = \frac{N_G}{N_T} \times 100\%$$

where : N_G is the number of good (functional) sites

N_T is the total number of sites.

A yield of Y = 85% means that 85% of the chips operate as they should and can be sold to customers. High yield values are, of course, desirable to help insure the economic stability of the company. However, yield enhancement is a complex problem that requires countless hours of thinking and experimenting with the process line.

Yield analysis is based on predicting the yield of a particular process, and requires a through understanding of all aspects of the silicon processing sequence. One working in this area is faced with increasing the value of Y for a given design. One variable that is critically important to increasing the yield is the area A_{die} of the die. The number of total die sites N_T on a wafer of diameter d is estimated from :

$$N_T = \pi \frac{(d - d_e)^2}{4A_{die}}$$

where, d_e is the wasted edge distance that arises from placing rectangular sites onto a round wafer. Empirical analysis shows that large area die are plagued by smaller yields. A simple expression that describes this is :

$$Y = e^{-\sqrt{DA}}$$

where : A is the area of die.

D is the defect density in units cm^{-2}.

The defect density is the average number of defects per cm^2 on the wafer. D represents the "limits of perfection" that can be expected of the silicon wafer ; this is due to the fact that every crystal wafer has random imperfects that cannot be eliminated. In modern technology, $D = 1\ cm^{-2}$ is a reasonable value for the defect density limit.

Some physical defects tend to occur in clusters on the wafer. An estimate for the yield when these dominate is given by :

$$Y = \left(1 - \frac{A_{die}\,D}{C}\right)^{C}$$

where, C is an empirical parameter that accounts for the clusters. This idea has been used to a write a binominal equation of the form.

$$Y = \frac{1}{\left(1 + \frac{A_{die}\,D}{C}\right)^{2}}$$

Alternately, when several die fail in a large area A_{fail} of the wafer, then the yield has been approximated using the expression

$$Y \approx (1 - g)\, e^{-A_{die} D}$$

where, $$g = \frac{A_{fail}}{A_{wafer}}$$

is the fractional area, where the defects exist.

Yield analysis is a very specialized aspect of VLSI manufacturing persons working in the area tend to have strong background in physics, general and physical chemistry, mathematics, statistics or engineering (chemical, material, or electrical-electronics), and groups to maximize yield values. It is not possible or solve a problem until it is discovered and defined. The "design of experiments" that can be pinpoint problems and lead to solutions becomes very critical.

12.3 BASIC DEVICE FABRICATION METHODS

In the basic device fabrication, we are concerned primarily with silicon technology, because its development is more advanced than that of any other semiconductor.

Some device fabrication methods are as given below :

1. Alloyed to junction fabrication
2. Diffused mesa junction fabrication
3. Epitaxial substrate process
4. Ion implantation process.

12.3.1 Alloyed Junction Fabrication

In the alloy method (Fig. 12.2*a*) and small pellet of aluminium is placed on an *n*-type ⟨ 111 ⟩ oriented silicon wafer. The system is then heated to a temperature slightly higher than the eutectic temperature (~ 580°C for the Al-Si system) so that a small puddle of molten Al-Si mixture is formed. The temperature is then lowered and the puddle begins to solidify. A recrystallized portion, which is saturated with the acceptor impurities and with the same crystal orientation, forms the heavily doped *p*-type region (p^+) on the *n*-type substrate. The aluminium button on top can be used as an ohmic contact for the *p*-type region. For the ohmic contact on the wafer and alloyed at about 400°C to form a heavily doped *n*-type region (n^+). On a *p*-type wafer, the roles of the aluminium and the Au-Sb alloy can be interchanged to form an n^+ junction on top and p^+ ohmic contact on the bottom of the wafer. The junction location obtained by the alloy method depends critically on the temperature-time alloying cycle and is difficult to control precisely.

12.3.2 Diffusion Mesa Junction Fabrication

The solid-state diffusion method was developed in 1956 to give more precise control of the impurity profile. A diffused mesa junction method, where *p*-type impurities (*e.g.*, boron in the form of BBr_3) are diffused into the *n*-type substrate, is shown in (Fig. 12.2*b*). After, diffusion, portions of the surface are protected (*e.g.*, by wax or metal contacts) and the rest are etched out to form the mesa structures.

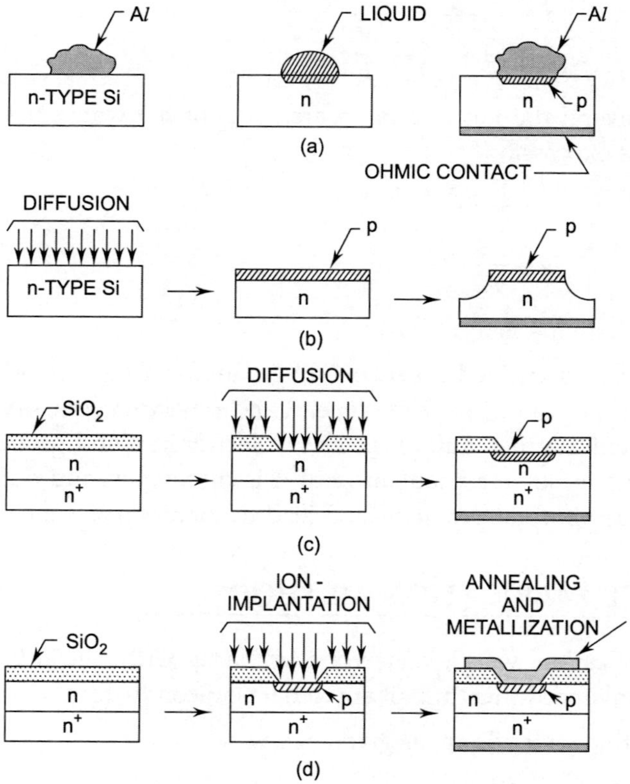

Fig. 12.2 (a) *Alloyed Junction*
Fig. 12.2 (b) *Diffused Mesa Junction*
Fig. 12.2 (c) *Diffused Planar Junction on Epitaxial Substrate*
Fig. 12.2 (d) *Ion Implantation.*

The simple one-dimensional diffusion process can be given by the Fick equation :

$$\frac{\partial C(x,t)}{\partial t} = D\frac{\partial^2 C(x,t)}{\partial x^2} \qquad ...(1)$$

where, C is the impurity concentration

D is the diffusion coefficient.

For a "limited source" condition, with the total amount of impurities S, the solution of equation (1) is given by the Gaussian function

$$C(x, t) = \frac{S}{\sqrt{\pi Dt}} \exp\left(-\frac{x^2}{4Dt}\right)$$

For the "constant surface concentration" condition with a surface concentration C_s, the solution of equation (1) is given by the error function complement.

$$C(x, t) = C_s \operatorname{erfc}\left(\frac{x}{2\sqrt{Dt}}\right)$$

The diffusion profiles of many impurities can indeed be approximated by the proceeding expression. Many, however, have more complicated profiles, for examples, As in Si with a diffusion process depending strongly on the impurity concentration.

12.3.3 Epitaxial Substrate Process

Fig. 12.2(*c*) is the epitaxial substrate, for example, an *n* layer on n^+ substrate. Such a substrate is generally used in the planar process to reduce series resistance. Epitaxy, derived from the Greek word epi meaning "on" and taxis, meaning "arrangement" describes a technique of crystal growth by chemical reaction used to form, on the surface of crystal, thin layers of semiconductor materials with lattice structures identical to those of the crystal. In this method, lightly doped high resistivity epitaxial layers are grown on and supported by a heavily doped low-resistivity substrate thus ensuring both the desired electrical properties and mechanical strength.

Epitaxial layers can be formed by the vapour-phase growth technique. The basic reaction that result in the growth of silicon layers is given below :

$$SiCl_4 + 2H_2 \rightarrow Si\ (solid) + 4HCl\ (gas)$$

Typically, the silicon layer is grown at a rate of about 1 μm/min at 1200°C or higher temperature with a mole fraction of $SiCl_4$ at 0.01% (ratio of $SiCl_4$ molecules to the total number of molecules in the gas). Epitaxial layers can also be formed by liquid-phase growth, with has been extensively used for compound semiconductors or by molecular beam epitaxial, which can give precise control of semiconductor compositions down to atomic dimensions.

12.3.4 Ion Implantation Process

Fig. 12.2(*d*) shows a *p-n* junction formed by ion implantation. This method gives the more precise control of an impurity profile. Ion implantation can be done at room temperature, and the implantation-induced lattice damages can be removed by a meaning at about 700°C or less.

Therefore, ion implantation is a relatively low-temperature process composed to diffusion, which is generally done at 1000°C or higher.

12.4 P-N JUNCTION DIODE FABRICATION

Whenever, *n*-type and *p*-type diffusion regions are placed next to each other, *p-n* junction diode results. A useful structure is the *n*-well diode shown in Fig. 12.3(*a*) below.

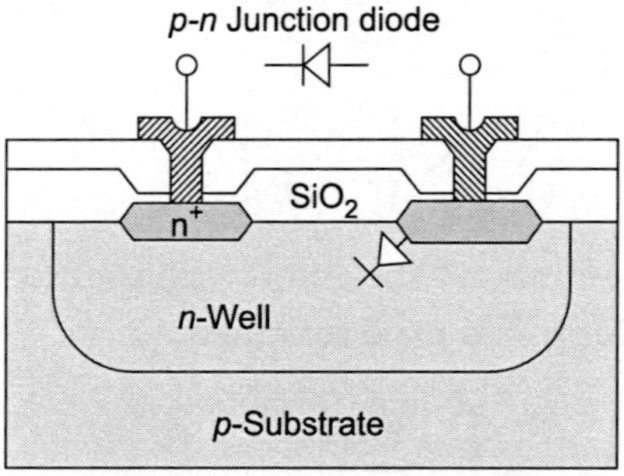

Fig. 12.3(a) *P-N Junction Diode*

The diode fabricated in an *n*-well can provide a high breakdown voltage. This diode is essential for the input clamping circuits for protection against electrostatic discharge. The diode is also very useful as an on-chip temperature sensor by monitoring the variation of its forward voltage drop. The more detail of *p-n* junction fabrication is discussed during basic fabrication process. A lateral P-N-P transistor is shown in Fig 12.3 (*b*).

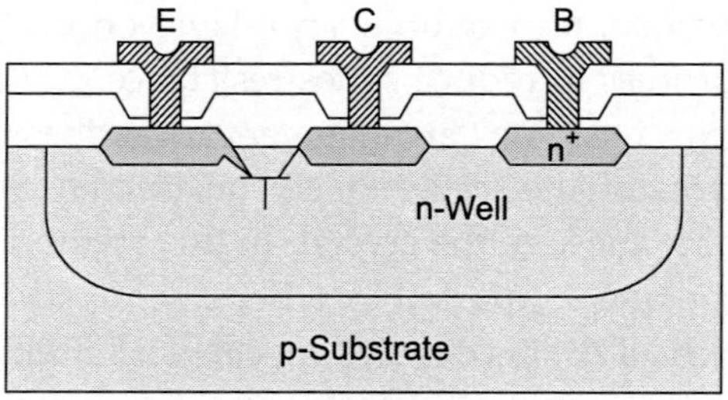

Fig. 12.3(b) *A Lateral P-N-P transistor*

12.5 BIPOLAR JUNCTION TRANSISTOR (BJT) FABRICATION

Early bipolar integrated circuits used junction isolation and diffusion for selective doping. One important difference between integrated transistors and the discrete transistors is that in the monolithic case of all three terminals must be available on the top surface of the chip. For example, interconnection of several isolated bipolar transistor current must pass along a high-resistance path in the lightly doped *n*-epitaxial material, while flowing from the active part of the collector to the contact. The resulting series collector resistance can be determined to the transistor properties and should be avoided if possible. One common method of decreasing the collector resistance is to include a heavily doped *n*-type layer below the collector. Usually, this n^+ burried layer is included by diffusion of n^+ regions into the *p*-type substrate before the epitaxial layer is grown. (See Fig. 12.4(*a*)).

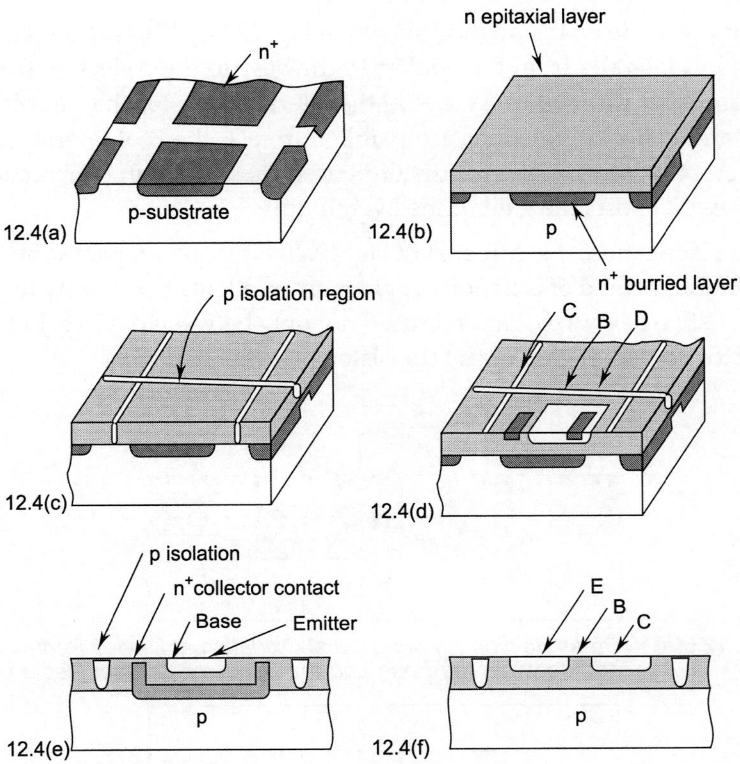

Fig. 12.4 (a) *Buried Layer Diffusion.*
Fig. 12.4 (b) *Growth of Epitaxial Layer.*
Fig. 12.4 (c) *Isolation Diffusion.*
Fig. 12.4 (d) *Double-Diffused Epitaxial Transistor.*
Fig. 12.4 (e) *Improved Collector Contact.*
Fig. 12.4 (f) *Lateral P-N-P Transistor.*

After the *n*-epitaxial layer is grown the *p*-type isolation diffusion is made, the wafer has a pattern typified by (Fig. 12.4(*b*)). During the isolation diffusion and in subsequent diffusions, there is some movement of the n^+ burried layer into the epitaxial region. To minimize this spreading of the burried layer, it is usually doped with Sb, which diffuses rather slowly in Si.

After the isolation pattern, has been diffused through the epitaxial layer, the wafer is masked for the base diffusion (boron) and then again for the emitter diffusion n^+ stripes above the collector areas to improve the collector contact. The resulting transistor structure is shown in Fig. 12.4(*d*). (The oxide and metallization layers are omitted for clarity). If very low collector resistance is required, deep n^+ diffusion can be made from the surface to the burried layer Fig. 12.4(*e*). This step is done before the base diffusion. After all diffusions have been completed, the surface is passivated and the interconnection pattern is included, as discussed below.

We have presented the processes for fabrication of *n-p-n* transistors, since this is the most common type of bipolar device used in IC". If *p-n-p* transistors are to be included as well, special techniques must be used. One way to make a *p-n-p* device in the *n*-type island of Fig. 12.4(*c*) would be to employ a triple diffusion : *a p*-layer, then *n*-layer, and finally another *p*-layer. This method is un-desirable because of the extra diffusion step required and because the ultimate emitter region has been compensated three times. When *p-n-p* transistors are required in an otherwise *n-p-n* circuit, they are usually made by the method shown in Fig. 12.4(*f*). This is called a lateral transistor, since the current flows laterally from the emitter to the base to the collector. The base width and other parameters are more difficult to control in the lateral transistor than in planar transistor, in which the emitter and collector junctions are parallel. In fact, the β of lateral transistors is often near unity ; therefore, the lateral *p-n-p* transistor is used most often in conjunction with a planar *n-p-n* such that the *n-p-n* boosts the gain of the overall unit.

In modern device fabrication, the diffusion of Fig. 12.4(*c*) is largely replaced by ion implantation. Also, the local oxidation method of isolation provides an excellent opportunity for size reduction of transistors. In (Fig. 12.5(*a*) below) oxidation is used not only between devices, but also between the base and the collector contact region of the transistor.

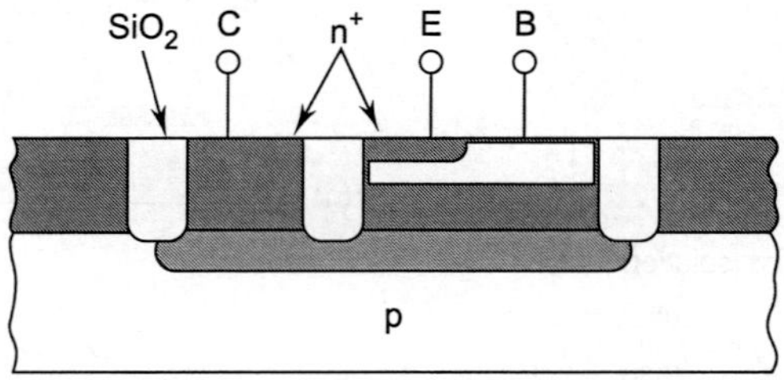

Fig. 12.5(a) *Cross Section Showing Local Oxidation at Device Perimeter and between the Base and the Collector Contact Region*

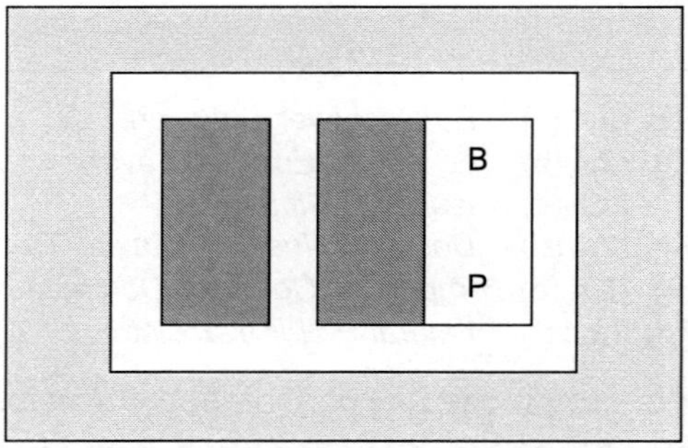

Fig. 12.5(b) *A Top View of the Device before Metallization*

As a result, the mask for the base implant or diffusion can extend over part of the oxide region on both sides without affecting the device geometry. Similarly, the n^+ mask for the emitter and the collector contact does not have the usual close registration tolerance. Each of these implant or diffusion areas can extend over the adjacent oxide region without affecting the geometry of the device. Since, space is not wasted adjacent to the isolation and mask registration is less critical, the dimensions of the device can be made small. This process, introduced by Fairchild and called Isoplanar, has important advantages in bipolar integrated circuits. The area of an Isoplanar transistor can be about one-half that of an equivalent junction isolated device. This reduction in area, along with the reduced sidewall capacitance provided by oxide isolation, produces a much faster device with high packing density on the chip.

12.6 CMOS FABRICATION

A schematic cross section of modern CMOS transistors, consisting of an *n*-channel MOSFET and a *p*-channel MOSFET integrated on the same chip, is shown in Fig. 12.6.

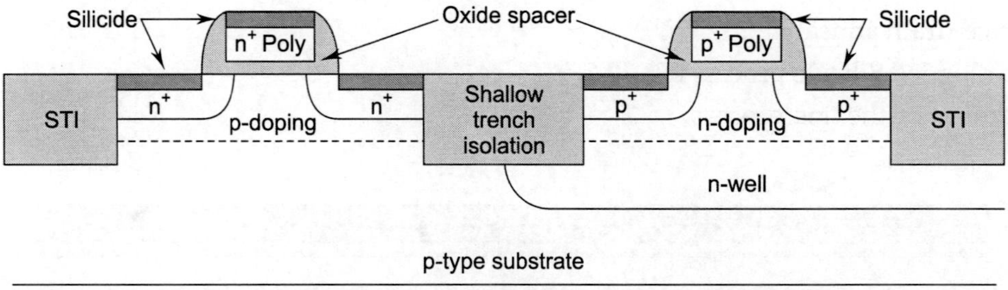

Fig. 12.6 *Schematic Device Cross Section for an Advanced CMOS Technology*

The key physical features of the modern CMOS technology as illustrated in Fig. 12.6, include: *p*-type polysilicon gate for the *p*-channel MOSFET and *n*-type polysilicon gate for the *n*-channel MOSFET, refractory metal silicide on the polysilicon gate as well as on the same source and drain diffusion regions and shallow-trench oxide isolation.

In the electrical design of the modern CMOS transistor, the power-supply voltage is reduced with physical dimensions in some coordinating manner. A great deal of design detail goes into determining the channel length, or separation between the source and drain, accurately, maximizing on current of the transistor while maintaining an adequately low off current, minimizing variation of the transistor characteristics with process tolerances, and minimizing the parasistic resistances and parasistic capacitances. A generic process flow for fabricating the CMOS transistors are outlined as below:

It features Shallow Trench Isolation (STI), dual n^+/p^+ polysilicon gates, and self aligned silicide. The front-end-of-the-line process consists of six or seven masking levels and is suitable for sub-0.5-μm generations of VLSI logic and SRAM technology.

CMOS Fabrication Steps :

1. Starting material *p*-type substrate or p^- epi on p^+ substrate for latch up prevention.
2. Grow pad oxide. Deposit CVD (chemical vapour deposition) nitride Fig. 12.7(*a*).
3. Lithography to cover the active region with photoresist.
4. Reactive ion etching (RIE) nitride and oxide in the field region.
5. Reactive ion etching (RIE) shallow trench in silicon Fig. 12.7(*b*).
6. Grow pad oxide. Deposit thick chemical vapour deposition (CVD) oxide. Fig. 12.7(*c*).
7. Chemical-mechanical polishing planarization. Fig. 12.7(*d*).
8. *n*-well lithography and implant (also channel doping).
9. *p*-well lithography and implant (also channel doping) Fig. 12.7(*e*)
10. Grow gate oxide.
11. Deposit polysilicon film. Fig. 12.7(*f*)
12. Gate lithography
13. Reactive ion etching (RIE) polysilicon gate. Fig. 12.7(*g*)
14. Sidewall reoxidation.
15. n^+ source-drain lithography and implant (also dope n^+ polysilicon gate).
16. p^+ source-drain lithography and implant (also dope p^+ polysilicon gate). Fig. 12.7(*h*)
17. Oxide (or nitride) spacer formation by chemical vapour deposition (CVD) and Reactive ion etching (RIE).

18. Source-drain anneal Fig. 12.7(*i*)
19. Self-alligned Silicide process Fig. 12.7(*j*)
20. Back-end-of-the line process.

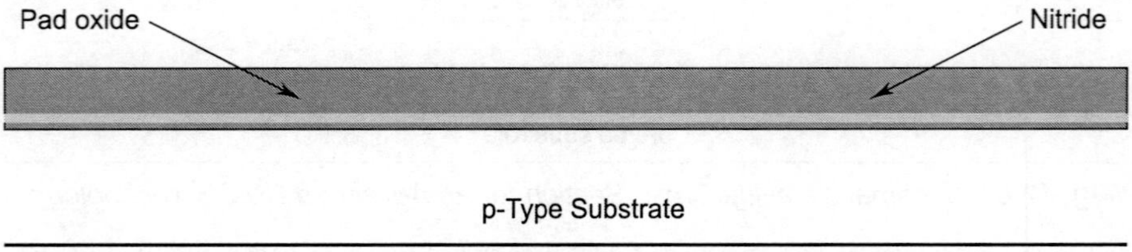

Fig. 12.7 (a)

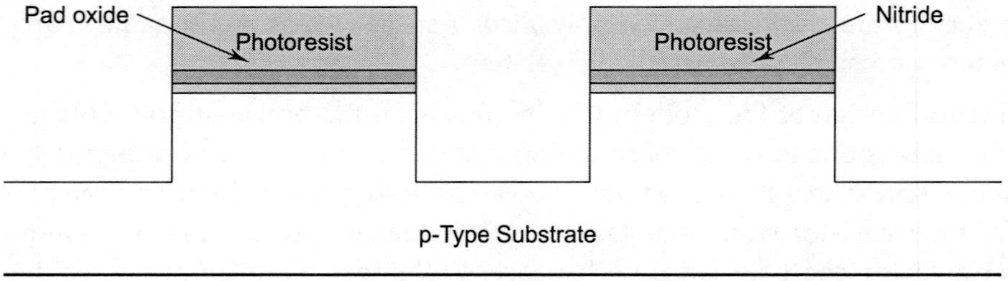

Fig. 12.7 (b)

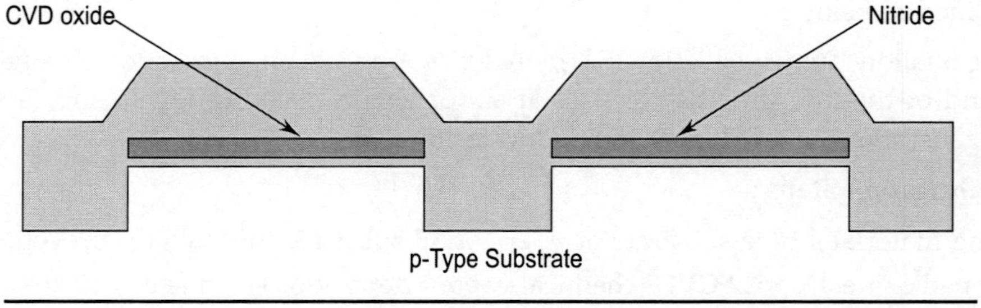

Fig. 12.7 (c)

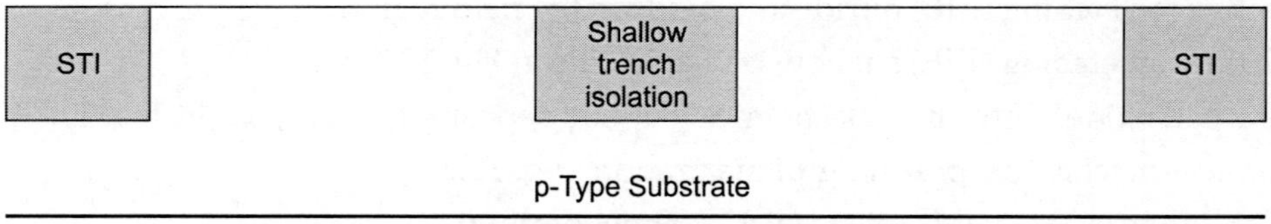

Fig. 12.7 (d)

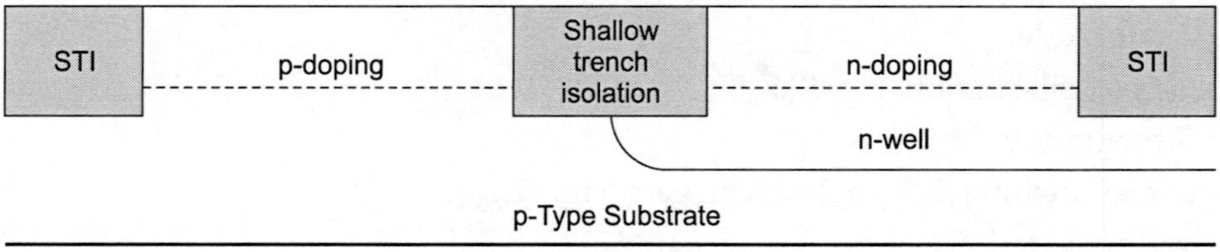

Fig. 12.7 (e)

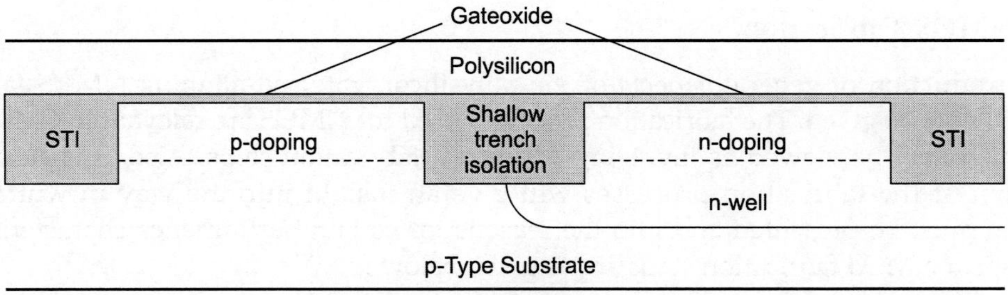

Fig. 12.7 (f)

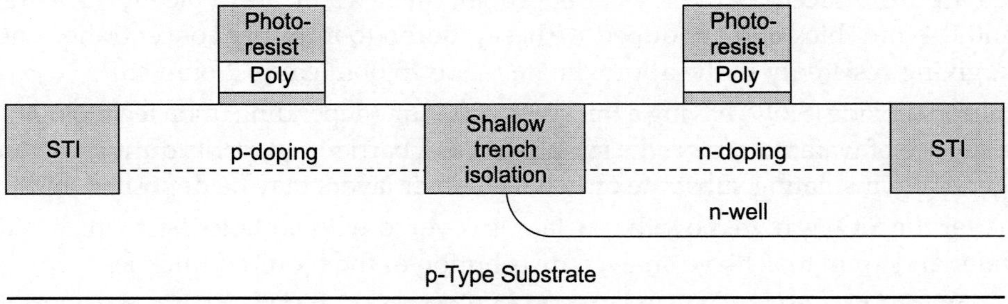

Fig. 12.7 (g)

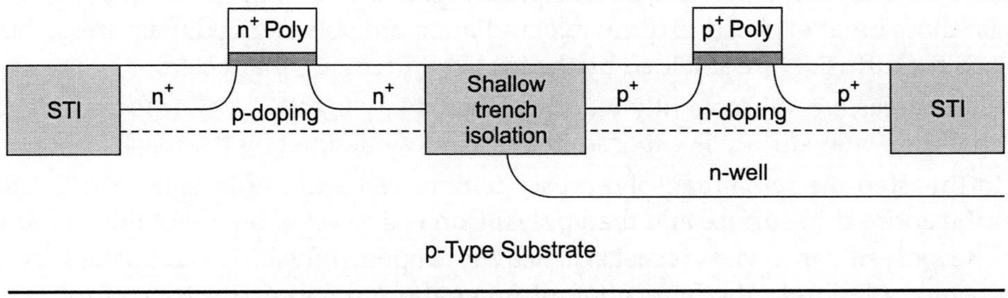

Fig. 12.7 (h)

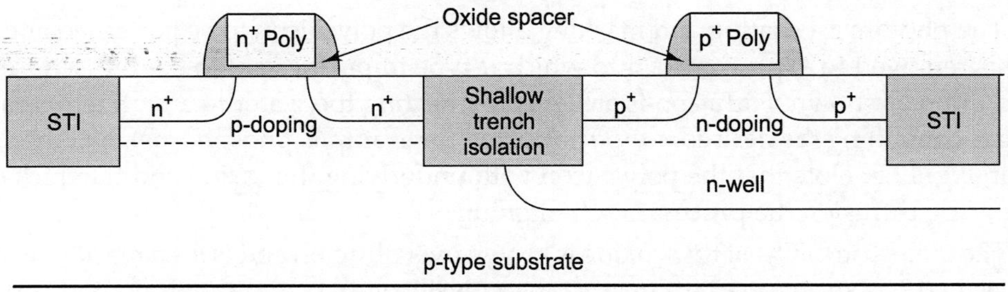

Fig. 12.7 (i)

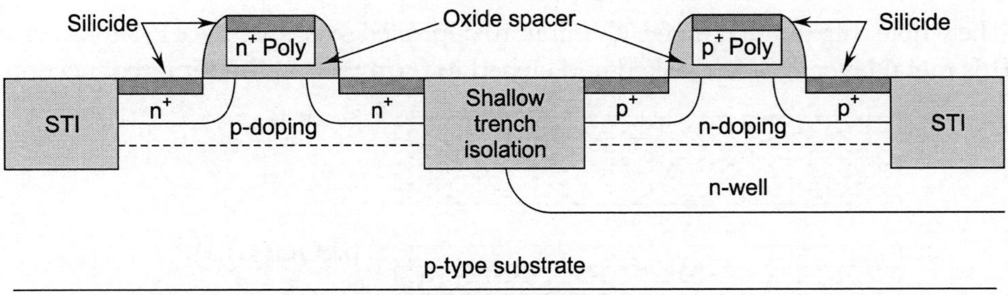

Fig. 12.7 (j)

Fig. 12.7 *CMOS Fabrication Steps (12.7 (a) to 12.7 (j))*

12.6.1 NMOS Fabrication

For the introduction of general aspects of the polysilicon gate self-aligning NMOS fabrication process will now be given. The fabrication processes used for NMOS are relevant to CMOS and Bi CMOS which may be viewed as involving additional fabrication steps. Also, it is clear that an appreciation of the fabrication processes will give an insight into the way in which design information must be presented and into the reasons for certain performance characteristics and limitations. An NMOS fabrication steps are as given below :

Step 1. A Thin wafer cut from a single crystal of silicon of high purity into which the required *p*-impurities are introduced as the crystal is grown. Such wafer are typically 75 to 150 mm in diameter and 0.4 mm thick and are doped with, say, boron to impurity concentrations at 10^{15} cm^3 to $10^{16}/cm^3$, giving resistivity in the approximate range 25 ohm cm to 2 ohm cm.

Step 2. Silicon dioxide (SiO_2), having a thickness of 0.1 μm (depending upon technology), is grown all over the surface of wafer to protect the surface, act as a barrier to dopants during processing, and provide a generally insulating substrate onto which other layers may be deposited and patterned.

Step 3. After the SiO_2 grown, now the surface is covered with a photoresist which is deposited onto the wafer and spun to achieve an even distribution of the required thickness.

Step 4. A layer of photoresist is then exposed to ultraviolet light through a mask which defines those regions into which diffusion is to be take place together with transistor channels. Assume, for example, that those areas exposed to ultraviolet radiation are polymerized (hardened), but that the areas required for diffusion are shielded by the mask and remain unaffected.

Step 5. These areas are subsequently readily etched away together with the underlying silicon dioxide so that the wafer surface is exposed in the window defined by the mask.

Step 6. In this step the remaining photoresist is removed and a thin layer of SiO_2 (0.1 μm) is grown over the entire chip surface and then polysilicon is deposited on top of this to form the gate structure. The polysilicon layer consists of heavily doped polysilicon deposited by chemical vapour deposition (CVD). In the fabrication of fine pattern devices, precise control of thickness, impurity concentration, and resistivity is necessary.

Step 7. The photoresist coating and masking allows the polysilicon to be patterned and then the thin oxide is removed to expose areas into which *n*-type impurities are to be diffused to form the source and drain as shown. Diffusion is achieved by heating the wafer to a high temperature and passing a gas containing the desired *n*-type impurity (for example, phosphorus) over the surface as indicated in Fig. 12.8. Note that the polysilicon with underlying thin oxide and the thick oxide act as masks during diffusion the process is self-aligning.

Step 8. The dielectric (SiO_2) of thick oxide is grown over all again and is then masked with photo-resist and etched to expose selected areas of the polysilicon gate and the drain and source areas where connections (*i.e.,* contact cuts) are to be made.

Step 9. The whole chip that has metal (aluminium) deposited over its surface to a thickness typically of 1 μm. This metal layer is then masked and etched to form the required interconnection pattern.

1.

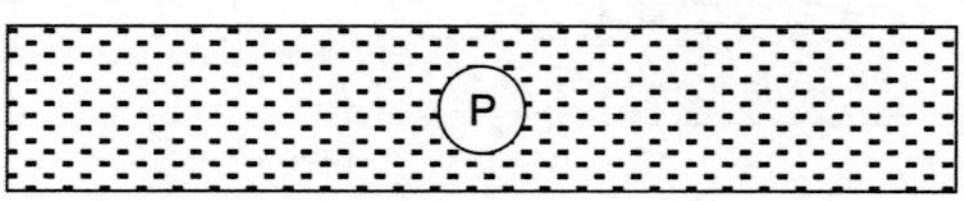

Substrate

2.

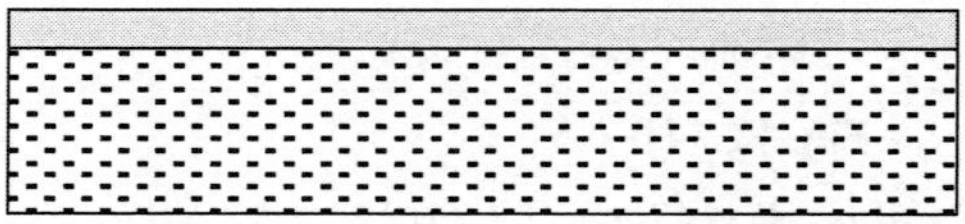

Thick oxide (1μm)

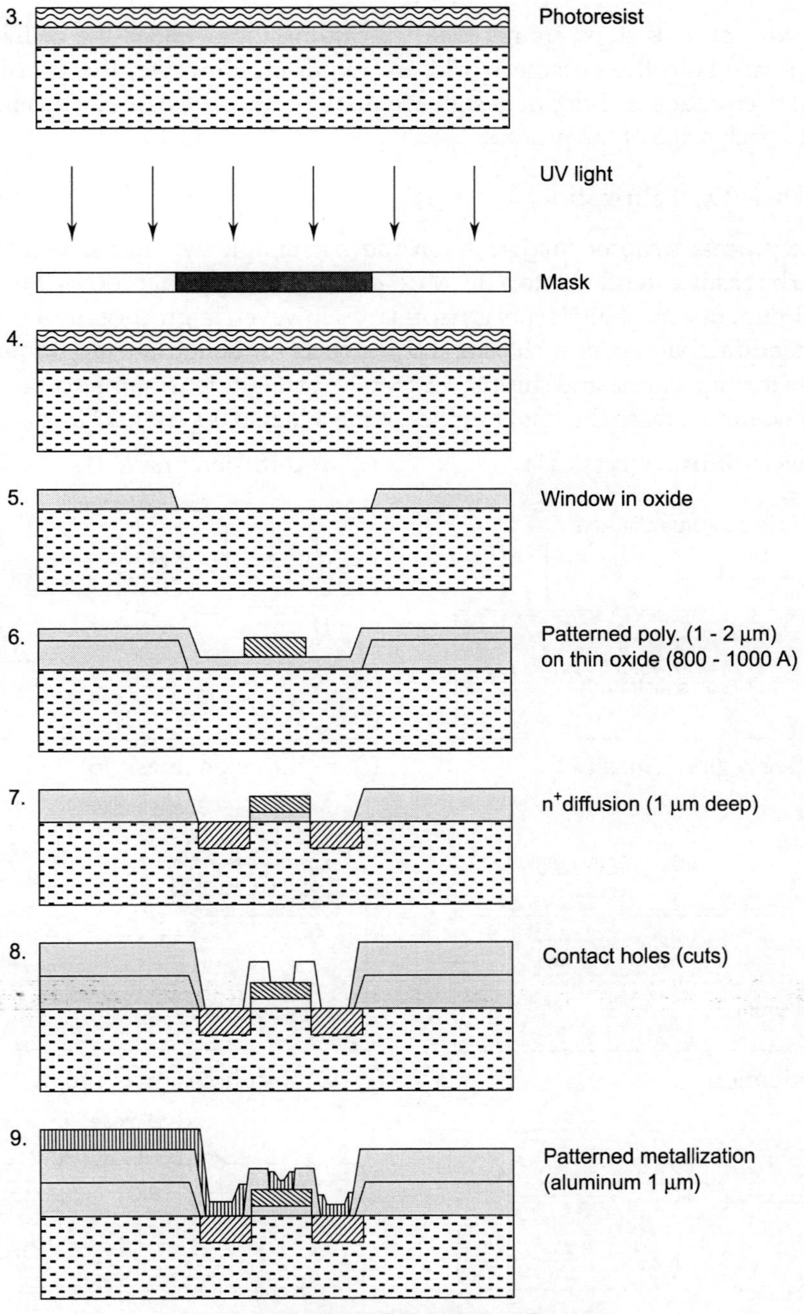

Fig. 12.8 *NMOS Fabrication Process*

It will be seen that the process revolves around the formation or deposition and patterning of three layers, separated by silicon dioxide insulation. The layers are diffusion within the substrate, polysilicon on oxide on the substrate, and metal insulated again by oxide. To form depletion mode devices it is only necessary to introduce a masked ion implantation step betwen step 5 and step 6 in Fig. 12.8. Again, the thick oxide acts as a mask and this process stage is also self-aligning.

Consideration of the processing steps will reveal that relatively few masks are needed and the self-aligning aspects of the masking processes greatly ease the problems of mask registration. In practice, some extra process steps are necessary, including the overglassing of the whole wafer, exccept where contacts to the outside world are required. However, the process is basically straight forward to envisage and circuit design eventually comes down to the business of delineating the masks for each stage of the process.

12.6.2 *n*-well CMOS Fabrication

An *n*-well CMOS process is chosen for discussion due to its simplicity. Another benefit of this choice is that it can also be easily extended into a BiCMOS process. The typical process flow is as shown in Fig. 12.9. A minimum of 7 masking layers is necessary. However, in practice most CMOS processes will also require additional layers such as *n* and *p* guards for better latchup immunity, a second polysilicon layer for capacitors, and multilayer metals for high-density interconnections. The inclusion of these would increase the total number of masking layers from 15 to 20.

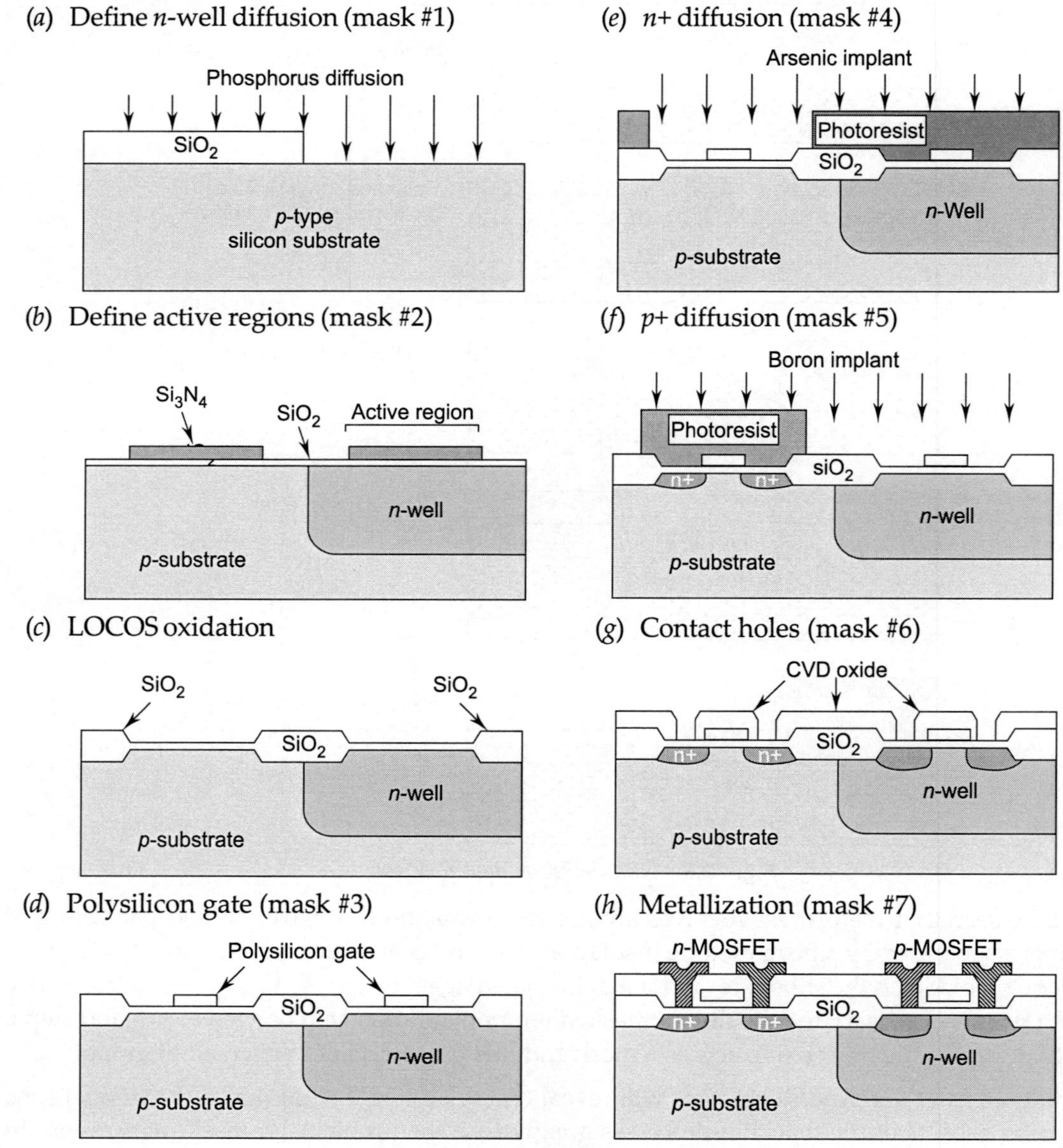

Fig. 12.9 *A Typical n-wellCMOS Process Flow*

The starting material for the n-well CMOS is a p-type substrate. The process begins with an n-well diffusion Fig. 12.9(a). The n-well is required wherever p-type MOSFET's are to be placed. A thick silicon dioxie layer is etched to expose the regions for n-well diffusion. The unexposed regions will be protected from the n-type phosphorus impurity. Phosphorus is usually used for deep diffusions because it has a large diffusion coefficient and can diffuse faster than arsenic into the substrate.

The second step is to define the active region (where, transistors are to be placed) using a technique called local oxidation (LOCOS). A silicon nitride (Si_3N_4) layer is deposited and patterned relative to the previous n-well regions Fig. 12.9(b). The nitride-covered regions between transistor (Fig. 12.9(c)). This thick-field oxide is necessary for isolating transistors. It also allow interconnection layers to be routed on top of the field oxide without inadvertently forming a conduction channel at the silicon surface.

The next step is the formation of polysilicon gate (Fig. 12.9 (d)). This is one of the most critical step in CMOS process. The thin oxide layer in the active region is first removed using wet etching followed by the growth of a high-quality thin gate oxide. Current 0.13 μm and 0.18 μm process routinely use oxide thickness as thin as 20 Å to 50 Å (1 a angstron = 10^{-8} cm). A polysilicon layer, usually arsenic doped (n-type), is then deposited and patterned. The photolithography is most demanding in this step, since the finest resolution is required to produce the shortest possible MOS channel length.

The polysilicon gate is a self-aligned structure and is preferred over the older type of metal gate structure. A heavy arsenic implant can be used to form the n^+ source and drain region of the n-MOSFET's. The polysilicon gate also acts as a barrier for this implant to protect the channel region. A layer of photoresist can be used to block the regions where p-MOSFET's are to be formed (Fig. 12.9(e)). The thick-field oxide stops the implant and prevents n^+ regions from forming outside the active regions. A reversed photolithography step can be used to protect the n-MOSFET's during the p^+ boron source and drain implant for the p-MOSFET's Fig. 12.9(f). In both cases the separation between the source and drain diffusions - the channel length is defined by the polysilicon gate mask alone, hence the self-alignment.

Before contact holes are opened, a thick layer of CVD oxide is deposited over the entire wafer. A photomask is used to define the contact window opening (Fig. 12.9(g)) followed by a wet or dry oxide etch. A thin aluminium layer is then evaporated or sputtered onto the wafer. A final masking and etching step is used to pattern the interconnection [Fig. 12.9(h)].

The final passivation step prior to packaging and wire bonding. A thick CVD oxide or pyrox glass is usually deposited on the wafer to serve as a protective layer.

The Berkeley *n*-well fabrication : N-well CMOS inverter is as shown in Fig. 12.10. There are number of n-well fabrication processes, and, in order to look more closely at typical fabrication steps. We will use the berkeley n-well process as an example. This process is as shown as a flow diagram (Fig. 12.11) below.

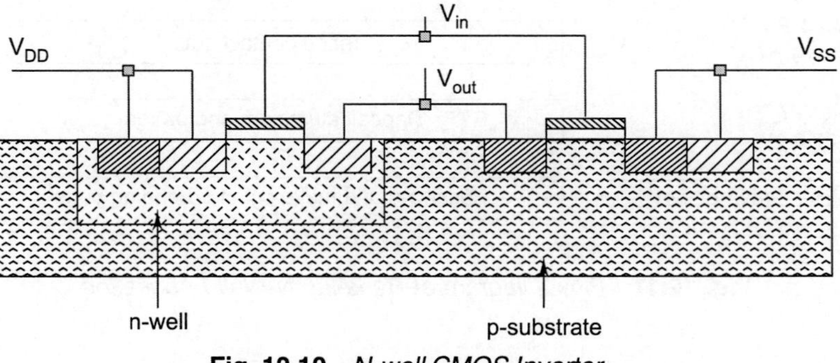

Fig. 12.10 *N-well CMOS Inverter*

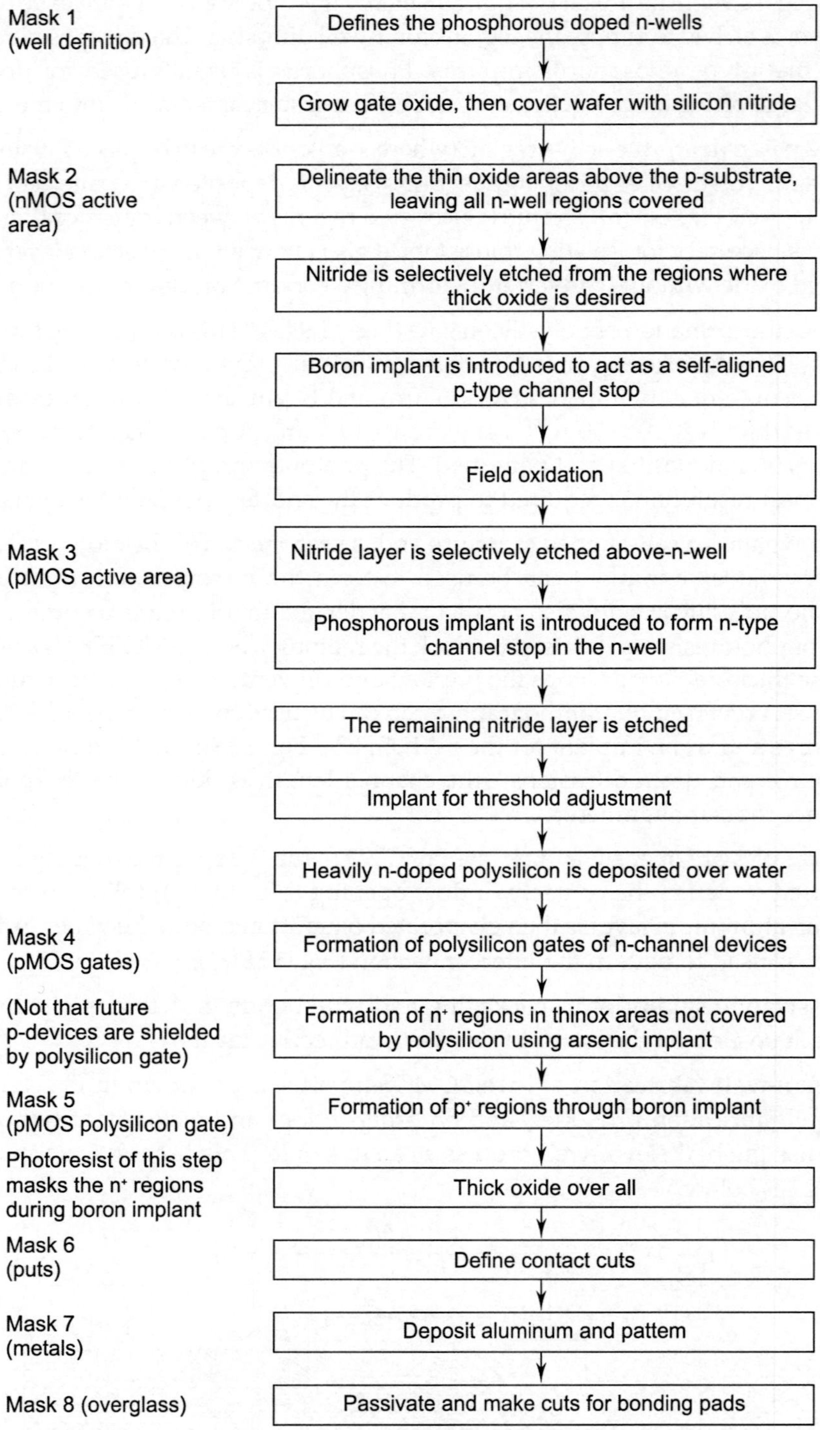

Fig. 12.11 *Flow Diagram of Berkelay N-Well Fabrication*

12.6.3 P-Well Fabrication Steps

A brief overview of the fabrication steps may be obtained with reference to Fig. 12.12(*a*) below, noting that the basic processing steps are of the same nature as those used for *n*MOS.

In premitive terms the structure consists of an *n*-type substrate in which *p*-device may be formed by suitable masking and diffusion and, in order to accommodate *n*-type devices, a deep *p*-well is diffused into the *n*-type substrate as shown in figure This diffusion must be carried out with special case the *p*-well doping concentration and depth will affect the threshold voltage as well as the breakdown voltages of the *n*-transistors.

To achieve low threshold voltage, we need either deep well diffusion or high well resistivity. However, deep wells require large spacing between the *n*- and *p*-type transistors and wires because of lateral diffusion and therefore a larger chip area. The *p*-well act as substrates for the *n*-devices within the parent *n*-substrate, and provided that voltage polarity restrictions are observed, the two areas are electrically isolated.

However, since there are now in effect two substrates, two substrate connections (V_{DD} to V_{SS}) are required as shown in Fig. 12.12.

All steps like, masking, pattering, and diffusion the process is similar to nMOS fabrication. In summary, the processing steps for the *p*-well CMOS device fabrication are as given below :

Step 1. Defines the areas in which the deep *p*-well diffusions are to be take place.

Step 2. Defines the thinox regions, namely areas where the thick oxide is to be stripped and oxide grown to accommodate *p*-and *n*-transistors and diffusion wires.

Step 3. Used to pattern the polysilicon layer which is deposited after the thin oxide.

Step 4. A *p*-plus mask is now used (to be in effect 'Anded' with Step 2) to define all areas where *p*-diffusion is to take place.

Step 5. This is usually performed using the negative form of the *p*-plus step and, with step 2, defines those areas where *n*-type diffusion is to take place.

Step 6. Contact cuts are now defined.

Step 7. The metal layer pattern is defined by this step.

Step 8. An overall passivation layer is now applied and steps is needed to define the opening for access to bonding pads.

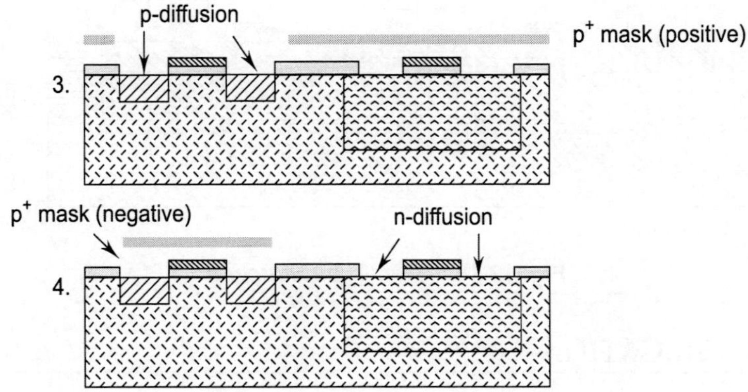

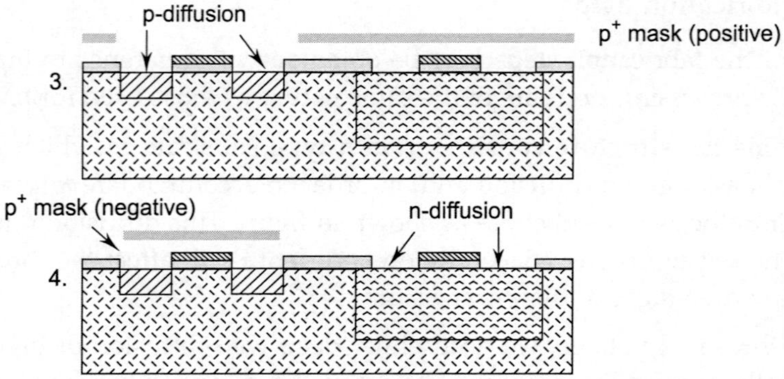

Fig. 12.12(a) *CMOS p-Well Process Steps*

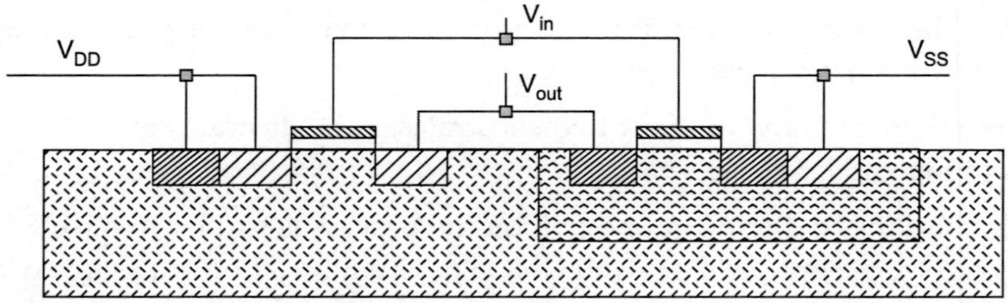

Fig. 12.12(b) *CMOS p-Well Inverter Showing V_{DD} and V_{SS} Substrate Connection*

12.6.4 Twin-Tube Fabrication Process

A logical extension of the *p*-well and *n*-well approach is the twin-tube fabrication process. The twin-tube fabrication process started with a substrate of high resistivity *n*-type material and then create both *n*-well and *p*-well regions.

This process of twin-tube fabrication is possible to preserve the performance of *n*-transistors without compromising the *p*-transistors. Doping control is more readily achieved and some relaxation in manufacturing tolerances results. This particularly important as far as latch-up is concerned. In general, the twin-tube fabrication process allows separate optimization of *n*-type transistor and *p*-type transistors. The arrangement of an inverter is shown in Fig. 12.13 for explaining the twin-tube fabrication process.

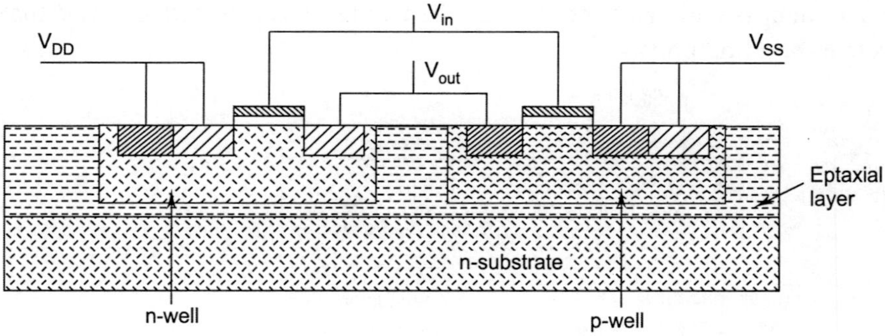

Fig. 12.13 *Twin-Tube Structure*

12.7 BiCMOS FABRICATION

BiCMOS is a modified cross technology that includes bipolar junction transistors as circuit element. In digital design, BiCMOS stages are used to drive high capacitance lines more efficiently than MOSFET only circuit. BiCMOS processing is more efficiently than standard CMOS, and bipolar transistors have an intrinsic voltage drop that cannot be avoided making them undesirable for low-voltage applications. A *n-p-n* vertical bipolar transistor can be integrated into the *n*-well CMOS process with the addition of a *p*-base diffusion region which is a combination of bi-polar transistor and unipolar transistor CMOS is known as biCMOS process. Fig. 12.14(*a*) below shows the BiCMOS devices.

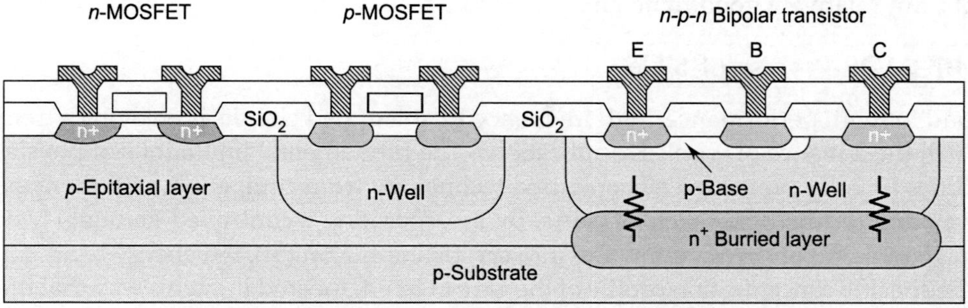

Fig. 12.14(a) *Cross-Sectional Diagram of a BiCMOS Process*

The characteristics of this device depend on the base width and the emitter area. The base width is determined by the difference in junction depth between the n^+ and the *p*-base diffusions. The emitter area is determined by the junction area of the n^+ diffusion at the emitter. The *n*-well serves as the collector for the *n-p-n* transistor. Typically, the *n-p-n* transistor has *a* β in the range 50 to 100 and a cut-off frequency greater than 10 GHz. Normally, an n^+ burried layer used to reduce the series resistance of the collector, since the *n*-well has a very high resistivity. However, this would further complicate the process with the introduction of p-type epitaxy and one more masking step.

Other variations on the bipolar transistor include the use of polyemitter and self-aligned base contact to minimize parasitic effects. The fact that most BiCMOS processes do not have optimized *p-n-p* transistors makes circuit design somewhat difficult. However, in non-critical situations, a prasistic lateral *p-n-p* transistor can be used as shown in Fig. 12.14(*b*) below.

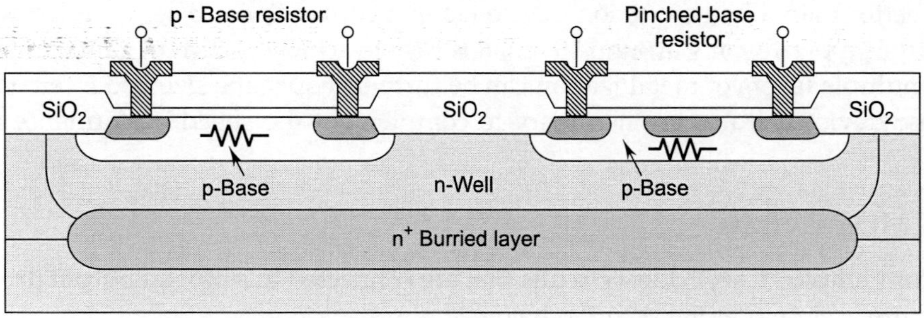

Fig. 12.14(b) *p-Base and Pinched p-Base Resistor*

In this case, the *n*-well serves as the *n*-base region, with p^+ diffusions as the emitter and the collector. The base width is determined by the separation between the p^+ diffusions. Since the doping profile is not optimized for the base-collector junction, and because the base width is limited by the minimum photolithographic resolution, the performance of this device is not very good - typically β of around 10, with a low cut-off frequency.

12.7.1 *p*-Base Resistor in BiCMOS

With the additional *p*-base diffusion in the BiCMOS process, two additional resistor structures are available. The base diffusion can be used to form a straightforward *p*-base resistor.

Since, the base region is usually of a relatively low doping level and with a moderate junction depth, it is suitable for medium-value resistors (a few kilo ohm). If a large resistor value is required, the pinched-base resistor can be used. In this structure, the *p*-base region is encroached by the n^+ diffusion, restricting the conduction path. Resistor values in the range of 10 kΩ to 100 kΩ can be obtained. As with the diffusion resistors. These resistors exhibit poor tolerance and temperature coefficients, but relatively good matching.

12.7.2 BiCMOS Process of SiGe

The demand for high performance, high frequency RF integrated circuits is enjoying a tremendous growth, with the advance of wireless applications. The fundamental limitations of physical material properties initially prevented silicon-based technology from competing with more expensive III-IV compound technologies such as GaAs. By incorporating a controlled amount (typically no more than 15 mole %) of germanium (Ge) into crystalline silicon (Si), the energy band gap can be altered. The specific concentration profile of the Ge can be engineered in such a way that the energy bandgap can be gradually reduced from that in pure Si region to a lower value in the region. This energy-band reduction produces a built-in electrical field that can assist the movement of carriers, hence giving faster operating speed.

Therefore, SiGe bipolar transistor can achieve significant higher cut-off frequency (*e.g.*, in the 50-70 GHz). Another benefit is that the SiGe process is compatible with existing Si-based fabrication technology, ensuring a very favorable cost/performance ratio. To take the advantage of the SiGe material characteristics, basic bipolar transistor structure must also be modify to further reduce parasitic capacitance (for higher speed) and to improve the injection efficiency.

The device makes use of trench isolation to reduce the collector sidewall capacitance between the *n*-well/*n*+ burried layer and the *p*-substrate. The emitter size and the p^+ base contact size are defined by a self-alignment process to minimize the base-collector junction (Miller) capacitance. This type of device is called a heterojunction bipolar transistor (HBT). Since, the emitter-base junction is from two different types of material, polysilicon emitter and SiGe base. The injection efficiency is significantly better than a homojunction device (as in a conventional BJT). Coupled with the fact that base width is typically only around 50 nm, it is easy to achieve a current gain of more than 100. In addition, multiple layers of metallization can be further reduce the size and interconnect resistance. All these device features are necessary to complement the speed performance of the SiGe material.

12.7.3 BiCMOS Circuits

BiCMOS circuits employ CMOS logic circuits that are connected to a bipolar output driver stage. A general structure is shown in Fig. 12.15(*a*) below.

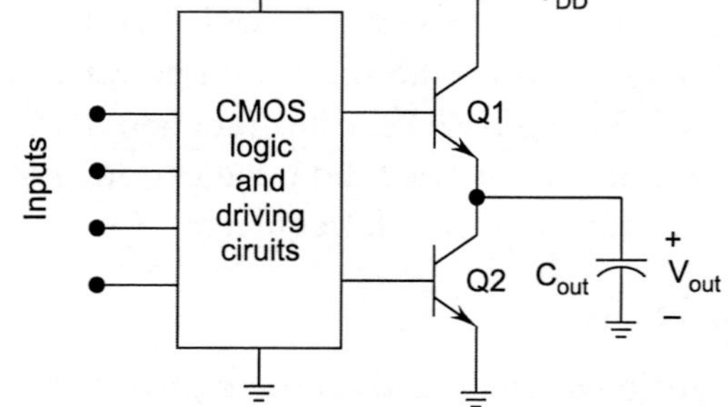

Fig. 12.15(a) *General Form of a BiCMOS Circuit*

The CMOS network is used to provide logic operations and drive the output bipolar transistor Q_1 and Q_2. Only one BJT is active at a time. Transistor Q_1 provides the high output voltage while Q_1 discharges the output capacitance and gives the low output state. The inverting circuit in Fig. 12.15(*b*) below gives an example of the operational detail.

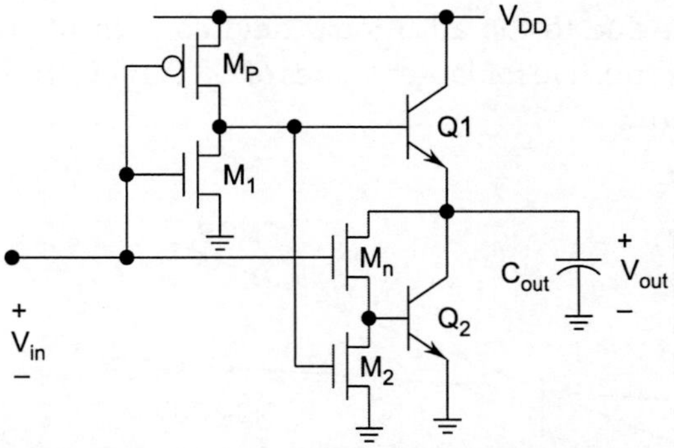

Fig. 12.15(b) *An Inverting BiCMOS Driver Circuit*

The NOT logic operation is performed by FET's M_p and M_n, even though they are separated from each other. The other two FET's M_1 and M_2 are used to provide paths to remove charge from the base terminals of Q_1 and Q_2 respectively. This speeds up the switching of the circuit, enhancing its use as an output driver.

12.7.3.1 *BiCMOS NAND2 Circuit*

The CMOS circuitry can be modified to provide logic functions. A NAND2 gate based on this design is shown in Fig. 12.16(*a*) below :

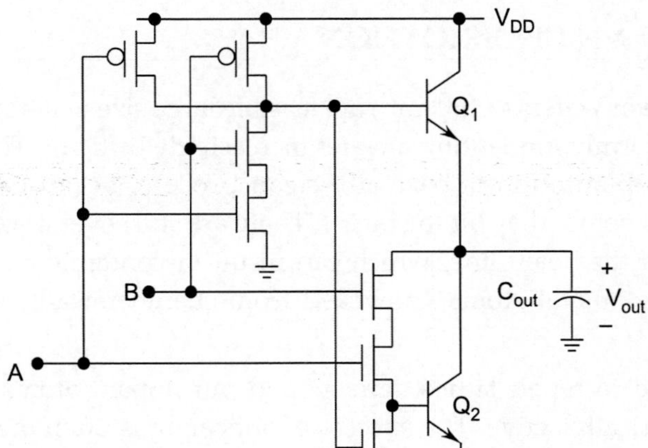

Fig. 12.16(a) *A BiCMOS NAND 2 Circuit*

A careful examination of the circuit shows that the logic is formed by the parallel pFET driving Q_1, and the series *n*FET's between the collector and the base of Q_2. The other FET's are used as pull-down devices to turn off the output transistors. Other logic functions can be designed using this as a basis. In general, the upper output transistor uses a standard-design CMOS circuit as a driver. The *n*FET section is replicated and placed in between the collector and base of the lower output transistor; adding a pull-down *n*FET to the base completes the design.

It is apparent that BiCMOS circuits are more complicated than their CMOS equivalents. If we write the total output capacitance as :

$$C_{out} = C_{transistors} + C_L$$

Where, C_L is the external load, we see that the parasitic transistor capacitance $C_{transistor}$ will be larger in a BiCMOS circuit due to the additional devices present. This leads to an important conclusion : BiCMOS is only effective for larger values of C_L. A typical plot of time delay t_d versus C_L is shown in Fig. 12.16(*b*) below.

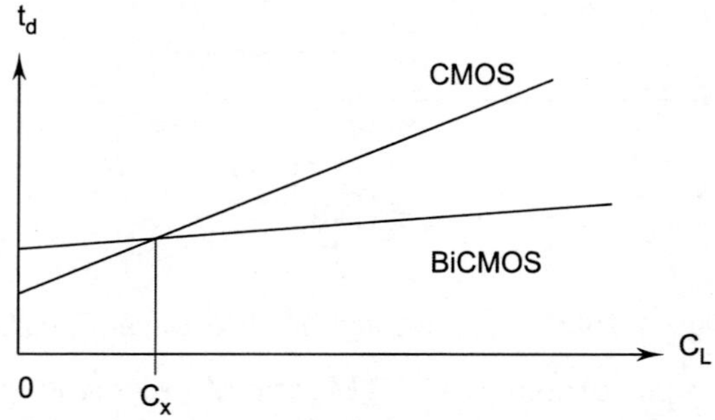

Fig. 12.16(b) *Gate Delay Versus External Load Capacitance*

Due to higher parasitic device capacitance, the CMOS and BiCMOS behaviors cross at a value $C_L = C_x$. For $C_L < C_x$, a standard CMOS design provides faster switching than a BiCMOS circuit. The speed increase in seen only for loads where C_L is much larger than C_x. This restricts the application of BiCMOS circuits to applications such as driving long date buses. Moreover, the cost and problem of V_{BE} drops are important factor in using the technology in digital VLSI.

12.8 PROBLEMS IN VLSI FABRICATION

Now-a-days large diameter wafers (> 150 mm) are feasible large size wafers are necessary for VLSI. This makes the task uniformly implanting a wafer increasingly difficult. This in turn has effect on sheet resistance. Ion-implantation is basically clean process, because contaminant ions are separated from the beam before they bit the target. There are still several sources of contamination possible near the end of the beam line, which can result in contaminant dose upto 10% of the intended in dose, for *e.g.*, metal atoms knowcked from chamber walls, wafer holder, masking aperture and so on.

Annealing, is required to repair lattice damage and put dopant atoms on substitutional site, where they will be electrically active. The success of annealing is often measured in terms of the fraction of the dopant that is electrically active, as found experimentally using a Hall effect technique. For VLSI, the challange in annealing is not simply to repair damage and activate dopant (possible by high-temperature annealing). But to do so while minimizing diffusion so that shallow implants remains shallow. This has motivated much work in rapid thermal annealing (RTA), where annealing times are on the order of seconds. RTA use tungsten-halogen lamp or graphite resistive strips to heat the wafer from one or both sides as against conventional furnace annealing where times or on the order of minutes.

Modern device structures, such as the lightly-doped drain (LDD) for MOSFET require precise control of dopant distribution vertically and laterally on a very find shallow n^+ layers with implantation energies within the reach of standard machines. As stated earlier, the ion velocity, perpendicular to the surface, determines the projected range of an implanted ion distribution. If the wafer is tilted at a large angle to the ion beam then the effective ion energy is greatly reduced. Tolted ion beam, thus, make it possible to achieve extremely shallow dopant distributions using comparatively high implantation energies.

We can circumvent the problem of implanting a shallow layer in silicon completely if instead we implant entirely into a surface layer and then diffuse the dopant into the substrate. This is most often done when the surface film is to be used as a conductor making contact to the substrate. This is most often done when the surface film is to be used as a conductor making contact to the substrate. Diffusion results in steep dopant profiles without damage to the silicon lattice. Dopant diffusion in silicides and polysilicon is generally much faster than in single silicon, so the implanted atoms soon become uniformly distributed in the film.

SUMMARY

In this chapter "process integration (Device fabrication)", technologies have been described that require hundreds of processing steps. The general trend for each individual technology has been toward smaller dimensions and more complicated processing. The shrinking of device dimensions has allowed improvement of device performance as well as an increase in the number of devices. The basic trend for NMOS and CMOS development will continue to be driven by reducing the device dimensions. Guidelines have been established for miniturizing the dimensions of the MOS devices to maintain ideal transistor characteristics.

These guidelines were derived with the assumption that the internal electric fields would be maintained constant and the power supply voltage would be reduced. In practice, the supply voltages have been not scaled with the many IC's that need to run at the same voltage. In addition, the simple scaling rules have not taken into account some of the parasitic effects that become dominant at the smaller dimensions, such as fringing capacitances and series resistances. Therefore, the continuing trend in MOS devices will be to reduce the dimensions as much as possible and to continue to modify the structure to make devices that can sustain as high a supply voltage as possible. However, at some point, there will be a fundamental limit to how much device dimensions can be reduced. Some recent estimates have set this limit at 90 to 65 nm dimensions for the channel length of the transistors.

This may be the limit to simply scaling the devices in two-dimensional surface area, but other trend to increase the density of the devices is extend the structures into the third dimension. This means either changing the substrate configuration to improve isolation of the devices or making multiple layers of transistors so that the devices are stacked upon each other.

TUTORIAL SHEET

1. What are the basic device fabrication on methods?
2. Explain P-N junction diode fabrication.
3. Explain BJT fabrication.
4. What are the various steps involved in CMOS fabrication?
5. Explain BICMOS working and fabrication.
6. What are the various problems involved in VLSI fabrication?

13

PACKAGING AND ASSEMBLY

13.1 INTRODUCTION

Packaging of electronics circuits is the science and art of establishing interconnections and a suitable operating environment and to predominantly for electrical circuits. It supplies the chips with wires to distribute signal and power, remove the heat generated by the circuits, and also provide them with physical support and environmental protection. It plays an important role in determining the performance, cost, and reliability of the system. With the decrease in feature size and increase in the scale of integration, the delay on the chip circuitry is smaller than that introduced by package.

The ideal package would be one that is compact, and should supply the chip with a required number of signal and power connections, which have minute capacitance, inductance and resistance. The package should remove the heat generated by the circuits. Its thermal properties should match well with semiconductor chip to avoid stress-introduced cracks and failures. The package should be reliable and it should cost much less than the chip it carries. Packaged electronics is the embodiment of all electronic equipments such as calculators, personal computers (PCs), mainframe computers, telephones, televisions and so forth.

It consists of active components, such as integrated circuit (IC) chips, flat-panel and cathode-ray tube displays, loudspeakers and so on.

The active components are interconnected supplied with electric power and housed in packaging. Success in a very competitive marketplace hinges on superior performance and cost performance is a broad term which encompasses the ability to perform a multitude of functions, airline reservations, banking transactions, and so-on-within a certain time, preferably without imposing an unacceptable wait on the users. It is task dependent, so only the general trends will be discussed throughout this book.

13.2 PURPOSE OF PACKAGING

Electronic packages contain electrical circuit components—up to several millions or even tens of million—mainly transistors assembled in integrated circuit (IC) chips, but also resistors, diodes, capacitors, and other components. To form circuits, these components need interconnections.

Individual circuits must also be connected with each other to form functional entities. Many of the intercircuit connections are steadily migrating into the IC chips as the levels of integration continue to increase. Mechanical support and environmental protection are required for the IC's and their interconnections. To function, electrical circuits must be supplied with the electrical energy which is consumed and converted into the thermal energy (Heat). Because all circuits operate best within a limited temperature , packaging must provide an adequate means for removal of heat.

13.2.1 Functions of Packaging

The packaging has four major functions:

1. **Signal distribution:** The main function of packaging is to distribute the signal to different components involving mainly topological and electromagnetic consideration.
2. **Power distribution:** Which involves electromagnetic, structural and material aspects.
3. **Heat dissipation:** The cooling is also function of packaging which involving heat dissipation by structural and material consideration.
4. **Protection:** Protection means mechanical, chemical and electromagnetic saving of components and interconnections.

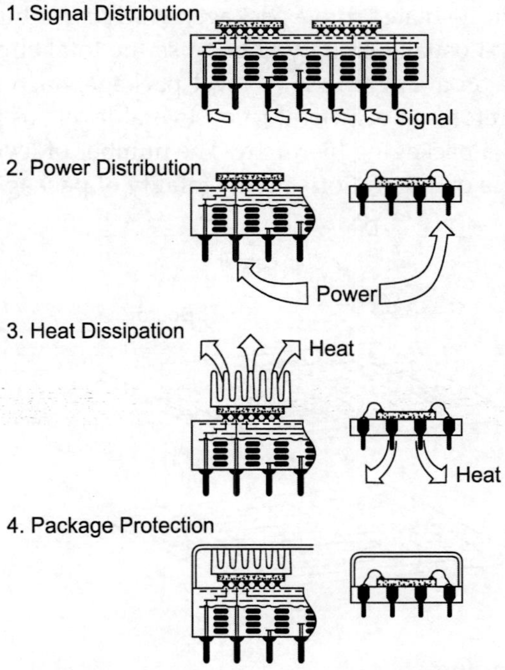

Fig. 13.1 *Four Major Functions of a Package*

13.2.2 Packaging Requirement

1. Speed:
    - Large bandwidth.
    - Short inter-chip propagation delay.
2. Thermal and Methanical:
    - High heat removal rate.
    - A good match between the thermal coefficients of the dice and the chip carrier.

3. Size:
   - Compact size.
4. Test and reliability:
   - Easy to test.
   - Easy to modify.
   - Highly reliable.
   - Low cost.
5. Pin count and wireability:
   - Large I/O count per chip.
   - Large I/O between the first and second level package.
6. Noise:
   - Low noise coupling among wires.
   - Good quality transmission lines.
   - Good power distribution.

13.2.3 Hierarchy of Packaging

The semiconductor chip is encapsulated into a package, which constitutes the first level of packaging. A printed circuit board is usually employed because the total circuit and bit count required might exceed that available on a single first – level package, such as capacitors, high power resistors, inductors etc. Therefore, as a general rule, several levels of packaging will be present. They are often referred to as a packaging hierarchy. The number of levels within a hierarchy may vary, depending on the degree of integration and the totality of packaging needs.

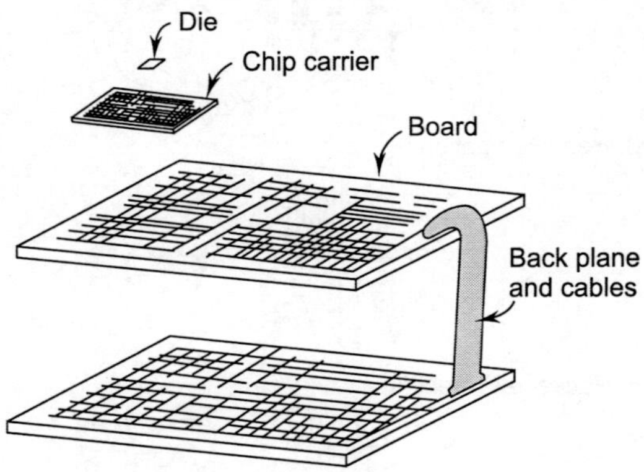

Fig. 13.2

13.2.3.1 First-level Packaging

The interfaces may be the exterior surfaces of an IC chip (die) itself-the common term being a base chip-often in the context of a "direct chip attach" or as it is still often the case, the IC chip has an enclosure, commonly referred to as a single chip module (SCM) or a chip carrier, or a header. In either it is considered a first level package. This is an extremely important and very diversified level of packaging which fills the entire part 2. Semiconductor packaging.

13.2.3.2 Second-level Packaging

The groups of chips (packaged in SCM's), along with other components such as capacitor, resistors, inductors, filters, switches and optical and RF components, are assembled on a second level packaging-covered in part-3 usually copper clad sheet of epoxy-glass laminates. After patterning of the metallic sheet followed by mechanical drilling and lamination to form interconnections, such structures are commonly called "printed wiring boards" (PWB) or card as in Fig. 13.3 below:

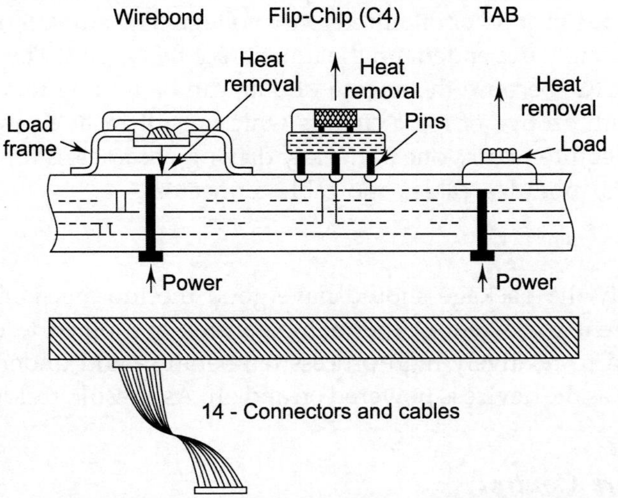

Fig. 13.3 *Packaging Level, Inter Connections and Technologies*

13.2.3.3 Third or Higher Level Packaging

The next packaging may be the outer shell of a small piece of equipment (*e.g.*, a hand-held calculator a portable phone or a pluggable unit to supplement or to enhance functions of a larger equipment a pluggable PC card with a hard disk or with a wireless modem). In a larger equipment (desktop or floor standing), several cards are plugged into a PWB. Such cards are sometimes referred to as "daughter cards" and the PWB as "mother boards". There may be several boards within a box/ frame. As the continuing increases in the number of circuits and or bits per IC chip tend to simplify the packaging hierarchy.

In addition to the relatively small and light IC chips, many electronics applications require bulkier and heavier camponents. Such as power transformers large electrolytic capacitors, disk drivers, display tubes, and so on. They are mounted separately and are electrically connected to the rest of the electronics via various cable. Cables also are used to connect boards either to each other to the rest of the equipment, including its external terminals (*i.e.*, a PC to its monitor, key-board, mouse, printer and via modem to the house telephone line.

13.2.4 Package Parameters

A successful package design will satisfy all given application requirements at an acceptable design, manufacturing and operating expense. As a rule, application requirements prescribe the number of logic circuits and or bits of storage that must be packaged to support electric power, kept within a proper temperature range, mechanically supported, and protected against the environment. Thus, IC packaging are designed to accomplish the following three basic functions : enclose the chip within a protective envelopment, provide electrical connection, and dissipate heat generated by the chip.

For the desiging the package for an application, the following parameters are considered:

13.2.4.1 Number of Terminals

The total number of terminals at packaging interface is a major cost factor. The terminals and signal inter connections constitutes the majority of conducting element. The other conductors supply power and provide ground or other reference voltage. The number of terminals supporting a group of circuits is strongly dependent on the function of this group. The smallest pinout can be obtained with memory ICs because the stream of data can be limited to a signal bit. Exactly the opposite is the case with groups of logic circuits which result from a random partitioning of a computer. The pinout requirement is one of the key driving parameters for all levels of packaging crips, chip carries, cards, modules, cables and cable connectors.

13.2.4.2 Reliability

For the better reliability the package should have good thermo mechanical performance. The materials having different coefficient of thermal expansion are joined to create interface. These interfaces are subjected to relatively high process temperature and undergo many temperature cycles in their useful life as the device is powered on and off. As a result, residual stresses are created in the interfaces.

13.2.4.3 Temperature Control

The operating junction temperature of a silicon chip should be low enough to prevent triggering the temperature-activity failure mechanisms. The package should provide a good medium for heat transfer from junction to the ambient/heat sink.

It is generally recommended to keep the Junction temperature below 50°C to ensure proper electrical performance and to contain the propensity to fail. The thermal expansion caused by heating up the packaging structure is not uniform it varies in accordance with the temperature gradient at any point in time and with the mismatches in the thermal coefficient of expansion.

The heat is transferred from the chip to the surface of the package by conduction and from the package surface to the ambient by convection and radiation.

13.2.4.4 Electrical Performance

The electrical performance at the IC package level is of great importance for microwave designs and has gained considerable attention recently for silicon digital devices due to ever increasing speed of today's circuits and their potentially reduced noise margins.

As signal propagates through the package, it is degraded due to reflections and line resistance. Controlling the resistance and the inductance associated with the power and ground distribution paths to combat ground bounce and the simultaneous switching noise has now becomes essential controlling the impedance environment of the signal distribution path in the package to mitigate the reflection related noise is becoming important. Reflections in addition cause an increase in the transition time and may split the signal into two or more pulses with the potential of causing erroneous switching in the subsequent circuit and thus malfunction of the system. Controlling the capacitive coupling between signal traces in the signal distribution path to reduce cross talk is gaining importance.

Increased speed of the devices demands that package bandwidth be increased to reduce undue distortion of the signal. All these criteria are related through geometric variables, such as conductor cross-section and length, dielectric thickness, and the dielectric constant of the packaging body. These problems are usually handled with transmission line theory.

13.2.4.5 Testability

Implicit in reliability considerations is the assumption of a flawless product function after its initial assembly—a defect manufacturing. Although, feasible in principal, it is rarely practiced because of the high costs and possible loss of competive edge due to conservative dimensions, tolerances, materials and process choices. So, several tests are implied to assess the reliability of the packages.

13.2.4.6 Rent Rule

The number of terminals required have reconfirmed findings of rent. The rent rule is used for finding the number of terminals required in a chip for the input, and output of data and signal required for chip operation.

The relationship is known as Rent's rule, which can be generally expressed as.

$$N = KM^P$$

Where, N is the number of required input-output terminals, *k* is the average number of terminals used for an individual logic circuit,

M is the total number of circuits,

P ($o \leq p \leq 1$) is a constant called Rent's constant.

Generally, the "dictate" of Rent's rule applies to information being conveyed to the assembly of circuits. Not necessary to the number of actual lines or terminals. Information can be coded on several lines, some what analogous to other devices, to trade the total number of terminals against the need to code-decode and to buffer data, with the resultant performance penalty. Such a penality is often small when many logic blocks are encountered by the data between entering and exiting the enclosure (chip, card, board) as it occurs in a microprocessor. Therefore, significant pinout reduction can be utilized with an acceptable performance degradation. In the case of high performance logic, only very few logic blocks are present between the input and output terminals. Consequently, to ultilize most circuits and to avoid severe performance degradation, Rent's rule has to be followed.

13.3 TYPES OF PACKAGE

To meet the requirement of high speed and density the IC packages have been developed over time to time. The history of IC package development has been the continuous battle to miniaturization. The different types of packages are as given below:

1. Through Hole Package.
2. Surface Mounted Package.
3. Flip-chip Package.
4. 3-D VLSI Package.

13.3.1 Through Hole Package

(Through the board hole mounting technology uses precision hide drilled through the board and plated with copper). This copper plating forms the connection between separate layers. These layers consist of thin copper sheet stacked together and insulated by epoxy fiberglass. These are no dedicated via structures to make connections between wiring levels, through hole serve that purpose. Though holes form a sturdy support for the chip carrier and resist thermal and mechanical stress caused by the variations in the expansions of the components at raised temperatures. Different types of through hole package can be further classified as:

13.3.1.1 *Dual in Line Package*

This is a rectangular package with two rows of pins in its two sides. Here, first the die is bonded on the read frame and in the next step, chip input-output and power/ground pads are wire-bonded to the lead frame and the package is molded in plastic. DIP's are the workhorse of the high-volume and general-purpose logic products.

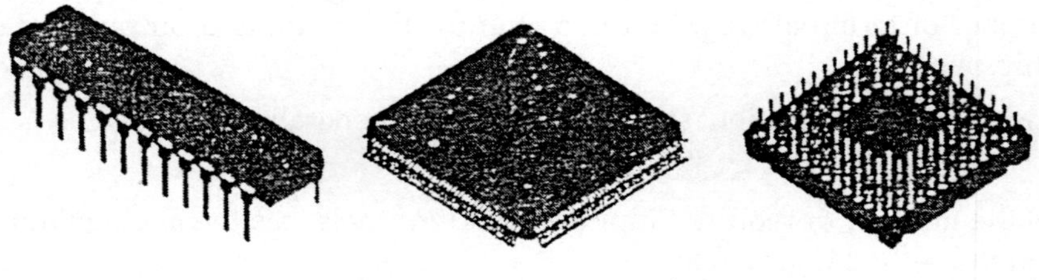

(a) Dual in-line package (DIP). (b) Quad flat package (QFP). (c) Pin grid array package (PGA).

Fig. 13.4 *Different Through Mount Packages*

13.3.1.2 *Quad Flat Package (QFPs)*

With the advances in VLSI technology, the lower available pin counts of the rectangular DIP becomes a limiting factor. With pins spaced 2.2 mm apart and analyan two sides of the package, the physical size of the DIP has become too large. On the other hand, the physical size of an un package micro electronic circuit has been reduced to a few millimeters. As a result, the DIP package has become up to so times larger than the bare die size itself, thus defeating the objective of shrinking the size of the integrated circuits. So, one solution is to provide pins all around. I QFP's pins are provided on all four sides. Thin QFP's are developed to reduce the weight of the package.

13.3.1.3 *Pin Grid Array (PGA)*

PGA can offer a much larger pin count. A pin grid array has leads on its entire bottom surface rather than only at its periphery. It has cavity up and cavity down versions. In a cavity down version a die is mounted on the same side as the pins facing toward the PC board, and a heat sink can be mounted on its backside to improve the heat flow. When the cavity and the pins are on the same side, the total number of pins is reduced because the area occupied by the cavity is not available for brazed pins. The mounting and wire bonding of the dice are also more difficult because of the existence of the pins next to the cavity. High pin count and larger power dissipation capability of PGAs make them attractive for different types of packaging.

13.3.2 Surface-Mounted Packages

The shortcomings of through the board mounting are solved by the surface mounted packages. In surface mounted package technology, a chip carrier is soldered to the pads on the surface of a board without requiring any through holes. The smaller component sizes, absence of through holes and the possibility of mounting chip on the both sides of the PC board improve the board density. This reduces package parasitic capacitances and inductances associated with the package pins and board wiring. The various types of surface mount package are available in the market and can be divided into the following categories:

Small-outline Package (SOP)

The small-outline package has gull wing shaped leads. It requires less pin spacing than through hole mounted DIPs and PGAs. SOP package usually have small lead counts and are used for discrete, analog and SSI/MSI logic parts.

Plastic-leaded Chip Carriers (PLCCs)

(PLCC's) such as gull-wing and leaded chip carriers, offer higher pin counts than SOP. Plastic leaded chip carriers pack denser and are more suitable for automation than gull-wing leaded carriers because their leads do not extend beyond the package.

Leadless Ceramic Chip Carriers (LCCC's)

LCCC's take advantage of multilayer ceramic technology. The conductors are left exposed around the package periphery to provide contacts for surface mounting. Dice in leadless chip carriers are mounted in cavity-down position, and the back side of the chip faces away from the board, providing a good heat removal path. The ceramic substrate also has a high thermal conductivity. LCCC's are thermetically sealed.

13.3.3 Flip Chip Packages

The length of the electrical connections between the chip and the substrate can be minimized by placing solder bumps on the dice, flipping the chips over, aligning them with the contacts pads on the substrate, and reflowing the solder balls in a furnace to establish the bonding between the chip and the package. This method provides electrical connections with minute parasitic inductance and capacitance. In addition contact pads are distributed over the entire chip surface. This save silicon area, increase the minimum input-output and power/ground terminals available with a given die size, and provides more efficiently routed signal and power/ground interconnects on the chip.

Chip Size Packages (CSP)

To combine the advantages of both chip and bare chip in one solution, a variety of CSPs have been eveloped. CSPs can be divided into two categories:

(*i*) Fan-in type (*ii*) Fan-out type

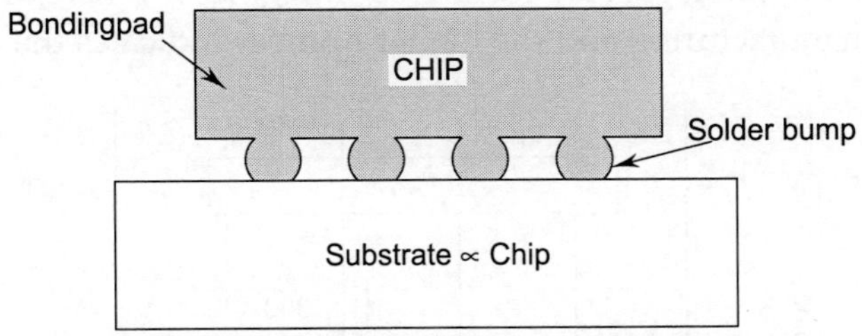

Fig. 13.5 *Flip-Chip Using Solder Bumps*

(*i*) *Fan-in type:* Fan-in type CSPs are suitable for memory applications that have relatively low pin counts. This type is further divided into two types, depending on the location of bonding pads on the chip surface, these are the center pad type and the peripheral pad type. This type of CSP keeps all the solder bumps within the chip area by arranging bumps in area array format on the chip.

(*ii*) *Fan-out type:* The fan out CSPs are used mainly for logic applications, because of the die size to pin count ratio, the solder bumps cannot be designed within the chip area.

13.3.4 3-D VLSI Packaging

The driving forces behind the development of 3-D packaging technology are the multi-chip module technology. Although the requirements for the 3-D technology are more aggressive. These requirements include the need for significant size and weight reductions, high performance and small delay, higher reliability and potentially reduced power consumption.

Silicon-on-silicon Hybrid

A silicon substrate can also be used as an interconnection medium to hold multi-chips as an alternative to ceramic substrates. This is called silicon-on-silicon package, or sometime, hybrid wafer-scale integration.

Thin film interconnections are fabricated on a wafer and separately processed, and test dice are mounted on this silicon substrate via wire bonding. Using this technique, chip fabricated in different technologies can be placed on the same hybrid package. The silicon substrate can also potentially contain active devices that serve as chip-to-chip drivers, bus and input-output multiplexers, and built-in-test circuitry.

13.4 DIE ATTACHMENT TECHNIQUES

Connections between chip and package are commonly performed by die attachment techniques. These die attachment or interconnects are to provide electrical signal between the chip pads and package, different bonding techniques are used.

These die attachment or connections techniques are given below:

1. Wire bonding.
2. Controlled collapse chip connection (C-4).
3. Tape automated bonding (TAB).

Depending on the number and spacing of input-output connections on the chip and the substrate as well as permissible cost these techniques are differentiated. The possible number of connections per chip with each of the technologies are illustrated in Fig. 13.6. Where, the arrow indicates the current technology in manufacturing and the higher number indicates the technology feasibility.

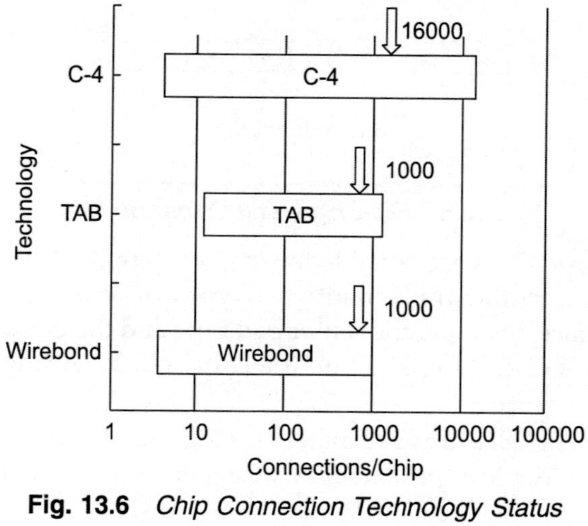

Fig. 13.6 *Chip Connection Technology Status*

13.4.1 Wirebonding Technique

Wirebonding is the most common chip-bonding technology, spanning the needs from consumer electronics to main frames. Wirebonding (Fig 13.7) is a method to connect a fine wire between an on chip pad and a substrate pad. This substrate may simply be the ceramic base of a package or another chip. The common material used are gold and aluminum.

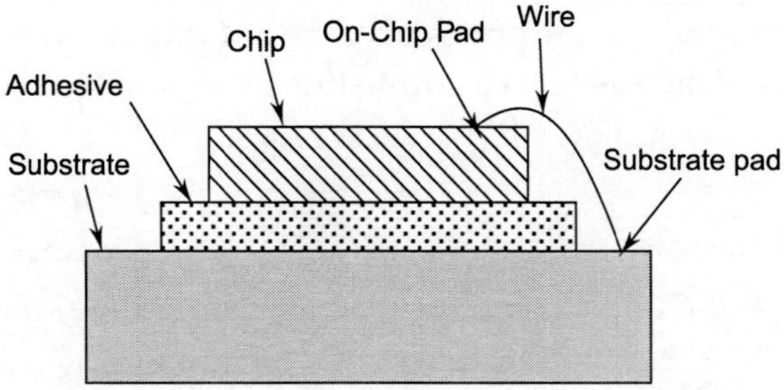

Fig. 13.7 *Wirebonding Assembly*

Its widespread use is primarily based on the fact that the maximum number of chip connections in products in use can be accomplished with this technology in addition to provide the low cost per connection. It should be pointed out that wire bonding concurrently provides for thermal dissipation by backbonding or diebonding the chip to the substrate.

The *disadvantages of wirebonding are*:

(*a*) The connection density (input/ output/ chip size in mm) increases by a factor of 2 (from 750/400) to 4 (5000/1250). Thus, if leading-edge wire bonding is to meet today's requirement. It needs to be enhanced by a factor of 4 in pitch a major challenge.

(*b*) The peripheral pad connections on the edge of the chip particularly on large chips do not provide optimal power distribution to all the transistors on that chip.

(*c*) The most important limitation of wire bond has to with the economics of chip usage itself. Because wire bonding cannot be performed easily in manufacturing over the active area of chip compared to area array solder flip-chip connections.

(*d*) The size of the chip larger by about one third with the wirebond chip. This difference will account for tens to hundreds of dollars, as it directly relates to the number of chips per wafer.

(*e*) In addition to the above reasons, the reliability of the flip chip has been shown three order of magnitude better than wirebond.

13.4.2 Controlled Collapse Chip Connection (C-4)

(C-4) technique of chip connecting is also called solder bumps. The solder bumps are small spheres of solder (solder balls) that are bonded to contact area or pads of semiconductors devices and subsequently used for face down bonding. The length of the electrical connections between the chip and the substrate can be minimized by placing solder bumps on the die, flipping the die over, aligning the solder bumps with the contacts on the substrate and re-flowing the solder balls in a furnace to establish the bonding between the die and the substrate.

This technology provides electrical connections with minute parasitic inductances and capacitances. In addition, the contact pads are distributed over the entire chip surface rather than being confined to the periphery. As a result, the silicon area is used more efficiently, the maximum number of interconnects is increased and signal interconnections are shortened. But, this technique results in poor thermal conduction, difficult inspection of the solder bumps, and possible thermal expansion mismatch between the semiconductor chips and the substrate.

13.4.3 Tape Automated Bonding (TAB)

The various advancement in tape automated bonding technology are being pursued by Japanese companies as Shinko-Denshi, NEC, Fujinitso, Mitsui-kinzoku, Oki and Nitto Denko. One particular enhancement of TAB being pursued by Nitto Denko is illustrated in Figure below, comparing the new two layer direct copper bonding process with a conventional three layer process. The new process coats polyamide onto copper, the opposite of the 3M process in the united states that coats copper on kapton or other polyamides by electroplating.

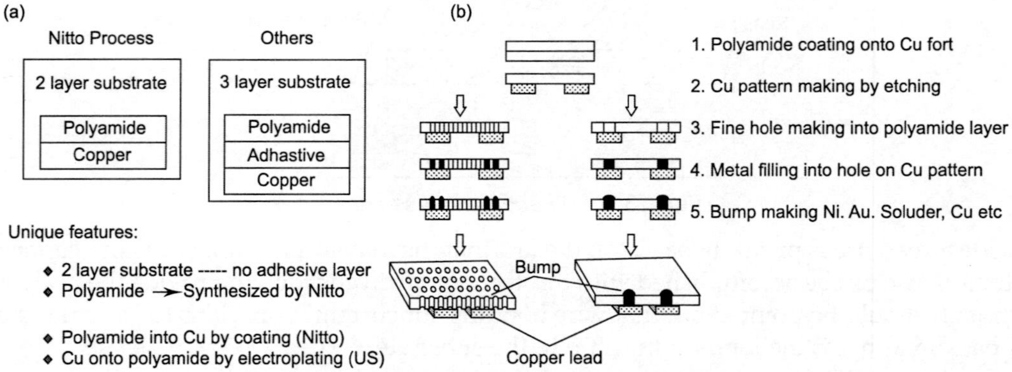

Fig. 13.8 *TAB (Tape Automated Bonding)*

The advantages claimed for new process are:

1. High Heat Resistance.
2. No Curling.
3. Low Moisture Absorption.
4. Strong Peel Strength.

In tap-automated bonding technology, a chip with its attached metal film is placed on a multilayer polymer tape. The interconnections are patterned on a multilayer polymer tape. The tape is positioned above the bare die, so that the metal tracks correspond to the bonding sites on the die.

TAB technology provides several advantages over wire bonding technology. It requires a small bonding pad, smaller on chip bonding pitch, and a decrease in the quantity of gold used for bonding. It has better electrical performance, lower labour costs, higher I/O counts and lighter weight, greater densities and the chip can be attached in a face up-down configuration. TAB technology includes time and cost of designing and fabricating the tape and the capital expense of the TAB bonding equipment. In addition each die must have its own tape patterned for its bonding configuration. Thus, TAB technology has typically been limited to high volume application.

13.5 PACKAGING FABRICATION TECHNIQUES

The single chip packages are generally based either on refractory-ceramic technology or on metal lead frames and molded plastics. Ceramic packages are usually used for state-of-art devices, where maximum reliability is required. For more mature products where low cost is critical, but which still require a hermetic seal, CERDIP technology is employed. This technology uses a combination of lead frames, dry-pressed refractory ceramic parts and glass sealing. Plastic packaging is usually better for more mature products where low cost is paramount and hermeticity is not required. Until recently plastic had implied lower reliability, now however, plastic technology has improved, and is highly reliable when made with proper process control. This section discusses processing technologies that are generally applicable to a variety of package type. The different three types of packaging fabrication technologies are described here:

1. Hermetic Vs Non-Hermetic Packaging.
2. Ceramic Packaging.
3. Plastic Packaging.

13.5.1 Hermetic Vs Non-Hermetic Packaging

There is a natural division of definitions between the two most common packaging approaches for integrated circuits:

Hermetic packaging which presents a sealed, impermeable package to the operating environment, and what we term non-thematic, which is permeable. The hermetic package usually consists of a cavity package made of ceramic, metal, glass, so forth, that are, for the most part, resistant to permeantion by fluids and gases. On the other hand, non-hermetic packages are generally associated with organic coatings or seals such as epoxies, silicones, polyamides, and so forth, which are not capable of preventing penetration of fluides and gases.

These are also new technologies that are designated "hermetic-equivalent" packages that use novel approaches to obtain hermeticity through depositions of metal films to seal the electronics package. Both hermetic and non-hermetic packages are able to house single device or multichip systems.

Hermetic Packaging

The moisture on a device's surface during operation has historically been shown to lead to the principle causes of failure in the field. The extremely small geometries involved in integrated circuits, differing galvanic potentials between metal structures, and the presence of high electric fields all make the device susceptible to interactions with moisture. Once packaging method designed to prevent performance degradation due to moisture's deleterious effects is the hermetic package. Ceramic or metal is used to form an enclosure to isolate the electronic devices from the ambient operating environment. By eliminating condensible moisture from the cavity of the package during the sealing process and preventing the ingress and egress of moisture at the package perimeter during its operating life, excellent long-term reliability can be achieved.

The word hermetic is defined as completely sealed by fusion, solder, and so on, so as to keep air or gas from getting in or out, in other words, airtight. In termetic packaging practice, such seals are non-existent. Small gas molecules will enter the package over time through diffusion and permeation. Eventually, these gases will reach equilibrium within the cavity of the package. In light of this permeation, long life can still be obtained in the field because of the extremely slow nature of this activity. Accelerated tests and leak testing for screening and qualification are specified in the military standed MIL-STD-883 and represent the foundation and industry accepted approach to testing for reliability in hermetic packages.

Non-hermetic Package

The widespread availability of non-hermetic packaging followed that hermetic packaging by many years due to problems associated with the initial polymer material used for encapsulated circuits fell into this category. The inability of these early polymers to sufficiently retard the deleterious effects of moisture once it reached the delicate surface of the circuit, inter connect, and substrate led to poor performance both in testing and in the field. In adequate adhesion, contaminants within the material itself, incompatible thermal expansion, and stress-related problems, in addition to a relatively immature knowledge of filler technology, all combined to prevent plastic packaging immediate acceptance. With the significant efforts in the areas of resins, fillers, material formulations, and process development work, polymer packaging finally began to make it presence felt in the early 1970's although finally began to make its presence felt in the early 1970's although they had been used in some form as early as the late 1950's. During this time, significant progress was made in improving the quality of the glass passivation layer that is deposited on the active area of the device as a first line of defense against moisture-related problems.

The combination of these technological advances provides the essential increase in reliability that was needed for polymer packaging to begin seeing widespread use. During plastic packaging's infancy period, many failure modes were identified and materials concenies addressed that resulted in different polymers and polymer formulations and its applications. This has led to recent demonstrations of plastic packages that approach the reliability of hermetic packages in many applications and environments. Certainly, plastic-molded packages are the most dominant packaging method in use today for commercial and industrial-grade electronics. It is estimated that over 90% of all integrated circuit are marketed in this form.

Non-hermetic packaging encompasses not only the aforementioned plastic-molded package but also plastic and ceramic cavity packages that are sealed with a polymer, and the more recent chip on board (COB) and multichip module (MCM) approaches to packaging when a hermetic seal is not used. Most of the high-performance MCM's being produced, MCM-C (Ceramic) and MCM-D (deposited), are currently packaged in hermetic packages, but the lower-cost MCM-L(laminated) which is a wirebonded or tape automated bonding (TAB) device on a printed wiring board, tends to be encapsulated with organic polymers in an overcoat fashion. Much remains to be understood in the use of polymers for protection of semiconductor devices. Generalities of the required material properties are known, and how to avoid the failures associated with them, but specific degradation process and activities at the material surface interface remain the object of many research dollars.

As mentioned earlier, corrosion was the primary cause of failure in early polymer packaging. New formulations with better filler technology resulted in materials that do not impart stress-related failures on devices and their associated interconnect, until the recent emergence of large dies, are attempted. In the newer technologies of COB and MCM's, corrosion is once again a primary cause of failure with the existing encapsulation materials, especially with wirebonded devices as in most plastic-molded packages where the thin aluminum bond pad on the device is exposed to the coating interface. Moisture can also have deleterious effects on the long term adhesion between organic material and the substrate being protected. Moisture acts as a debonding agent through one or a combination of the following mechanisms:

1. Attach of the metal surface by moisture can form a weak, hydrated oxide surface.
2. Moisture-assisted chemical bond breakdown.
3. Moisture-related degradation or depolymerization.

The natural protection of metals, surface oxides, can actually be detrimental to the durability of the polymers adhesion in wet environments. Both metals oxides are relatively polar. Water (H_2O) will preferentially adsorb onto the oxide surface and create a weak boundary layer at the metal-polymer interface that can lead to adhesion problem. In wet environments, long-term corrosion prevention properties of a polymer can be adversely effected by this weakening interface region.

13.5.2 Ceramic Packaging

Ceramic and glasses defined as inorganic and non-metallic materials, have been an integral part of the information-processing industry. Glasses and ceramic are used for different purposes as defined below:

1. For data processing by the use of semiconductor device interconnected onto packages.

2. Ceramic are also used for storing information by the use of such magnetic materials as iron and chromium oxide and ferrites.
3. For displaying information by the use of low temperature solder glasses sealed to transparent faceplate glasses.
4. For printing information by use of ceramic-metal print heads and corrosion resistant glass.
5. For transferring information by the use of very low loss glass fibers.

Integrated circuit (IC) packaging is one of most important applications of ceramic in microelectronics. Ceramic materials in the form of dual-in-line package chip carriers and pin-grid arrays are used throughout the industry, from consumer electronics to main frame computers. Ceramic possess a combination of electrical, thermal, mechanical and dimensional stability properties unmatched by any group of materials.

For example, ceramics have dielectric constants from 4 to 20,000, thermal expansion coefficients matching silicon (3×10^{-6}/ °C) to matching copper (17×10^{-6}/°C), and thermal conductivity from that of insulated brick (0.1 w/m k) to better than aluminum metal (220 w/m k), which is one of the best metallic thermal conductors known. Dimensional stability as measured by either change of dimensions as a function of temperature and time or shrinkage control during substrate processing has been achieved at better than ± 0.05% nominal shrinkage, allowing as many as 100 layers of ceramics and metal to be cofired into one substrate. A number of manufacturers, such as IBM, are routinely manufacturing these type of substrates but with a lower number of layers with nominal zero shrinkage and tolerance better than ± 0.1%. Such as illustrated in Fig. 13.9 below. Recent technology breakthroughs have led to the use of ceramics as superconductors, offering no electrical resistance to the flow of current at low temperatures.

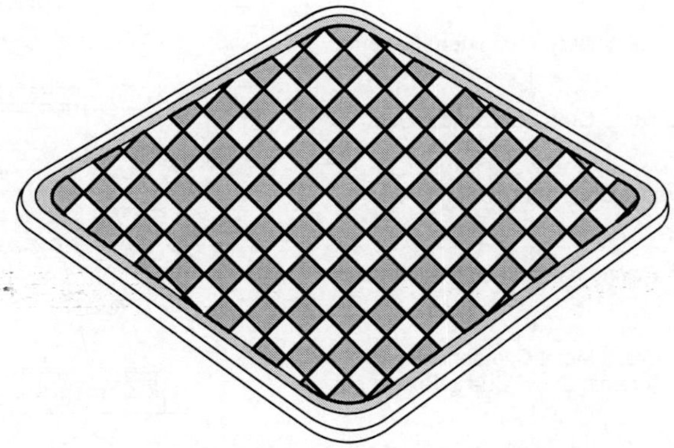

Fig. 13.9 *Ceramic Substrate*

As a result, a number of commercial applications in medical, electronics and transportation industries are being developed. If the temperature could be pushed closer to room temperature, a direct and profound application for power distribution of electronic package could be developed.

Types of Ceramic Packages

Ceramic Packages offer two types of technologies for design, which are as described below:

1. Ceramic Solid Logic Technology.
2. Ceramic Dual-in-line Package.

1. **Ceramic Solid Logic Technology:** During 1963, IBM introduced for its own use a ceramic package called SLT (Solid Logic Technology), 12 × 12 mm in size, made of 96% dry pressed alumina on which conductor and resistor pastes were screened and fired at about 800°C.

Sixteen connector pins of copper were swaged into the holes in the substrate, and the entire assembly was immersed in a solder bath to coat the conductors and pins, forming electrical contacts between pins and conductor lines. Semiconductor chips were soldered in place and the cap sealed to the module with epoxy as shown in Fig. 13.10 below.

Fig. 13.10 *Solid Logic Technology Package*

2. **Ceramic Dual-in-line Packages:** The availability of dual-in-line package (DIP's) and the ease with which they can be transported and automatically inserted into the holes of printed circuit boards and soldered in place made them an industry standard. Ceramic DIP's were fabricated using either ceramic green sheets or dry-pressing processes. Figure. 13.11 below illustrates the tape processes used in fabricating DIP's.

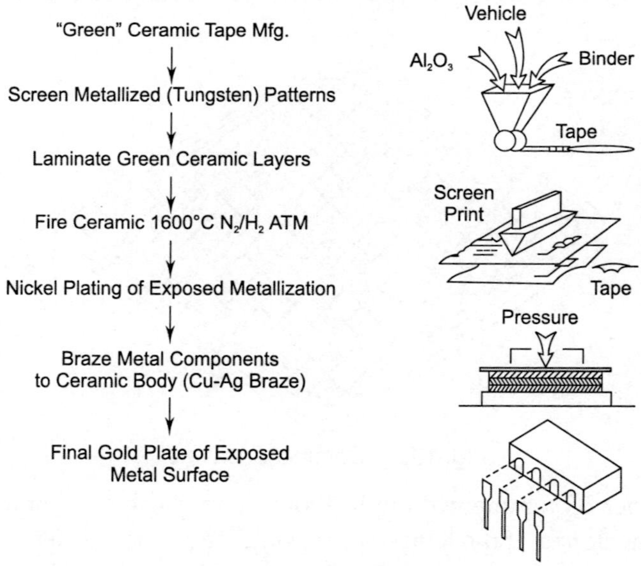

Tape Processes Used in Fabrication DIP

Fig. 13.11 *DIP*

With either process, the maximum number of leads provided with DIP packages has been 64. This limit is due to fabrication difficulties, the question of the reliability of solder joints to the board, and excessive lead inductance, among other reasons. The tape process involves forming a slurry of well-dispersed alumina and glass in a suitable organic vehicle, and doctor blading to form tape, which is metallized with tungsten metal paste.

The desired number of layers is laminated and sintered in a controlled atmosphere between 1500°C and 1600°C. Exposed tungsten metallizations are nickel and gold plated and then brazed with a copper-silver alloy. Pressed ceramic as the name implies, involves molding a mixture of alumina lubricants and binders to form the desired shape, which is then sintered to form a monolithic body. Metal frames are bonded to this body by using low temperature glasses. Hermetic sealing posed problems early in the development of DIPs. These problems were traced to the incomplete removal of water prior to sealing. Glasses in the system $PbO - B_2O_3 - SiO_2 - Al_2O_3 - ZnO$, Capable of meeting all requirements, have been successfully developed and used to solve the hermeticity problem.

Advantages of Ceramic Package

1. Ceramic substrate provides the highest wiring density of all substrates technology. Even though the line and via dimensions as well as their pitch are coarser and further apart compared to thin film technology.
2. Ceramics is the only packages that achieved the highest reliability, and as consequence, the packages that are in use today for the very stringent reliability requirement are ceramic packages.
3. Ceramics by their nature are hermetic.
4. Their dimensional stability during and after high temperature processing is expectional.
5. Chemical inertness of most of the ceramic to water, acids, solvents and other chemicals is outstanding.

Disadvantages of Ceramics

1. Ceramics are brittle making them prone to catastrophic failure.
2. Ceramics sensitive to stress unless designed carefully for each applications they are used.
3. In very-high performance applications requiring very low dielectric constant and very high packaging densities, ceramics cannot compete with thin film organic package.

13.5.3 Plastic Packaging

The first evidence of plastic encapsulation was seen in the early 1950's using compression molding of phenolics. The phenolics materials was compressed around electric devices, leaving the electrical connectors protruding. The problem with this method was that the delicate connections were often severed due to the high pressure. By the early 1960's plastic encapsulation emerged, as an inexpensive simple alternative to ceramics and metal encasing, and during 1970's virtually all high-volume Integrated Circuits (IC's) were encapsulated in plastic. For example, one cost comparison of a 14-pin package in 1979 for quantities of 5,00,000 packages and a gold price of $ 200 per ounce sets the cost per plastic package at $ 0.063 and cost of a ceramic dual-in-line package (DIP) at $ 0.82.

Although, epoxy novolac was the first material used for plastic encapsulation, phenolics and silicones were the dominant plastics of the 1960's. At that time, plastic encapsulated microelectronics (PEM's) were plagued by numerous reliability problems owing largely to the poor quality of the encapsulation system. Over the decade, numerous formulations of expoxies were developed which had lower curing strinkages and contamination levels and thus took over as the main encapsulating material in the early 1970's. Although, silicones are still sometimes used, a typical encapsulant used today is a complex mixture of cross-linkers, accelerators, flame retardants fillers, coupling agent, mold release agents and flexibilizers in an epoxy resin matrix.

Plastic Package Construction

A plastic package or plastic encapsulated microelectronics (PEM's) consists of a silicon chip, a metal support or lead france, wires that electrically attach the chip circuits to a lead frame with incorporates external electrical connections and a plastic epoxy encapsulating material to protect the chip and the wire interconnects. The lead frame is made of a copper alloy, alloy 42 or alloy 50 and is plated with gold and silver or palladium, either completely or in selected areas over nickel or nickel/cobalt. The silicon chip is usually mounted to the lead frame with an organic conductive formulation of epoxy. Wires, generally of gold but also of aluminum or copper, are bonded to the aluminum bonding pads on the chips and to the fingers of the lead frame. The assembly is then typically transfer molded in epoxy. Following the molding operations, the external leads are plated with a lead-tin alloy, cut away from the strip, and formed into desired shapes. Plastic packages can be premolded or postmolded. In the former, a plastic base is molded; the chip is then placed on it and connected to an I/O fan-out pattern with wire.

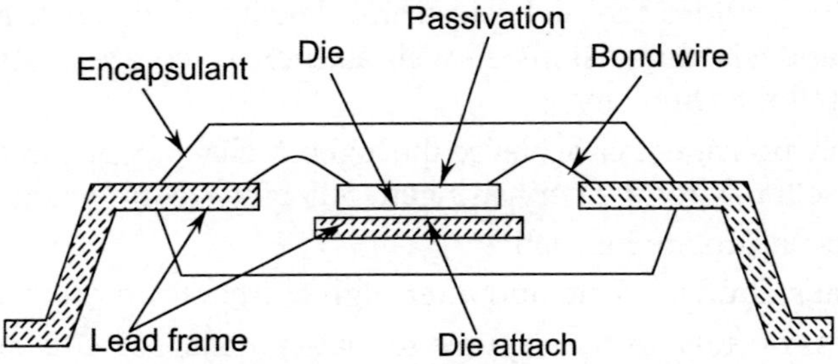

Fig. 13.12 *Plastic Package Construction*

The die and wire bonds are usually protected by an epoxy attached lead which forms a cavity. The two types of plastic package are:

1. **Premolded plastic packages:** Premolded plastic packages are most often used for high-pin-count devices or pin grid arrays that are not amenable to flat lead frames and simple fan-out patterns.
2. **Postmolded plastic packages:** The postmolded plastic packages are that in which, the die is attached to a lead frame, which is then loaded into a multicavity molding tool and encapsulated in a thermoset molding compound via the transfer molding process. Postmolded packages are less expensive than premolded one because there are fewer parts and assembly steps. In the 1990's about 90% of plastic packages were made using postmolding techniques.

Plastic encapsulation has come a long way. It offers attractive benefits over ceramic technology in terms of form factors, weight, performance, cost and availability. The reliability of plastic packages is no longer a stumbling block to their widespread applications. Plastic encapsulated microelectronics have generally encountered formidable challenges in gaining acceptance for use in government and military applications. In fact, it was only in the early 1990's that the industry dispelled the nurtron that hermetic packages were superior in reliability to plastic packages, in spite of the fact that hermetic packages typically had low production and procurement volumes, along with outdated government and defense department standards and handbooks associated with their manufacture and use. Prior to the 1980's moisture induced failure mechanisms, such as corrosion, cracking and interfacial delamination were significant.

However, improvement in encapsulants, die passivation, metallization technology, and assembly automation have made plastics the encapsulation technology of the future. Currently, the best endorsement for plastic encapsulated microelectronics (PEM)'s is from automotive manufactures, which consume plastic integrated circuits at a rate of over 2.7 million a day. PEM's will continue to account for the vast share of the integrated circuits market in coming year, but hermetic packages, with their special characteristics will continue to have a unique market in the electronics industry.

13.6 PACKAGE SEALING AND ENCAPSULATION

Package sealing is performed to protect the device and package metallization from corroding environments and from mechanical damage due to handling. Moisture is one of the major source of corrosion of the semiconductor devices. Electro-oxidation and metal migration are associated with the presence of moisture. The diffusion rate of moisture also depends on the encapsulant material as shown in Fig. 13.13 below, which compares the permeability

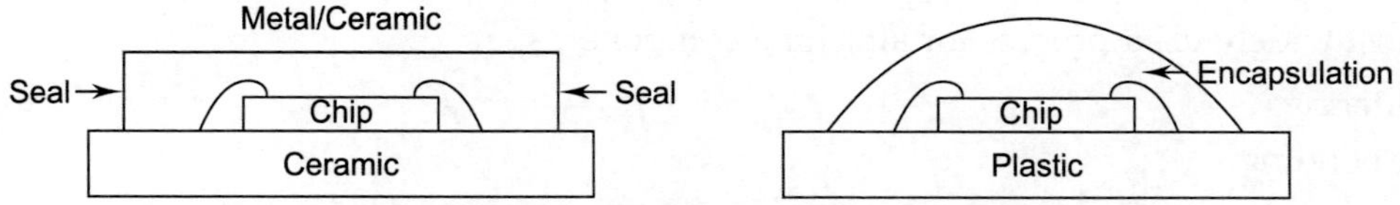

Fig. 13.13 *Basic Concept of Sealing and Encapsulation*

of water for various organic and inorganic materials. In spite of the poor permeability of organic and inorganic materials in the figure, a number of encapsulants, particularly silicones certain epoxies, polyamides, and imide-siloxanes have been shown to have great promise. In plastic packaging, a variety of plastic encapsulants have been developed to minimize the diffusion of water. This results in significant improvements in reliability of plastic encapsulated packages. The diffusion coefficients of these materials range from 1×10^{-8} cm^2/s for epoxy resins, 12×10^{-8} cm^2/s for typical silicone 305, to 0.1×10^{-8} cm^2/s for NYSOL transfer molding material. The developments with silicone rubber have recently led to further improvements with silicon gel, which is soft and can be deposited to thinner dimensions with minimal stress problems.

13.7 PACKAGING IN WIRELESS APPLICATIONS

Wireless applications typically involve RF, high-frequency digital and mixed-mode circuits. Wireless packaging requires minimal electrical parasitic effects that need to be well characterized. In wireless applications, the trend is to integrate multiple modules on a single chip, so the thermal management of the whole chip becomes crucial. The IC package must have good thermal properties. Metal as a material shows optimal properties concerning thermal conductivity, electromagnetic shielding, mechanical and thermal stability.

For thermal expansion, best match to semiconductors and ceramic material can be achieved with molybdenum, tungstem, or special composities like kovar. Ceramic materials are applied, both as parts of the package as well as for subsystem carring RF transmission lines. To this end, and to provide electromagnetic shielding, these materials partly have to be metallized. Aluminum nitride, beryllia, aluminum silicon carbide, and CVD diamond show best thermal conductivity and are therefore applied in high power applications, while alumina is well known for standard microwave applications. Integration of passive components is a major challenge in wireless packages. More and more efforts are being made to integrate passive components, power devices on a chip with the other mixed signal circuits. The size of the package becomes an issue.

Micromachining technology provides a way to make miniature packages that conform to RF circuits, while providing physical and electrical shielding. Conformal packages made by applying micromachining technology provide the capability to isolate individual circuit elements and improve circuit performance by eliminating the radiation and cross-coupling between the adjacent circuits. At high frequencies, interconnections need to be carefully designed. Microstrip interconnects, coplanar waveguide are mostly used for microwave packaging. Flip-chip packaging has tremendous potential for future RF packaging.

13.8 ASSEMBLY PROCESSES

The thick-film substrate provides an assembly vehicle for the addition of components, other than the inductors, resistors and capacitors that were deposited during the screening processes. These components include transistors, diodes, integrated circuits, capacitors, diodes, inductors and so on. In establishing an assembly process, care must be taken to insure that each step is compatible with the preceding steps and will not adversely affect the performance of the circuit.

The most widely used process for attaching components are:

1. Soldering.
2. Die bonding.
3. Wire bonding.
4. Ultrasonic bonding.

13.8.1 Soldering Process

Soldering process is most widely used for attaching components of low temperature solder, usually of a lead-tin composition. With the proper care, the right kind of flux, and an appropriate solder, reliable solder joints can be made. A careful analysis of the materials used in the component, substrate and solder is required. The most common flux is white rosin, mildly activated forms of rosin and fluxes containing a small amount of weak organic acid can also be used.

Even though the metal is not alway easy to tin, only the mildest fluxes are permissible, because of the problem of removing the flux. There are several ways in which solder can be applied to the circuit

Point-to-Point Wiring

Point-to-Point wiring technique requires that the components be interconnected by soldering their leads to the connection pads with a soldering iron. The components must have leads or ribbon terminations which have been cut to length and preformed. This system is set up very fast and has a very low initial cost. However, obtaining reliable solder joints requires well-trained operators and is a relatively slow and expensive process. The solder joint can be reworked at least once, giving repairability to the circuit; however, reworked joints are of poor mechanical strength.

Dip Soldring

This technique makes use of solder fountain. The substrate must have holes which allow the component to be placed on the back with the leads extending through the holes to the solder pad. After all the components has been assembled, the substrate is passed over the surface of the solder pot or fountain.

Prefluxing of the substrate is required to obtain good solder joints. This technique is the same as that used for attaching components to printed circuit boards, and therefore requires lead type components which have had their leads prepared for inspection. The system is inexpensive and easy to set up, inspect, and rework. It also affords a built-in heat sink for the component, and gives a joint good mechanical strength and electrical connection. Because all of the components are soldered at one time, there is a minimum amount of localized thermal shock to the substrate. The main disadvantage of this system is that it requires relatively large-sized components and therefore results in a large micro-circuit package.

Solder Reflow

In this system, the substrate is prepared with an application of solder onto the pad areas. The component is then put in place, and the solder is reflowed by means of application of heat. This heat can be applied by conduction, in a convention oven, with an infrared strip heater, or even by induction. The main requirement is that sufficient heat be generated to raise the temperature of the solder to its mealting point, so that alloying can take place with the component pad. The component pads must have a solderable surface in the form of a metallized or solder-coated pad. This system is usually used on leadless micro miniature components, and results in a small circuit package. The solder can be applied to the circuit as a paste or a liquid; close dimensional control is required to get reworkable joints.

The system allows for good process control and minimum thermal shock to the substrate, and can be reworked only when caution is used. The resulting solder joints are very difficult to inspect and have poor mechanical strength. The reflow solder technique is used for the lead less inverted-device type component, and also by a leading computer manufacturer for flip-chip type components.

13.8.2 Die Bonding

The general requirements on a die bond are one or more of the following:

- Low ohmic contact.
- Electrical isolation.
- Adequate heat sinking capability.
- Absence of P-N Junctions.
- Adequate mechanical bond strength.

The method most commonly used for chip mounting is alloying. Two less prevalent methods are soldering and epoxy plastic mounting. Factors of primary importance are adhersion, thermal conduction, environmental stability, and cost. Alloying, as the term is used here, consists of combining the silicon of the back of the chip with another material to form a liquid phase which adheres to both the header base and the back of the chip. Gold is used in nearly all cases, because it has a eutectic formation of 370°C, which is high enough to with stand additional processing steps in most cases but low enough to preclude any damage to the chip and to promote easy processing.

The gold is usually supplied in the form of gold, gold-germanium, or gold-silicon perform, or a heavy gold layer on the substrate. In order to make the alloy formation, the thin layer of silicon dioxide on the back of the chip must be broken to allow the silicon and metal to combine. This is usually accomplished by mechanically scrubbing the chip on the gold material, whether it is preform or substrate metallization, from 390°C to 450°C. A protective atmosphere of nitrogen is often used to reduce the formation of oxide during the operation.

Once the alloy is formed, the substrate is removed from the heating block, and the alloy is allowed to solidify. The entire operation, requiring no more than from ten to twenty seconds, can be accomplished with a variety of equipment, from a pair of tweezers and a gas-shrouded hot block to a semiautomatic magazine-fed alloy utilizing vibrating, oscillatory, or ultrasonic scrubbing.

13.8.3 Wire Bonding

The process employed to achieve electrical continuity between the circuit and the package is that of wire bonding. The following characteristics are required of the joining process:

(*i*) It must be capable of bonding wires of small diameter to the metallized areas of the substrate.
(*ii*) It must introduce no contamination to the surface of the circuit.
(*iii*) It must produce a bond that will remain stable, both electrically and mechanically during subsequent processing of the package and exposure to the service environment.

A number of factors influence the joining of metals by the application of the pressure. It is clear that if the metals have perfectly smooth metal surfaces, free from contaminating films of organics and oxides, the probability of achieving a successful bond is greatly enhanced. In practice, it is not possible to achieve such surface. Placing two surfaces together produces only a few points of actual contact, the surfaces being held apart by small irregularities which flow under load.

The main factors which influence the pressure welding process are:

- Topography of the surface.
- Mechanical properties of the metal and its oxides.
- Presence of organic film.
- Applied load
- Temperature and time at temperature under load.

Wire bonding is the most expensive operation in hybrid circuit fabrication. In addition, more environmental failures are due to faulty bonding wires than to any other single cause. Two types of bonding are generally used for hybrid circuit fabrication:

(*i*) Thermal compression.

(*ii*) Ultrasonic.

Thermal compression is the older method.

Ultrasonic having become well established only in the past few years.

13.8.4 Ultrasonic Bonding

Ultrasonic techniques are suitable for lead bonding on transistor models where there is sufficient bonding area, which can be determined by looking at the geometry picture on the data sheet. Ultrasonic bonding is physically performed similar to wedge bonding, and energy is supplied as ultrasonic vibration rather than a heat.

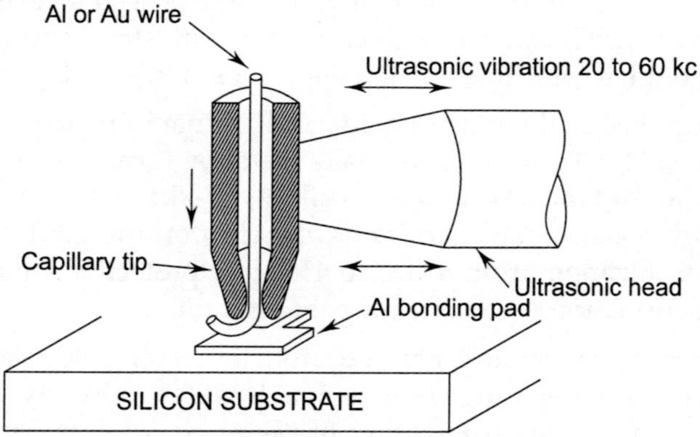

Fig. 13.14 *Ultrasonic Bonding*

Since, the system is not heated, the formation of aluminum oxide is less likely. Consequently aluminum wire bonding is easily accomplished. Ultrasonic bonding was originally developed in the semiconductor industry when analysis revealed that a major cause of failure in semiconductor devices was the formation of gold-aluminum intermetallic which exhibit low mechanical strength-commonly known as purple plague. Ultrasonic bonding has a further advantage in hybrid circuit. In that the process eliminates the need of elevated temperatures during bonding.

This is extremely important in hybrid circuit assembly, where fifty to one hundred bonds are not unusal. Exposure of the active devices to elevated temperatures for extended periods of time can lead to degradation of the device characteristics. In ultrasonic bonding, two similar or dissimilar metals are joined. A pulse of alternating current of suitable frequency is delivered to the transducer, which converts the AC power to acoustic power at the same frequency. A linear effect is obtained if a large polarizing field is used in conjunction with a small AC driving field. The acoustic power is transmitted through the coupling system into the materials to be bonded while these are held together by a relatively low static clamping force. The net acoustic power delivered to the bond zone is from 15 to 30 percent of that originally drawn from the power line.

13.9 ASSEMBLY TECHNIQUES

The different assembly techniques are as given below:

1. Surface Mount Assembly.
2. Chip-and-Wire Assembly.
3. Mixed Technologies.
4. Special Modules.

13.9.1 Surface Mount Assembly

The surface-mount assembly technique can be categorized under lowest-cost, fastest turnaround assembly method, using pre-packaged component soldered to glass-epoxy board, flex circuit, or thick film ceramic substrate. Many package styles are available, such as SIP, DIP. Pins may be attached as in line leads 90° leads etc.

13.9.2 Chip-and-Wire Assembly

In order to minimize interconnect electronic circuit parasitics, a high-density layout is one of the assembly techniques recommended as a solution. The highlight of this technique is epoxy attachment or wire bonding of integrated circuits and components to a glass-epoxy board (chip-on board) of course, another way is attachment to a thick or thin film ceramic substrate. Currently, many package styles are available and can be listed as follows:

- Epoxy seal, using a B-stage ceramic lid epoxies to the substrate (quesic-hermetic seal).
- Encapsulation, in which bare semiconductor die are covered with a "glob top" (low-cost seal).
- Metal (typically kovar) package with kovar lid either welded or soldered to the package (hermetic seal).
- Leads are typically plug in type, SIP, DIP, PGA etc.

13.9.3 Mixed Technologies

Another category of assembly technique recognized as "mixed technologies" combines chip and wire with surface mount assembly techniques on a single substrate, which may be a glass-epoxy board or ceramic substrate. Heat sink/heat spreaders are available in a variety of materials. The Maxtek module includes a 4-layer, 20 mil-thick glass epoxy board mounted to a beryllium copper heat spreader. Selectively gold plated for wire bonding pads at each end for use with elastomeric connectors:

- Three methods of IC die attach.
- Epoxied directly to glass-epoxy board.
- Epoxied directly to the BeCu heat spreader through a cutout in the board.
- Epoxied to the head spreader, throught a cutout, via a thermally conductive submount to electrically isolate the die from the heat spreader.
- Solder-mounted resistors and capacitors.
- 50 cm differential signal lines.
- IC die may be "glob topped" or covered in either a ceramic or plastic lid for protection.

13.9.4 Special Modules

Under special modules we can emphasize a highlight technologies of complex assemblies of mixed technology substrates, flexible circuits and electromechanical components complex assemblies of mixed technology substrates often utilize a B-stage epoxy lid or glob top over chip-and wire circuitry. These technologies enable designers to provide an integrated solution complex system problems like in a module, which is CRT Driver System capable of displaying 4 million pixels with 1.5 cm rise and fall times. The circuit incorporates a thin film MCM connected by a special high frequency connector to an FR-4 board with thick-film ceramic low inductance load resistor and flex circuit with an integral impedance-matching inductor, connecting directly to the CRT.

SUMMARY

The rapid development of IC technology has created a critical need for packaging sealing and encapsulation using advanced polymeric materials for encapsulant, interlayer dielectrics and passivation layers. Recent advances in high-performance polymeric materials, such as improved silicone elastomers, ultra soft silicone gels, low-stress epoxies, low thermal expansion coefficient and photode-finable polyimides and cyclobutenes have begun to provide polymers to meet reliability requirements of VLSI technology. In packaging pin counts will undoubtedly continue to increase with increasing silicon capability. Alumina ceramic packages will dominate the high-performance VLSI packaging technology until factors such as its high dielectric constant, modest thermal conductivity, or cost force a change to other contending materials.

Higher packaging density on the PWB level will drive the package design toward smaller lead pitches. If chip integration keeps pace with decreasing feature size, today's chip-interconnection technologies, especially wire bonding, will be challenged and could be unstable at I/O > 200. If wire bonding is to survive, it will need improvements in placement accuracy to meet demands of tighter bond-pad pitches. The TAB and flip-chip will be revisited, to assess their future in VLSI packaging. Both technologies lead themselves to hybrids, especially TAB with its pretestability features. Flip-chip has the advantages of a very-low inductance connection to the device, compared to wire or TAB.

TUTORIAL SHEET

1. What is the purpose of packaging?
2. What are the functions and requirements for packaging?
3. Explain various package parameters.
4. Define Rent's rule. What is its significance?
5. What are the different types of packages?
6. What is C-4 technique and what are its disadvantages?
7. What is TAB?
8. What is the difference between Hermetic and non-herimetic package?
9. What is Ceramic packaging? Explain with its types? What are its advantages and disadvantages?
10. Differentiate between package sealing and encapsulation.
11. What are the uses of packaging in wireless applications?
12. What are the most widely used process for attaching components?
13. What are the different assembly techniques?

14

INTRODUCTION TO MEMS

14.1 INTRODUCTION

Micro-electro-mechanical-system or MEMS represents an extraordinary technology that premises to transform whole industries and drive the next technological revolution. These devices can replace bulky actuators, and sensors with micro-scale equivalent that can be produced in large quantities by fabrication process used in integrated circuit. This reduces cost, bulky size, weight and power consumption while increasing performance, production volume and functionality by order of magnitude. *For example,* one well known MEMS device is the accelerometer for a car airbag chip that replaced a system of conventional mechanical sensors.

Furthermore, it is clear that current MEMS products are simply precursors to greater and more pervasive applications to come including genetic and disease testing, guidance and navigation system, power generation, RF devices (especially for cell phone technology), weapons systems, biological and chemical agent detection and data storage. Micromirror based optical switches have already proven their value ; several start-up companies specializing in their development have already been sold to large network companies for hundreds of millions of dollars. A MEMS silicon motor together with a strand of human hair is as shown in Fig. 14.1. The promise of MEMS is increasingly capturing the attention of new and old industries alike as more and more of their challenges are solved with MEMS.

14.1.1 What is MEMS?

Micro-electro-mechanical-system (MEMS) is a process technology used to create tiny integrated devices or system that combines mechanical and electrical components. They are fabricated using integrated circuit (IC) batch processing techniques and can range in size from few micrometer to millimeters. These devices have the ability to sense, control and actuate on the micro-scale and generate effects on the micro-scale.

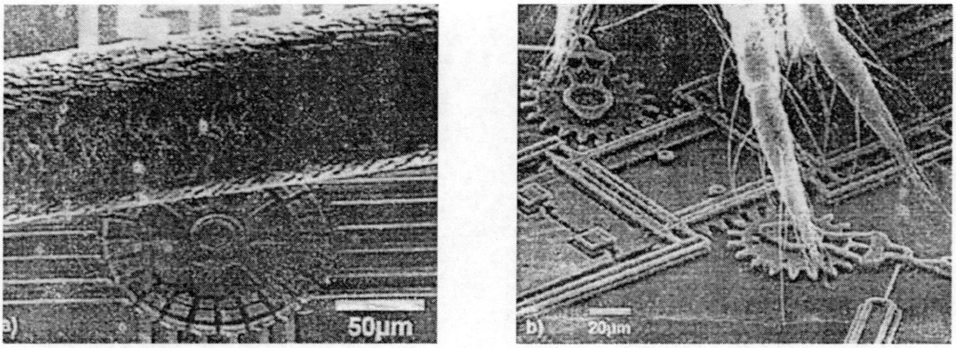

Fig. 14.1 *A MEMS Silicon Motor Together with a Strand of Human Hair*

Overview:

In the past decade, new fabrication techniques have developed which helped popularize micro-electro-mechanical-system (MEMS), and various noval devices have been reported in diverse areas of engineering and science. One such area is microwave and millimeter systems.

MEMS for VLSI applications should solve many problems of high frequency technology for wireless communications. The advanced developments of communication devices forced the market to acquire miniaturized efficient devices, which is possible only by the development of radio frequency (RF) MEMS.

The term MEMS refers to the design and fabrication of VLSI for integrated circuits. Traditional MEMS can be divided into two classes : MEMS actuators and MEMS sensors. The first one is a kind of moving mechanism activated by an electrical signal like micromotor. Micro-sensors are currently available for a large number of applications. Due to their case of fabrication, these were the earliest microsystems. The reason for the actuators not becoming popular is that the amount of energy generated by such tiny system does not cause much impact in the associated system.

The passive MEMS devices include bulk micromachined transmission lines, filters and couplers. Active MEMS devices include switches, tuners and variable capacitors. They include, in addition to the conventional microelectronics packaging, integrating antenna structures for command signals into micro electromechanical structures for desired sensing and actuating functions. The system also may need micropower supply, micro relay and microsignal processing units. Microcomponents makes the system faster, more reliable, cheaper and capable of incorporating more complex functions. In the beginning of the 1990'^{s}, MEMS emerged with the aid of the development at integrated circuit (IC) fabrication processes. Where sensors, actuators and control functions are co-fabricated in silicon. The remarkable research progresses have been achieved in MEMS under strong capital promotions from both government and industry.

In addition to the commercialization of some less-integrated MEMS devices, such as microaccelerometers inkjet printer heads, micromirrors for projection, etc., the concepts and feasibility of more complex MEMS devices have been proposed and demonstrated for the applications in such varied fields as microfluidics, aerospace, biomedicine, chemical analysis, wireless communication, data storage, display, optics etc.

Some branches of MEMS, such as micro-opto-electromechanical system (MOEMS), micro-total analysis system (µTAS), etc., have attracted a great deal of research interest since their potential application market.

14.1.2 History of MEMS

The history of MEMS is useful to illustrate its diversity challenges and applications. The following list summarizes some of the key MEMS milestones.

1950′s

1958 : Silicon strain gauges commercially available.

1959 : "There' Plenty of Room at Bottom" – Richard Feynman gives a milestone presentation at California of Technology. He issues a Public Challenge by offering $ 1000 to first person to create an electrical motor smaller than $\frac{1}{64}$ th of an inch.

1960′s

1961 : First silicon pressure sensor demonstrates.

1967 : Invention of surface micromachining. Westinghouse creates the Resonant Gate Field Effect Transistor, (RGT). Description of use of sacrificial material to free micromechanical devices from the silicon substrate.

1970′s

1970 : First Silicon accelerometer demonstrated.

1979 : First micromachined inkjet nozzle.

1980′s

1980 : Early 1980's : First experiments in surface micromachined silicon. Late 1980's : micromachining leverages microelectronics industry and widespread experimentation and documentation increases public interest.

1982 : Disposable blood pressure transducer

1982 : "Silicon as a Mechanical Material". Instrumental paper to entice the scientific community reference for material properties and etching data for silicon.

1982 : LIGA process.

1988 : First MEMS conference.

1990′s : Methods of micromachining aimed towards improving sensors.

1992 : MCNC starts the multi-user MEMS process (MUMPS) sponsored by Defense Advanced Research Projects Agency (DARPA)

1992 : First micromachined hinge.

1993 : First surface micromachined accelerometer sold (Analog Devices, ADXLSO)

1994 : Deep Reactive Ion Etching is patented.

1995 : Bio MEMS rapidly develops

2000 : MEMS optical - networking components become big business.

14.2 MEMS TYPES AND CHARACTERISTICS

14.2.1 MEMS Types

To understand the basic principles of MEMS reliability, it must first be noted that very little is understood about mechanics and materials at the microscopic level. At this scale, many common physical principles, such as inertia and gravity have insignificant effects, and atomic and surface forces dominate micromachine behaviour. Even when the material is known to have fully-understood properties at the micro scale, such as friction and wear almost completely new tribological data must be catalogued when dealing with micron sized features.

Categories Two Major of MEMS:

1. Sensors
2. Actuators.

14.2.1.1 Sensors

Sensor is a device that measure information from a surrounding environment and provide an electrical output signal in response to the parameters it measured.

Sensors, such as accelerometers, and gyroscopes tends to have no moving parts that operate under the conditions of sliding friction. As such, many of their performance characteristics may be measured with conventional tools used in the microelectronics industry.

14.2.1.2 Actuators

Actuator is a device that convert an electrical signal into action. It can create a force to manipulate itself, other mechanical devices or the surrounding environment to perform some useful function.

For example, Actuators such as micromirrors, microengines and gears represent an entirely different class of devices that requires the characterization of their electrical and their mechanical properties. Therefore, custom tools are required to analyze and measure their performance and reliability. Laboratories and companies have written proprietary software and developed test stands for this purpose.

14.2.1.3 Failure Issues

The primary obstacle to overcome in MEMS is adhension—that is, one moving part sticking to another part or to the ground plane. Also known as stiction, adhension occurs under various conditions : *e.g.,* two clean surfaces in contact, the surface tension of an etchant (pulling two structures together) during the etching release of a MEMS device, or an electrostatic attraction due to a trapped charge. Some solutions to the adhension problem are relatively straightforward (such as increasing the number of layers of a poly spring to minimize out-of-plane movement) and some require more efforts (such as coating a device with a self-assembling monolayer to eliminate capillary action).

Another concern is operational wear in actuators. Although some MEMS have been shown to operate for billions of operating cycles, the erosion of sliding surfaces—and the particulate contamination that results from such action—must be controlled for MEMS to realize their commercial potential.

Various Solutions:

1. Using harder structural materials (*e.g.,* diamond instead of poly).
2. Using liquid or solid lubricants.

3. Optimizing device design.
4. Optimizing the frequency of the operational cycle.

A major issue is packaging, a problem that appears to be the focus of numerous failure mode. Normal IC packaging is hermetically sealed to prevent contamination. Few MEMS devices (such as accelerometers) may be packaged with similar standards, but others, require interaction with the outside world—*e.g.*, gas detectors, medical 'Lab-on-a-chip' system and optical switches (requiring optical windows), and are therefore subject to intrusion by unwanted factors through those openings.

A dust mote can wreak serious havoc on an actuator of comparable size, and stray radiation can similarly damage sensitive components ; solution must maintain functionality while minimizing vulnerability. Similarly, a micromachine that requires free space to function is in additional danger to contamination during manufacture.

Also, a MEMS device must contend with heat dissipation, power supply, input and output signals, and other factors. Although these challenges in adhension, wear, and packaging have led to an abundance of research, companies have met with very limited commercial success. Institutional researches are making significant headway into these issues, but the momentum towards commercialization is still slow ; MEMS will require time to gain more substantial interest—and therefore resources—from industry.

14.2.2 Monolithic Integration of MEMS with CMOS

One sought—after potential of MEMS is the ability to integrate the functions of sensing, actuating, control and processing on one piece of real estate-reducing cost, size, and possible points of failure. However, MEMS require processing steps that are incompatible with some CMOS fabrication processes when combined on the same chip. If the MEMS were fabricated first, the processing would make the remaining surface finish unacceptable for CMOS photolithography ; on the other hand, if the CMOS were fabricated first, then the heat from the MEMS processing would cause the aluminium to contaminate the stoichiometry of the doped silicon and cause it to the fail. These problems make 'system-on-chip" integration a difficult fact.

Solution of these Problems:

(*a*) One is to fabricate the CMOS design using tungsten instead of aluminium. Thereby allowing the chip to undergo a subsequent MEMS fabrication without adversely affecting the CMOS portion.

(*b*) Fabricate the MEMS portion, placing the assembly in a 'trench' slightly below the surface of the water. Once the MEMS device are finished, the entire trench is filled and sealed with an oxide. The wafer is then planarized by chemical-mechanical polishing, thus making the wafer surface acceptable for conventional CMOS photolithography. After the fabrication of the circuit, the oxide is etched away, freeing the MEMS devices.

(*c*) Another solution involves an analysis of the MEMS and CMOS architecture to develop a complex, interleaved, custom process that allows the manufacture to fabricate both simultaneously. Although innovative, each of these solutions creates their own unique set of manufacturing challenges ; the benifits, of any approach must be weighted against its costs and complexity.

14.2.3 MEMS Components

MEMS an acronym that originated in United States is also referred to as Micro-Systems Technology (MST) in Europe and Micromachines in Japan. Regardless of terminology, the uniting factor of a MEMS device is in the way it is made. While the device electronics are fabricated using 'computer chip'. IC technology, the micromechanical components are fabricated by sophisticated manipulations of silicon and other substrates using micromachining processes.

MEMS Consists of:

1. Microsensors
2. Microelectronics
3. Microactuator
3. Microstructure.

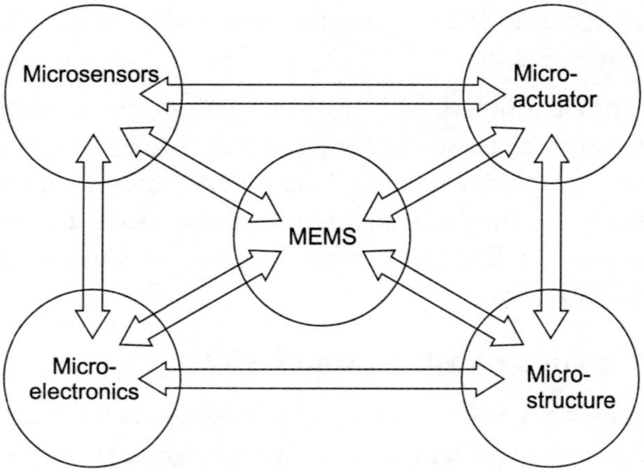

Fig. 14.2 *Schematic Illustration of MEMS Components*

14.2.3.1 Microsensors

Detect changes in the system's environment by measuring mechanical, thermal, magnetic, chemical or electromagnetic information or phenomena.

14.2.3.2 Microelectronics

Process this information and signal the microactuators to react and create some form of changes to the environment.

14.2.3.3 Microstructure

MEMS devices are very small : Their components are usually microscopic. Levers, gears, pistons, as well as motors and even steam engines have all been fabricated by MEMS. However, MEMS is not just about the miniaturization of mechanical components or making things out of silicon. MEMS is a manufacturing technology ; a paradigm for designing and creating complex mechanical devices and systems as well as their integrated electronics using batch fabrication techniques.

14.2.3.4 Microactuator

Microactuator is a device that converts an electrical signal into an action. It can create a force to manuplates itself, other mechanical devices or the surrounding environment to perform some useful functions.

From a very early vision in the early 1950's, MEMS has gradually made its way out of research laboratories and into everyday products. In the Mid-1990's, MEMS components began appearing in numerous commercial products and applications including:

(*a*) Accelerometers used to control airbag deployment in vehicles.
(*b*) Pressure sensors for medical applications.
(*c*) Inkjet printer heads.
(*d*) In projection displays and for micropositioners in the data storage system.
(*e*) The greatest potential for MEMS devices lies in new applications within telecommunications (optical be wireless), biomedical and process control areas.

→ MEMS has several distinct *advantages* as a manufacturing technology:

(*a*) Interdisciplinary nature of MEMS technology and its micromaching techniques.
(*b*) Unprecedented range of devices and synergies across previously unrelated fields (for *e.g.*, biology and microelectronics).
(*c*) MEMS with its batch fabrication techniques enables components and devices to be manufactured with increased performance.
(*d*) High reliability, combined with the obvious advantages.
(*e*) Reduced size, volume, weight and cost.

14.2.4 Definitions and Classifications

This section defines some of the key terminology and classifications associated with MEMS. It is intended to help the reader and newcomers to the field of micromaching become familiar with some of the more common terms.

Figure 14.3 illustrates the classifications of Micro Systems Technology (MST). Although, MEMS is also referred to as MST, strictly speaking, MEMS is a process technology used to create these tiny mechanical devices or systems, and as a result, it is a subset of MST.

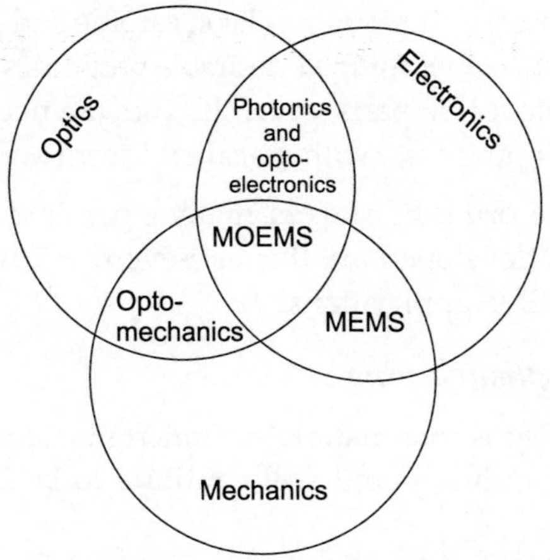

Fig. 14.3 *Classifications of Microsystems Technology*

Micro-optoelectro-mechanical system (MOEMS) is also a subset of MST and together with MEMS forms the specialized technology fields using miniaturized combinations of optics, electronics and mechanics. Both their microsystems incorporate the use of microelectronics batch processing techniques for their design and fabrication. There are considerable overlaps between fields in terms of their integrating technology and their applications and hence it is extremely difficult to categorise MEMS devices in terms of sensing domain and / or their subset of MST.

The real difference between MEMS and MST is that MEMS tends to use semiconductor processes to create a mechanical part. In constrast, the deposition of a material on silicon for *e.g.*, does not constitute MEMS, but is an application of MST.

→ **Transducer :** A transducer is a device that transforms one form of signal or energy into another form. The term transducer can therefore be used to include both sensors and actuators and is the most generic and widely used term in MEMS.

→ **MEMS** devices generally overlap several domains or do not even belong in anyone category. These energy domains include:

- Mechanical : force, pressure, velocity, acceleration, position
- Thermal : temperature, entropy, heat, heat flow.
- Chemical : concentration, composition, reaction rate.
- Radiant : electromagnetic wave intensity, phase, wavelength, polarization reflectance, refractive index, transmittance
- Magnetic : field intensity, flux density, magnetic moment, permeability
- Electrical : voltage, current, charge, resistance, capacitance, polarization.

14.2.5 Material for MEMS

MEMS devices are fabricated using a number of materials depending on the application requirement. One popular material is polycrystalline silicon, also called "polysilicon or poly'. This material is sculpted with techniques such as bulk or surface micromachining, and Deep Reactive Ion Etching (DRIE), proving to be fairly durable for many mechanical operations. Another is nickel, which can be shaped by PMMA (a form of plexiglass) mask plating, as well as by conventional photolithographic techniques.

Other materials—such as diamond, aluminium, silicon carbide and gallium arsenide - are currently being evaluated for use in micromachines for their desirable properties ; *e.g.*, The hardness of diamond and silicon carbide. To create moveable parts, several layers are needed for structural and electrical interconnect (ground planes purposes, with so-called "sacrificial" oxide layers in between.

The current manufacturing record is five layers, making possible a variety of complex mechanical systems. These capabilities, developed are the last several years, are beginning to unlock the almost unlimited possible of MEMS applications.

14.2.5.1 Materials for Micromachining

Substrates: The most common substrate material for micromachining is silicon. It has been successful in the microelectronics industry and will continue to be in area of miniaturization for several reasons:

(*i*) Silicon is abundant, inexpensive, and can be processed to unparallel purity.

(*ii*) Silicon's ability to be deposited in thin films is very amenable to MEMS.

In addition, it is surprisingly resistant to mechanical stress, having a higher elastic limit than steel in both tension and compression. Single crystal silicon also remains strong under repeated cycles of tension and compression.

The crystalline orientation of silicon is important in the fabrication of MEMS devices, because some of the etchants used attack the crystal at different rates in different directions.

Silicon is dominant as a substrate for MEMS, but research and development is ongoing with other non-semiconductor substrate material including metal glasses, quartz, crystalline insulators, ceramics and polymers. The ability is integrate circuitry directly onto the substrate is currently the underlying issue with today's MEMS substrate materials, hence the success of silicon.

14.2.5.2 *Additive Films and Materials*

The range of additive films and materials for MEMS devices is much larger than the types of possible substrates and including conductors. Semiconductors and insulators such as :

- Silicon-Single Crystal, polycrystalline and amorphous.
- Silicon compound ($Si_x N_y$, SiO_2 SiC etc.)
- Metals and metallic compounds (Au, Cu, Al, ZnO, GaAs, IrO_x, Cds)
- Ceramics (Al_2O_3 and more complex ceramic compounds)
- Organics (diamond, polymers, enzymes, antibodies, DNA etc.)
  - (*i*) High definition and reproduction of silicon device shapes using photography are perfect for high levels of MEMS precision.
  - (*ii*) Silicon microelectronics circuits are fabricated (a silicon wafer contains hundreds of identical chip not just one).

Other crystalline Semiconductors including germanium (Ge) and gallium arsenide (GaAs) are used as substrate materials due to similar inherent feature, but silicon is distinguished from other semiconductor in that it can be readily oxidized to form a chemically inert and electrically insulating surface layer of SiO_2 on exposure to steam.

The homogeneous crystal structure of silicon gives to it the electrical properties needed in microelectronic circuit but in this from silicon also has desirable mechanical properties. Silicon froms the same type of crystal structure as diamond, and although the interatomic bonds are much weaker, it is harder that most metals as shown in Fig. 14.4.

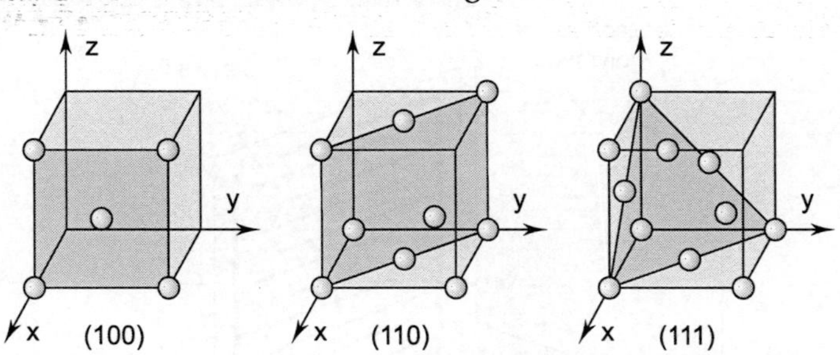

Fig. 14.4 *Low Crystallographic Index Planes of Silicon*

14.3 MICROSENSING FOR MEMS

Various microsensing and microactuations mechanisms have been developed to MEMS for diverse applications. The most commonly used sensing and actuating principles are introduced in this section. Many microsensors based on different sensing principles for MEMS have been developed including chemical sensors, gas sensors, optical sensors, biosensors, thermal sensors and mechanical sensors.

The major sensing mechanisms for mechanical microsensors are introduced in the following sections.

14.3.1 Piezoresistive Sensing

Piezoresistive sensing utilizes resistors where the resistance is varied through external pressure, to measure such physical parameters as pressure, force and flow rate or to be used as accelerometer.

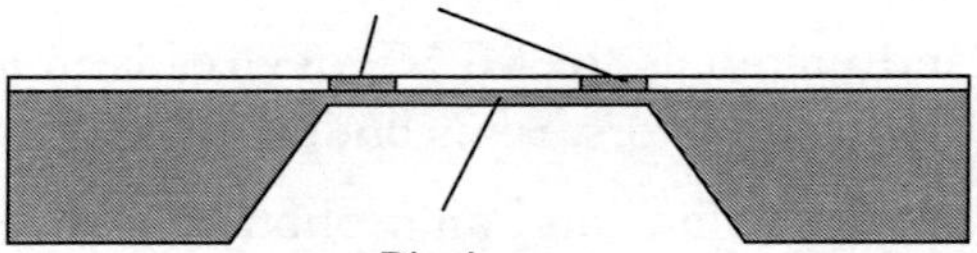

Fig. 14.5 (a) *Piezoresistive Sensing Structure*

A typical structure for piezoresistive microsensors is shown in Fig. 14.5 (*a*). The resistors are usually built on a silicon diaphragm. The deflection of the diaphragm leads to the dimension change of the resistors, resulting in the resistance changing as a result of the piezoresistive effect in silicon.

$$\frac{\Delta R}{R} = \left(1 + 2v\frac{\Delta l}{l} + \frac{\Delta \rho}{\rho}\right)$$

where, ΔR is the change of resistance, R is the original resistance, v is the poisson ratio, Dl is the length change of the resistor, l is the original length of the resistor, and Dρ and ρ represent the resistivity change and resistivity of the resistor, respectively.

It is easily found that the resistance of the resistors used for this type of piezoresistive microsensor is proportional to the external pressure when the resistivity change is ignored. Since the dimension change is proportional to the applied pressor.

Another piezoresistive type microsensor is schematically shown in Fig. 14.5 (*b*), where a semiconductor polymer foil is put on interdigital transducter (IDT) electrodes. If a voltage is applied to the

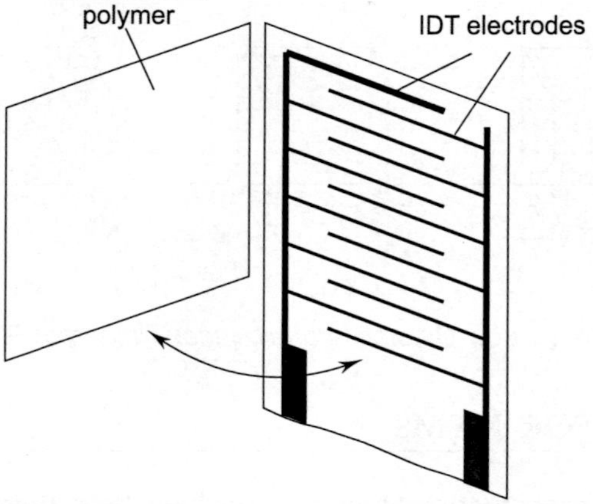

Fig. 14.5 (b) *Schematic of Piezoresistive Type Microsensor*

electrodes and there is no pressure applied, the resistance is at the level of mega ohms. When a force is applied, the resistance decreases as a result of current that flows across the shunting polymer foil. Here, the sensing resistance is inversely proportional to the pressure applied.

The performance of the piezoresistive microsensors varies with temperature and the pressure. The sensitivity of the sensors decreases as temperature increases. Any residual stress generated during the fabrication will also influence the sensitivity of the sensors. The non-linear deflection of the diaphragm occurs when the high-pressure-induced deflection is over 10% of the diaphragm thickness.

14.3.2 Capacitive Sensing

Induced capacitive sensing utilizes the diaphragm deformation capacitance change to convert the information of pressure, force etc., into electrical signal such as changes of oscillation frequency, time, charge and voltage. The structure of a typical capacitive microsensor is shown in Fig. 14.6.

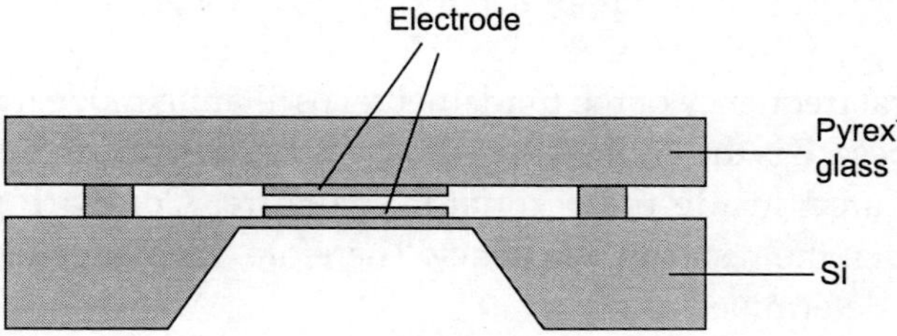

Fig. 14.6 *Capacitive Sensing Structure*

An electrode on the flexible diaphragm and the other one on the substrate construct the sensing capacitor. The capacitive microsensors can be used for pressure, force, acceleration, flow rate, displacement, position and orientation measurement, etc.

For capacitive microsensor, the capacitance change is not linear with respect to the diaphragm deformation, and also, the small capacitance (generally to 1 to 3 pF) requires the measurement circuit to be integrated on the chip. But, the capacitive sensing was found to have potential for higher performance than piezoresistive sensing in applications requiring high sensitivity, low power range and high stability.

14.3.3 Piezoelectric Sensing

Piezoelectric sensing is based on the piezoelectric effect of piezoelectric material. The electrical charge change is generated when a force is applied across the face of a piezoelectric film. For piezoelectric disc of a given thickness t, the voltage (V) generated across the electrode disc (Fig. 14.7) when subjected to a stress (T) would be $V = gt\,T$.

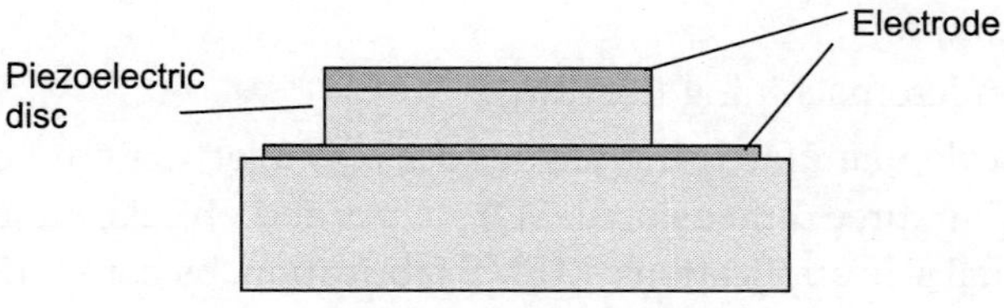

Fig. 14.7 *Piezoelectric Sensing*

Where, g is the piezoelectric voltage coefficient. Piezoelectric sensing is mostly used in sensors such as pressure sensors, force sensors, speedometers, and accelerometers, hydrophones and microphones.

14.3.4 Resonant Sensing

Resonant sensing is easily understood as the natural frequency of a string changing as a result of tensile force. In the development resonant microsensor, strain caused by pressure on the diaphragm leads to the natural frequency of a resonator varying. By picking up the natural frequency variation of the resonator, the physical information that caused the strain will be sensed.

For an example : The natural resonant frequency of a flexure resonator with both ends fixed can be obtained from :

$$f = \frac{4.73\,h^2}{2\pi d^2}\left[\frac{E}{12\rho}\left[1 + 0.2366\left(\frac{l}{h}\right)^2 \varepsilon\right]\right]^{\frac{1}{2}}$$

where f, is the natural frequency of the fundamental oscillating mode it is the resonator length, h is the resonator thickness, E is the Young's modulus, ρ is the density of the diaphragm material and ε is the strain generated inside the resonator structure. Comparing resonant sensing with piezoresistive sensing, the resonant frequency. Therefore, the gauge factor of the above resonant strain gauge can be determined as :

$$K_{gf} = \frac{1}{2}\left[0.2366\left(\frac{l}{h}\right)^2\right]\left[1 + 0.2366\left(\frac{l}{h}\right)^2 \varepsilon\right]^{-1}$$

$$\frac{\Delta f}{f} = K_{gf}\,\varepsilon$$

If a strain is 100 ppm, for a 1.2 mm long, 20 micron wide and 5 micron thick resonator strain gauge, the gauge factor can be as high as 3000, whereas the piezoresistive strain gauge factor is only about 2. Since, the gauge factor relates directly to the sensitivity of the sensor, the resonant sensing can be used to obtain highly sensitive microsensors. However, resonant sensing usually requires a more complex sensor structure than piezoresistive sensing.

14.4 MEMS FABRICATION METHODS

MEMS fall into three classifications:

1. Bulk micromachining.
2. Surface micromachining
3. High-aspect-ratio micromachining (HARM).

Conventional macroscale manufacturing techniques *e.g.*, injection moulding, turning, drilling, etc., are good for producing three dimensional (3D) shapes and objects, but can be limited in terms of low complexity for small size applications. MEMS fabrication, by comparison, uses high volume IC Style batch processing that involves the addition or substraction of two dimensional layers on a substrate (usually silicon) based on photolithography and chemical etching. As a result, the 3D aspect of MEMS devices is due to patterning and interaction of the 2D layers.

Additional layers can be added using a variety of thin-film and bonding techniques as well as by etching through sacrificial 'spacer layers'. Figure 14.8 the potential complexity of a MEMS system by the addition of independent structural layers.

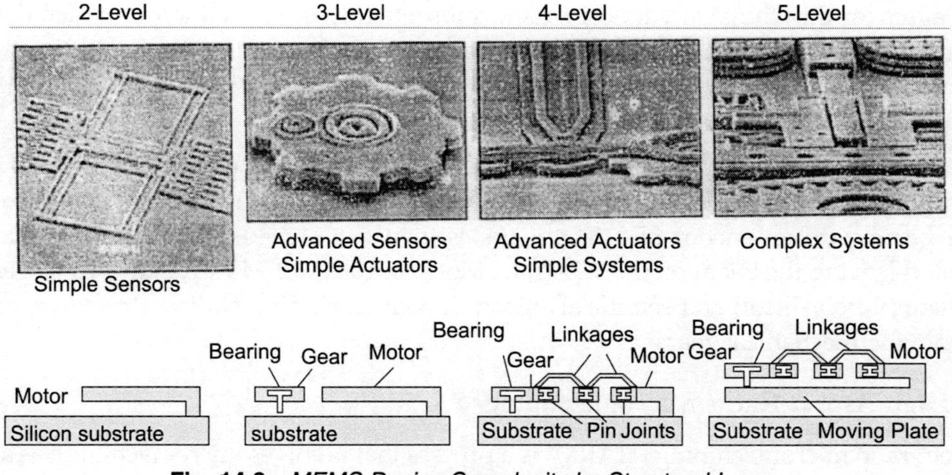

Fig. 14.8 *MEMS Device Complexity by Structural Layers*

14.4.1 Bulk Micromachining

Bulk micromachining involves the removal of part of the bulk substrate. It is a substractive process that uses wet anisotropic etching or a dry etching method such as reactive ion etching (RIE) to create large pits, grooves and channels. Materials typically used for wet etching including silicon and quartz, while dry etching is typically used with silicon, metals, plastics and ceramics.

14.4.2 Surface Micromachining

Surface micromachining involves processing above the substrate, mainly using it as a foundation layer on which to build. It was initiated in the 1980's and is the newest MEMS production technology. Material is added to the substrate in the form of layers of thin films on the surface (typically a silicon wafer). These layers can either by structural layers or act as spacers, later to be removed, when they are known as sacrificial layers. Hence the process usually involves films of two different materials : a structural material out of which the free standing structure is made (generally polycrystalline silicon or polysilicon, silicon nitride and aluminium) and a sacrificial material, deposited wherever either an open area or a free standing mechanical structure is required (usually an oxide).

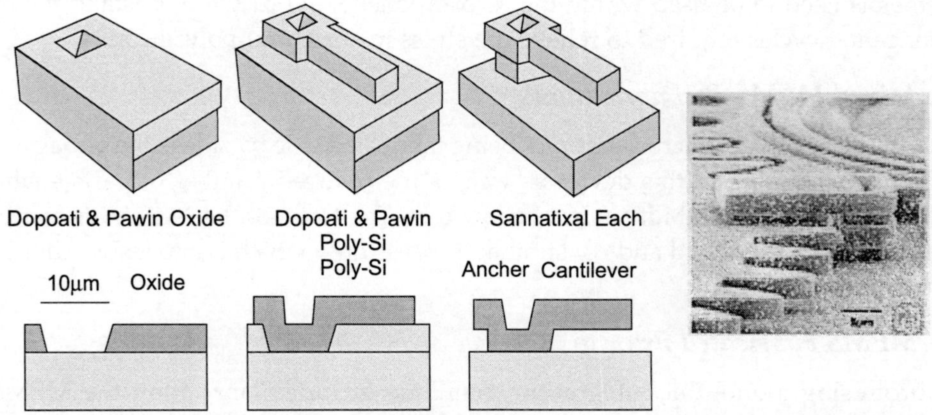

Fig. 14.9 *Surface Micromachining of a Cantilever Beam Using a Sacrificial Layer*

These layers (or thin films) are deposited and subsequently dry in sequence. Which the sacrificial material being finally wet etched away to release the final structure. Each additional layer is accompanied by an increasing level of complexity and a resulting difficulty in fabrication. A typical surface micromachined cantilever beam is shown in Fig. 14.9.

Here, a sacrificial layer of oxide is deposed on the silicon substrate using a pattern and photolithography. A polysilicon layer is then deposited and patterned using RIE processes to form a cantilever beam with an anchor pad. The wafer is then wet etched by remove the oxide layer releasing and leaving the beam on the substrate. More complex MEMS structures can be achieved using structural polysilicon and sacrificial silicon dioxide, including sliding structures, actuators and free moving mechanical gears.

14.4.3 High-Aspect-Ratio Micromachining

High-aspect-ratio micromachining (HARM) is a process that involves micromachining as a tooling step followed by injection moulding or embossing and if required, by electroforming to replicate microstructures in metal from moulded parts. It is one of the most attractive technologies for replicating microstructures at a high performance to cost ratio and includes techniques known as LIGA. LIGA is an important tooling and replication method for high-aspect-ratio microstructure. Products micromachined with this technique include high-aspect ratio fluidic structure such as moulded nozzle plates for inkjet printing and materials that can be used are microtitreplates in medical diagnostic applications. The polycarbonate polyamide and styrene.

→ Over the years different approaches have been developed for the integration of electronics interface. These include hybride integration using conventional wire bonding and flip-chips and monolithic integration. Monolithic integration offers superior system integration performance to hybrid systems but at an overall higher price in terms of involved technology and processing. Monolithic integration can be carried out in three ways.

14.4.3.1 IC before MEMS

Monolithic integration by IC first has proved to be successful and relatively cheap ; an example is the technology is Texas Instrument's DMD (Digital Micromirror Device). The process relies heavily on bulk micromachining and the addition of new layers through electroplating. It is a relatively simple integrated system but suffers from residual stresses within the device material. To date, refractory metals need to be used within the IC components in order to withstand the high temperature ennealing cycles required to relieve the stress in structural polysilicon.

14.4.3.2 Mixed MEMS-IC Fabrication

A typical example of MEMS microelectronics being fabricated side by side is the airbag accelerometer. Monolithic processing of this device as well as the reduced number of parts enable a very compact device with high reliability at a very low cost. The trade-off lies within its complexity as this process leads to a very rigid and constrained process flow which is expensive, thus requiring very high volumes.

14.4.3.3 MEMS Fabricated Prior to IC

The most promising monolithic integration technique includes fabricating the MEMS device prior to the microelectronics. Using technology known as *i*MEMS (Integrated Micro-electromechanical system) patented by Sandia National Laboratories, USA, MEMS components are fabricated in trench on a silicon substrate and then the standard electronics are processed onto the same substrate.

14.5 MEMS PACKAGING

MEMS Packaging : The proper operation of MEMS devices depends critically upon the 'clean' environment provided by the package and is considered on enabler for the commercialisation of MEMS. Packaging of microsensors presents special problems as a part of the sensor requires environmental access while the rest may require protection from environmental conditions and handling.

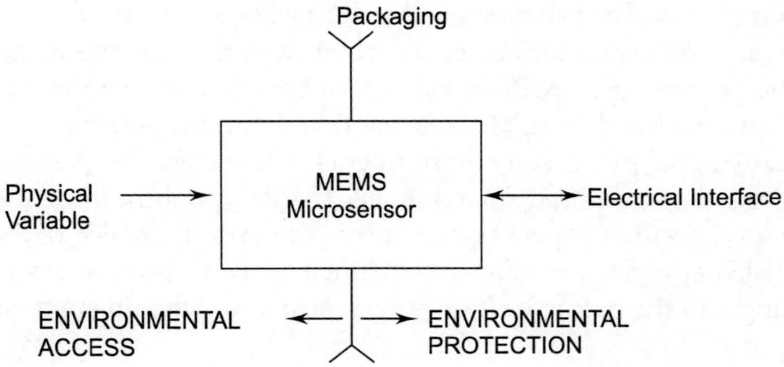

14.10

Although, there is no generic package for a MEMS device, the package should:

1. Provide protection and be robust enough to withstand its operating environment.
2. Allow for environmental access and connections to physical domain.
3. Minimum electrical interference effects from inside and outside the device.
4. Dissipate generated heat and withstand high operating temperature.
5. Minimum stress from external loading.
6. Handle power from electrical connection loads without signal disruption.

14.5.1 Types of MEMS Packaging

The most commonly used packages for sensors are usually based on derivatives of conventional semiconductor packages including plastic ceramic and metal can packages as shown in Fig. 14.11.

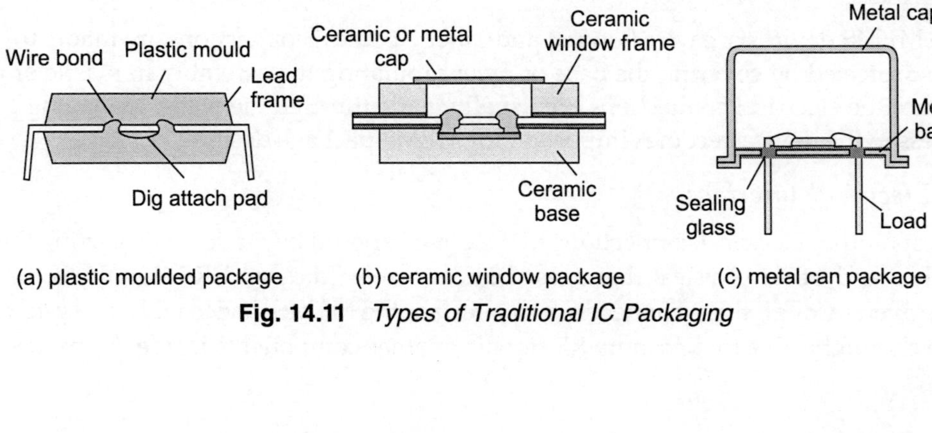

Fig. 14.11 *Types of Traditional IC Packaging*

Each of these packages has been adapted in one form or another for the packaging of silicon sensors. Since, the substrate on many integrated circuit requires an electrical connection to bias it. Sensor dies are usually mounted to a die attach pad in the package using a conductive bond. The die attach pad is typically joined to a metal lead frame with wire bonds providing the electrical connections to the lead frame.

Various bonding media include AuSi eutectic bonding, epoxy bonding and glass usually loaded with silver. The package is subsequently formed by plastic moulding, sealed ceramic or metal caps or with a brazed metal cap to the base of the metal package.

14.5.2 Role of MEMS Packages

The aim of MEMS package is to facilitate the integration of all components such that it minimizes the cost, mass and complexity. The function of a MEMS package can be summarized as providing : mechanical support, an electrical interface to the other system components and protection from the environment. The packaging provides an interface between the chip and the physical world. The package should protect the device, at the same time letting it perform its intended functions with less attenuation of signal in a given environments at low cost. The packaging becomes more expensive when protection is required for relatively fragile structure integrated into the device. For a standard integrated circuit, the packaging process can take up to 95% of the total manufacturing cost. Issues in MEMS packaging are much more difficult to solve because of stringent requirement in processing, handling and the nature of fragile microstructure ; the diversity also complicates the packaging problem.

Many MEMS sensors often require a sensing medium interface with a sensing area. For example, a pressure sensor packaging requires incorporation of a pressure port to transmit fluid pressure to the sensor. This makes a major difference between standard semiconductor device packages and MEMS package.

14.5.2.1 Mechanical Support

The fundamental nature of MEMS as a mechanical device, the protection and isolation of the device from thermal and mechanical shock, vibration, acceleration and other physical damage during operation because critical. The mechanical stress affecting a system depends on the application. For example, for the same space-borne application, the device package for a military aircraft is different from those used in communication satellite because the operating environment are different.

The coefficient of thermal expansion of a package should be equal to that of silicon for reliability because the thermal cycle may cause cracking or determination if the materials are unmatched. If the packaging solution is creating excessive stress in the sensing structure, it can cause a change of device performance.

Once the MEMS devices are wire bonded and other electrical connections are made, the assembly must be protected by covering the base or by encapulating the assembly in plastic or ceramic materials since the electrical connections are usually made through the walls. Managing package-induced stress in the device becomes important for MEMS package design.

14.5.2.2 Electrical Interface

Wire bonds and other electrical connections to the device should be made by protecting the device from scratches and other physical damages. Direct current (d.c.) and RF signals to the MEMS systems are given through these connections to interface the MEMS device with the systems. Also, the Package should be able to distribute RF signals to other components inside the package. High frequency RF signals can be introduced into the package by metal transmission lines or coaxial lines or the function can be electromagnetically coupled into the device. The final connection between the MEMS and the RF lines is usually made with wire bonds or flip-chip die attachments and multiplayer interconnections.

14.5.2.3 *Protection From the Environment*

Many of the MEMS and sensors are designed to measure outside variables from the surrounding environment. The hermetic packaging generally applicable to microelectronic devices is not suitable in many cases of MEMS devices. These devices might be integrated with the circuits or mounted to a circuit board and protected from mechanical damage. Only special attention to packaging will protect a micromachined device from aggressive surroundings. Protection starts at the dice level. Elements that cause corrosion or physical damage to the metal lines as well as other components such as moisture remains a concern for many MEMS devices. The moisture that may be introduced into the package during fabrication and before sealing can damage the materials. For example, aluminium lines can corrode quickly in presence of moisture, and gold lines degrade slowly in moisture. Junctions of dissimilar metals can also corrode in the presence of moisture. MEMS packages need to be hermetic, with good barriers against liquid and gases.

14.5.2.4 *Thermal Considerations*

The MEMS devices used for current applications do not have a high power dissipation requirement. The thermal dissipation from MEMS devices is not a serious problem since the temperature of the MEMS devices usually does not increase substantially during the operation. However, as the integration of MEMS with other high-power devices such as amplifiers in a single package increases, the need for heat dissipation will have to be addressed to protect the MEMS device from high temperatures. This thermal management can place a high design consideration on package design.

14.5.3 Computer Aided Design

Computer Aided Design (CAD) is generally used in MEMS for the design of photolithographic masks. This is a straightword process as MEMS structures are relatively large in comparison to the sub-micron structures usually associated with silicon chip components. As well as using CAD for mask design, CAD and finite element analysis (FEA) are important simulation tools for the design of MEMS applications. Unfortunately, to date there is a lack of adequate advanced software based tools to fully model, analyse and simulate MEMS microstructures as well as integrated MEMS/IC devices. This has acted as a barrier to the development of MEMS devices and systems.

One of the most successful and commercially available software design tools today is MEMCAD, a package from microcosm technologies in North Carolina, USA. The MEMCAD system defines device layout and process, constructs the three dimensional geometry of the device, assembles a detailed 3D model and analyses device performance as well as device sensitivity to manufacturing and design variations. MEMS pro, a package from Tanner Research in California, enable design of MEMS to simulate the growth / deposition, implementation / diffusion and etchs in a MEMS fabrication process.

14.6 MEMS APPLICATIONS

High volume MEMS can be found in a diversity of applications across multiple markets.

Applications of MEMS are shown in tabular form

Automotive	*Electronics*	*Medical*	*Communications*	*Defence*
Internal navigation sensors	Disk drive heads	Blood pressure sensor	Fibre-optic network components	Munitions guidance
Air conditioning compressor sensor	Inkjet printer heads	Muscle stimulators & drug delivery systems	RF Relays, switches and filters	Surveillance
Brake force sensors and suspension control accelerometers	Projection screen televisions	Implanted pressure sensors	Projection displays in portable communications devices and instrumentation	Arming systems
Fuel level and vapour pressure sensors	Earthquake sensors	Prosthetics	Voltage controlled oscillators (VCOs)	Embedded sensors
Airbag sensors	Avionics pressure sensors	Miniature analytical instruments	Splitters and couplers	Data storage
"Intelligent"	Mass data tyres	Pacemakers storage systems	Tuneable lasers	Aircraft control

MEMS Applications:

As an emerging technology MEMS products are centred around technology-product paradigns rather than product-market paradign. Consequently a MEMS device may find numerous applications across a diversity of industries. For example, the MEMS inkjet printer head nozzle in widespread use today has developed from a nozzle originally used in nuclear separation. The commercialisation of selected MEMS devices is illustrated in table below:

Commercialisation of selected MEMS devices

Product	*Discovery*	*Evolution*	*Cost Reduction/ Application Expansion*	*Full Commercialisation*
Pressure sensors	1954-1960	1960-1975	1975-1990	1990-present
Accelerometers	1974-1985	1985-1990	1990-1998	1998
Gas sensors	1986-1994	1994-1998	1998-2005	2005
Valves	1980-1988	1988-1996	1996-2002	2002
Nozzles	1972-1984	1984-1990	1990-1998	1998
Photonics/displays	1980-1986	1986-1998	1998-2004	2004
Bio/Chemical sensors	1980-1994	1994-1999	1999-2004	2004
RF switches	1994-1998	1998-2001	2001-2005	2005
Rate (rotation) sensors	1982-1990	1990-1996	1996-2002	2002
Micro relays	1977-1982	1993-1998	1998-2006	2006

14.6.1 New MEMS Applications

The experience gained from these early MEMS applications has made it an enabling technology for new biomedical applications and wireless communications comprised of both optical, also referred to as micro-optoelectromechanical system (MOEMS), and radio frequency (RF) MEMS.

14.6.1.1 BIOMEMS

Over the past few years some highly innovative products have emerged from bio-MEMS companies for revolutionary applications that support major societal issues including DNA sequencing, drug, discovery, and water and environmental monitoring. The technology focus on microfluidic system as well as chemical testing and sensors, flow controllers, micronozzles and microvalves to be produced.

Example of a new bio-MEMS device is the microtitreplate on which a number of cavities can be simultaneously filled accurately and repeatably by capillary force. This is a relatively single MEMS product in the form of piece of plastic with high-aspect-ratio micromachined mircochannels and is classified as a 'lab-on-a-chip' product.

14.6.1.2 MOEMS

(Micro-opto-electro-mechanical system) : Optical communications has emerged as the only practical means to address the network scaling issues created by the tremendous growth in data traffic caused by the rapid rise of the internet.

Current routing technology slows the information flow by transforming optical signal into electronics information and then back into light before redirecting it. An optical networks offer for superior througout capabilities and performance over traditional electronic systems.

→ The most significant MOEMS device products include waveguides, optical switches, cross connects, multiplexess, filters, modulators detectors, attenuators and equalizers.

14.6.1.3 RF MEMS

RF MEMS is one of the fastest growing areas in commercial MEMS technology. RF MEMS are designed specifically for electronics in mobile phones and other wireless communication applications such as radar, global positioning satellite system (GPS).

The technology includes circuit tuning elements (capacitors/inductors, resonators, filters, microphones and switches). These low-loss ultra-miniature and highly integrative RF functions can will eventully replace classical RF elements and enable a new generation of RF devices.

As it can be seen today, if RF MEMS components continue to replace traditional components in today's mobile phones, then phones could become extremely small (the size of wristwatch is not too far away), require little battery power and even be cheaper.

SUMMARY

MEMS have a tremendous future in replacing the components of many commercial products used today. The medical, wireless (including cellular and network technologies) ; biotechnology computer, automotive and aerospace industries are only a few that will benifit greatly from MEMS. Furthermore, this enabling technology promises to not only transform most major industries but to create entirely new categories of products. An almost infinite number of radical applications are possible because of its potential : nearly limitless functionality, infinitesimally small form factor, a phenomental price / performance ratio, and an architecture that lends itself to mass production, all of which are advantages that drove the success of the integrated circuit. MEMS will be indispensable factor for advancing technology in the 21th century.

TUTORIAL SHEET

1. Explain active and passive MEMS.
2. What are the different types of MEMS?
3. Explain classification of micro system technology.
4. What is the piezoresistive sensing?
5. How you categorise MEMS fabrication methods?
6. Explain MEMS packaging.

INDEX

V

W

X